Quantum Chemistry Aided Design of Organic Polymers

WORLD SCIENTIFIC LECTURE AND COURSE NOTES IN CHEMISTRY

Editor-in-charge: S. H. Lin

World Scientific Lecture and Course Notes in Chemistry — Vol. 2

Quantum Chemistry Aided Design of Organic Polymers

An Introduction to the Quantum Chemistry of Polymers and its Applications

Jean-Marie André & Joseph Delhalle
Facultés Universitaires ND de la Paix
Namur, Belgium

Jean-Luc Brédas
Université de Mons-Hainaut
Mons, Belgium

World Scientific
Singapore • New Jersey • London • Hong Kong

Published by

World Scientific Publishing Co. Pte. Ltd.
P O Box 128, Farrer Road, Singapore 9128
USA office: 687 Hartwell Street, Teaneck, NJ 07666
UK office: 73 Lynton Mead, Totteridge, London N20 8DH

QUANTUM CHEMISTRY AIDED DESIGN OF
ORGANIC POLYMERS – An Introduction to the Quantum
Chemistry of Polymers and its Applications

ISBN 981-02-0004-8

Printed in Singapore by JBW Printers and Binders Pte. Ltd.

TABLE OF CONTENTS

PREFACE

Until the end of the 1970's, polymer quantum chemistry calculations were mostly limited to academic exercises on simple and sometimes unrealistic model chain systems. The synthetic polymer chemists were, to say the least, rather doubtful as to the usefulness and interest of this new branch of quantum chemistry. However, the recent developments in methodological and computational aspects of polymer quantum chemistry and the rise of new polymers with appealing electroactive properties have immensely contributed to the emergence of the field.

The books already available on the subject such as proceedings of conferences or edited series of reviews covered by several authors on selected topics, are usually written for trained readers and thus mostly profitable to scientists who are already working in the field. In the four hundred or so pages of this book, we attempt to introduce the nonexperts to polymer quantum chemistry and its applications. Our aim is to be illustrative rather than comprehensive. The emphasis is not on the mathematics, but rather on showing how the "machinery" is working.

Illustrations are treated in detail within the framework of the simplest methods, free electron and Hückel theories, but similar applications obtained with more sophisticated techniques are presented and discussed as well. The last two chapters are devoted to topical questions on conducting polymers and organic chains for optoelectronics. The illustrative material comes solely from our own contributions which is in someway restrictive but also more convenient to provide a logical and unified presentation. Thus, we apologize in advance to the many authors who will not find their important contributions to the field of polymer quantum chemistry properly exposed. This book does not pretend at all to provide a complete overview of the field; rather, through examples chosen from our own work, we hope to give the reader the flavor of what can be the impact of quantum chemistry in the area of electroactive polymers. It is expected that the reader will find his way to the more general literature through the references provided at the end of the chapters.

This book is intended for those who are interested in understanding the electronic structure and properties of polymers. The scope of the book is to provide the nonspecialist reader with an introductory description of:

(i) quantum mechanical methods, mainly originating from quantum chemistry to calculate the electronic properties of polymers,

(ii) their use for interpreting and predicting results in areas where the electronic structure is playing an important role, for instance, the electrical conductivity and nonlinear optical properties of conjugated polymers.

The book is thus designed to be of interest primarily to university and industry scientists, research fellows and workers in the fields of theoretical chemistry and physics, materials science, polymer science, electronic device development, chemical physics, physical chemistry, and condensed matter physics. It should also serve as a reference book to lecture graduate students on the electronic structure of polymers or more generally of quasi-one dimensional materials. In this framework, it is worth stressing that the quantum theory of polymers bridges the gap between the electronic structure concepts in chemistry and physics. Since no book of this kind involving a strong interaction between theoretical and experimental aspects is available at the moment, it should meet a need for a timely monograph in a field of important and fastly growing interest and serve as an useful complement to four recent books:

J. LADIK, Quantum Theory of Polymers as Solids, Plenum Press (1988), 417 pp,

C. PISANI, R. DOVESI & C. ROETTI, Hartree-Fock Treatment of Crystalline Systems, Lecture Notes in Chemistry, Vol. 48, Springer-Verlag (1988), 193 pp,

R. HOFFMANN, Solids and Surfaces, A Chemist's View of Bonding in Extended Structures, VCH Publishers (1988), 142 pp,

J. BICERANO (Ed.), Molecular Level Calculations of the Structure and Properties of Non-Crystalline Polymers, Marcel Dekker (in press).

These books are of specific interest to workers involved in the field of polymer quantum chemistry and/or condensed matter physics; most of them are however not always of an easy access to readers not familiar with the practice of calculations.

Many people have contributed to the birth of our book.

This monography finds its roots in the Louvain school of chemical physics and quantum chemistry and it is a great honour for us to dedicate this work to the late Professor Maurice Van Meerssche and to Professor Georges Leroy. We have independently benefited from the remarkable sources of inspirations and human examples given during our respective postdoctoral educations by: Dr. Ron Chance, Professor Enrico Clementi, Professor Frank Harris, Professor Per-Olov Löwdin, and Professor Bob Silbey.

We would like to thank Professor Yu Lu from Academia Sinica who was instrumental in deciding us to start the endeavor of writing this book and persuading World Scientific to publish it.

This book owes much to Dr. Marie-Claude André-Roeland. Her efficient and careful editing of the text and figures have proven invaluable in the presentation of this manuscript.

We also thank the colleagues we have collaborated with for many years and who impacted so much on us from both personal and scientific standpoints : Ray Baughman, Jean-Louis Calais, Art Epstein, Alan Heeger, Alan MacDiarmid, John Morley, Lucjan Piela, Bill Salaneck, Bryan Street, Fred Wudl, Joseph Zyss.

We had the pleasure of working with the following collaborators who gave advice and deeply contributed to the materials presented here: Christian Barbier, Vincent Bodart, Luke Burke, Benoit Champagne, Simone Delhalle-Vercruyssen, Christian Demanet, Magdalena Dory, Joseph Fripiat, Geneviève Hennico-Grimée, Roberto Lazzaroni, Fabienne Meyers, Enrique Orti, Maria Cristina dos Santos, Sven Stafström, Bérengère Thémans-Etienne, Jean-Marc Toussaint, Daniel Vercauteren, Elie Younang.

Jean-Louis Calais, Joseph Fripiat, Barry Pickup, and Daniel Vercauteren carefully read the manuscript, and by their judicious comments, assisted and helped us to clarify our presentation.

The three of us have benefited from research positions as Research Assistant, Senior Research Assistant, Research Associate, and/or Senior Research Associate of the Belgian National Fund for Scientific Research (FNRS). We are most grateful to FNRS for the unvaluable help it has given in the development of our scientific research. One of us (JMA) is also indebted to the FNRS for a scientific mission during which a part of this book was initiated.

INTRODUCTION

Over the last two decades, the quantum theory of polymers and molecular crystals has witnessed a growing interest directly related to the appealing properties (electrical conductivity, polarizability and hyperpolarizabilities, phase transitions, etc.) displayed by quasi-one-dimensional unsaturated systems. Quantum chemistry is expected to play a major role in the development of novel materials with specific electronic and optical properties. The range of applications of this field is obviously not restricted to such challenging topics as the design of new types of conductors or superconductors (such as polysulfurnitride or possible organic superconductors at room temperature driven by an excitonic mechanism), or the rationalization of the role of biopolymers in biological processes. As a matter of fact, many interesting and puzzling questions are still open even in the case of the most common stereoregular polymers. For all these materials, and much in the same way as for simpler molecules, quantum chemistry is now starting to provide powerful instruments for interpreting and predicting the physical and chemical properties (structure, bonding, reactivity, etc.) for which a rather detailed knowledge of the electronic structure is required.

The concept of macromolecules is recent; it was introduced in 1921 by Herman STAUDINGER (1881-1965), the father of modern polymer chemistry (Nobel Prize in 1953) but the recognition of this concept has been relatively slow. An historical criticism against Staudinger has been frequently reported in the literature [1,2]: "Dear Colleague, leave the concept of large molecules well alone; organic molecules with a molecular weight above 5000 do not exist. Purify your products, such as rubber, then they will crystallize and prove to be lower molecular substances. Organic molecules with more than 40 carbon atoms do not exist. Molecules cannot be larger than the crystallographic unit cell, so there can be no such things as a macromolecule". This comment is only seventy years old. It was not until the middle of the 30's, that Staudinger's ideas were fully recognized. Nowadays, is not only the existence of macromolecules well established but also their importance in modern technology. It has been recently evaluated [3] that 25% of synthetic polymers are used in packaging, 21% in building and construction industry, 15% for electrical and electronic purposes, 10% as glues, cements, paints, coatings, 7% in car industry, 5% for home furniture and furnishings, 2.5% as cooking articles. On the average, a car contained 10 kg of plastics in 1960, 17 kg in 1966, 48 kg in 1972, 60 kg in 1979, and 85 kg in 1980. And this does not include at all biopolymers.

On a volume basis, the US annual production of plastics exceeds that of copper and aluminum and exceeded that of steel during the beginning of the 80's [4]. As, at the same time, the trend in plastics price over the last years has been consistently downward in contrast to the trend for most other materials, the plastics business has become far more competitive. Table I [5] details the US annual production of the main synthetic polymers over the last decade.

Table I   Production of polymeric materials by the US chemical industry (in millions of lb)

	1976	1980	1985	1987	1989
Man-made fibers	8130	8680	8122	8921	9096
Synthetic rubbers	2303	2015	1838	2182	2302
Plastics					
Thermosetting resins	3633	4093	5631	6263	6407
Thermoplastics resins	21810	26621	35202	41485	43543

If the oil crisis has evidently lowered the production increase during the last years, the 30 billion pounds of polymers produced in the United States correspond nevertheless to the enormous quantity of about 150 pounds per year and per person. Furthermore, we are also dealing with very important amounts of biopolymers. Each man has about 750 grams of hemoglobin; multiplied by 4.6 billions of individuals, this makes 3 million tons of hemoglobin. If polymers are mainly made from oil and are chemically built up from very simple elements: hydrogen, carbon, oxygen, chlorine, nitrogen, and sulfur, the resulting products are in some cases of very high social value, like for instance synthetic arteries and artificial hearts.

The word "polymer" embraces a large class of complex and diverse compounds among which quantum theory has to date considered only a rather restricted subset. The main limitations obviously come from the ability of the model to deal mostly with the physics and chemistry of "regular" polymers. There exists a number of systems whose structural and physical features allow a reasonable modeling by simply taking into account isolated infinite and regular chains. This legitimizes the application of quantum theory in the physics and chemistry of polymers in much the same way as quantum chemistry is applied to gaseous molecules and solid state physics to three-dimensional crystals. Using stereospecific catalysts, polymer chemists are able to synthesize chains of high stereoregularity (polyethylene, polypropylene, etc.) whereas well-organized polymers (polydiacetylene, polymethyleneoxide, etc.) are obtained through solid state polymerization. These examples do not exhaust the list of possible candidates for quantum investigations, but their well-defined structures illustrate the fact that in such cases theoretical models are not far removed from reality and should bring a valuable contribution to polymer science.

Polymers have given rise to fascinating discoveries in scientific research. An example is that of $(SN)_x$ (polysulfurnitride); the evolution of its electrical behaviour is listed in Table II as a function of time. It was synthesized around the beginning of this century and characterized as an explosive powder without any peculiar electrical property. Papers until the 1960's presented it as being a good insulator. In 1964, evidence was given that $(SN)_x$ is a semiconductor; in 1973, the proof was given that it is a metal. Finally in 1975, it was shown to become superconductor at very low temperature. Other examples concerning doped conjugated polymers will be developed later on.

Another reason for developing quantum chemical methods for polymers is the increasing use of refined techniques (XPS, IR, NMR, etc.) to solve fundamental polymer problems. For their interpretation, these techniques require highly developed theoretical procedures; for instance, XPS (X-Ray Photoelectron Spectroscopy or ESCA-Electron Spectroscopy for Chemical Analysis) and UPS (Ultraviolet Photoelectron Spectroscopy) spectra obtained with monochromatic radiation turn out to be powerful tools to study the electronic structure of polymers. In combination, they provide precise information on core level binding energies and line shapes and allow recording of the distributions of the valence electronic levels that constitute an actual "fingerprint" of the polymer. This illustrates the need for theoretical methods for polymers in order to supply the necessary background to interpret complex experimental data, such as XPS or UPS spectra.

Table II History of the electronic behaviour of $(SN)_x$ (polysulfurnitride)

1910	Insulator	A new sulphide of nitrogen
		F.P. Burt
		J. Chem. Soc., 1121 (1910)
		The sulfur nitrides $(SN)_2$ and $(SN)_x$
		M. Goehring and D. Voigt
		Naturwissenschaften, **40**, 482 (1953)
1964	Semiconductor	Spectra and the **semiconductivity** of
		the $(SN)_x$ polymer
		D. Chapman et al.
		Trans. Faraday Soc., **60,** 294 (1964)
1973	Metal	Polysulfurnitride, a one-dimensional chain
		with a **metallic** ground state
		V.V. Walatka, M.M. Labes, and J.H. Perlstein
		Phys. Rev. Lett., **31**, 1139 (1973)
1975	Superconductor	**Superconductivity** in polysulfurnitride $(SN)_x$
		R.L. Greene, G.B. Street, and L.J. Suter
		Phys. Rev. Lett., **34**, 557 (1975)

In view of the existence of quantum physics, quantum chemistry, quantum biology, and quantum pharmacology, on the one side, and of the role of polymers in everyday life and of their interesting properties, on the other side, our opinion was, fifteen years ago, that there was a timely need for developing a specific quantum methodology applied to polymer chemistry. The present book is intended to provide a comprehensive and unified description:

(i) of certain quantum mechanical methods, mainly originating from quantum chemistry, that are currently used to calculate the electronic properties of polymers (Chapters 1 & 2);

(ii) of their application for interpreting and predicting results in fields where the electronic structure is playing an important role like in the electrical conductivity (Chapter 3) and nonlinear optical properties (Chapter 4) of conjugated polymers.

At this stage, it is important to stress that the contributions described in this book mostly originate from our laboratories and are thus biased by our own background (heavily relying on the orbital approach). We firmly hope, however, to provide the reader with a general flavour of current achievements of quantum chemistry in the field of polymers.

References

[1]   R. Olby, J. Chem. Educ., **47**, 168 (1970).

[2]   F.W. Billmeyer, in "Textbook of Polymer Science", 2nd edition, J. Wiley (1971).

[3]   Chem. Eng. News, **65**, August 24, 27 (1987).

[4]   National Research Council, "Opportunities in Chemistry", Pimentel Report, p.49, National Academy Press, Washington DC (1985).

[5]   Chem. Eng. News, **68**, June 18, 40 (1990).

1

THE BAND THEORY OF POLYMERS

1.1 SUMMARY AND OBJECTIVES

The purpose of this chapter is:

(1) to state and justify the Hartree-Fock self-consistent approximation used throughout this book to study electronic properties of molecules and polymers,

(2) to define the periodic model of a polymer chain and sketch its limitations,

(3) to formulate the Bloch's theorem, explain the concept of the Brillouin zone, and describe its use in the free electron model and in the approximation based on linear combinations of atomic orbitals,

(4) to relate molecular orbital energies in a series of oligomers to the band structure of the polymer,

(5) to state the differences between one-dimensional solid state physics and quantum chemistry of polymers,

(6) to account for the shape of energy bands in the first Brillouin zone,

(7) to write down the Hückel expressions for the band structure of polyacetylene and analyze its electronic properties,

(8) to give rules to sketch the shape of electron bands in conjugated polymers.

1.2 THE ONE-ELECTRON HARTREE-FOCK SELF-CONSISTENT FIELD MODEL

For molecules and polymers, the standard quantum chemical approach is based on the Hartree-Fock (HF) theory. In this model, a single electron moves independently in the field of the fixed nuclei and in the averaged Coulomb and exchange field of all the other electrons. A set of molecular orbitals (MO's) describes the occupied and unoccupied one-electron wave functions. In molecular quantum chemistry, the relative energies of the molecular orbitals are drawn as single levels which are at most doubly occupied by a pair of electrons of opposite spins.

The basic postulate of quantum mechanics is to replace the concept of trajectories (full knowledge of positions and momenta of all particles at any time) by a wave function depending on the coordinates of all the particles in the system. The wave function has no physical significance by itself but its absolute square represents the probability density of the particles (electron density in the case of electrons). Since the wave function has no direct physical meaning, it can only be obtained by calculations. The rules were given by Schrödinger. In the case of stationary states, the wave function Ψ is the eigenfunction of the Schrödinger equation with the observable energy E as eigenvalue:

$$H \Psi = E \Psi$$

Here, the Hamiltonian operator H describes all the energy components of the system: usually, the kinetic energy of all particles and the electrostatic interactions (attraction between nuclei and electrons and repulsion between nuclei or between electrons). Inference from classical mechanics postulates that the kinetic energy operator T has the following form for a particle of mass m:

$$T = -\frac{h^2}{8\pi^2 m} \Delta$$

$$= -\frac{1}{2} \Delta = -\frac{1}{2} \nabla^2$$

in atomic units (Hartrees) and with the Laplacian:

$$\Delta = \nabla^2 = \frac{\delta^2}{\delta x^2} + \frac{\delta^2}{\delta y^2} + \frac{\delta^2}{\delta z^2}$$

and that the electrostatic interaction operator V between two charged particles q_i and q_j separated by a distance r_{ij} is:

$$V = \frac{q_i q_j}{4\pi\varepsilon_0 r_{ij}}$$

$$= \frac{q_i q_j}{r_{ij}} \text{ in atomic units}$$

Note that hereafter, atomic units will be generally used, except in sections where the concept of effective mass is necessary for understanding.

Unfortunately, as in classical mechanics, an exact solution of the Schrödinger equation cannot be found when the system contains three particles or more. That is already the case for the simplest many-electron atom He or of the simplest molecular ion H_2^+. Approximations must therefore be introduced. Historically, the most important of those are the Born-Oppenheimer approximation [1.1] to deal with polynuclear systems and the orbital approximation to deal with polyelectronic systems.

In the Born-Oppenheimer approximation, the fact that the mass of a nucleus is at least 1836 times heavier than the electron mass is used to neglect the kinetic operators of the nuclei with respect to the kinetic operators of the electrons. Physically, this corresponds to study the motion of the electrons in the field of fixed nuclei. This leads to an effective electronic energy $E(\mathbf{R})$ which is a parametric function of the nuclear coordinates $\mathbf{R}$ and is used as the potential for the nuclear motion. This approximation defines the important concept of potential energy surfaces, of primary importance in conformational analysis of macromolecules. The equilibrium structure corresponds to the absolute minimum of the energy surface. If there are multiple potential minima, isomers or conformers can be observed. A rigorous presentation of the Born-Oppenheimer approximation is very involved and goes beyond the scope of this book. The interested reader can consult the book by Born and Huang for more details [1.1].

In the orbital theory, it is assumed that a one-electron operator can be defined to represent the motion of a single electron in the field of all the fixed nuclei and the averaged field of interaction with all the other electrons. The logic of these principles is summarized in Chart 1.1.

Chart 1.1 Standard ideas of quantum chemical methods

Schrödinger many-electron equation: $H \Psi = E \Psi$

Not soluble exactly except for one-electron systems: H, He^+, Li^{++}, ...

$\Psi =$ Wave function of the whole electron-nuclear system

$=$ Function of 3n space-variables + n spin-variables

$H =$ Hamiltonian of the whole system
includes:
- . kinetic energy of all electrons
- . kinetic energy of all nuclei
- . attraction between all electrons and all nuclei
- . repulsion between all electrons
- . repulsion between all nuclei

$\Downarrow$ Born-Oppenheimer approximation
$\Downarrow$ Orbital theory

Hartree-Fock one-electron equation: $h \phi = \varepsilon \phi$

$\phi =$ One-electron spatial wave function (orbital)

$=$ Function of 3 space-variables

$h =$ One-electron Hamiltonian
includes:
- . kinetic energy of a single electron
- . attraction of a single electron with all nuclei
- . Coulomb repulsion and exchange interaction of
  a single electron with the averaged electron density

As indicated in Figure 1.1 in the case of the helium atom, it corresponds to describing the observable many-electron levels by the non-directly observable one-electron (orbital) levels which can be doubly occupied by electrons of spin α and β.

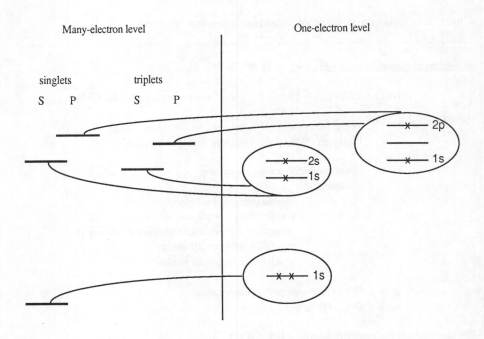

Figure 1.1   Many-electron versus orbital concepts as illustrated in the case of the helium atom

The transformation (approximation) of the many-electron equation to a one-electron equation is made through the Hartree-Fock (HF) methodology. This technique assumes an independent particle model in the form of an approximate many-electron wave function and utilizes the variation theorem. The simplest independent particle wave function for an n-electron system is a simple product of n orbitals. However, such a form does not include the spin properties and does not satisfy the Pauli exclusion principle. Thus, in order to take into account these requirements, an antisymmetrized product of n spin-orbitals, called a Slater determinant, is used. A four-variable spin-orbital $\Phi_i(\mathbf{r},\omega)$ is the product of a spatial orbital $\phi_i(\mathbf{r})$ and of a spin function $\sigma(\omega)$ which can be either $\alpha(\omega)$ or

$\beta(\omega)$. The HF equation is the one-electron equation which allows us to obtain the best orbitals, i.e., those spin-orbitals that, introduced in an antisymmetrized product of spin-orbitals give the best total energy of the system (the lowest and the closest to the experimental value in agreement with the variation principle) which can be obtained from such a wave function. These equations were suggested independently by the Englishman Hartree [1.2], the Russian Fock [1.3], and the American Slater [1.4]. Different versions of the Hartree-Fock theory have been proposed to deal more conveniently with the specific nature of the systems considered (closed shell, open shell). The closed-shell approximation forces the electrons to be paired in common spatial orbitals. Polymers are most often in such situation and therefore we consider only the closed-shell approximation. The basic equations are reviewed in Chart 1.2.

Chart 1.2   Main equations of closed-shell Hartree-Fock theory

---

<u>Wave function:</u>

$$\Psi(\mathbf{r}_1,\omega_1, \mathbf{r}_2,\omega_2,... \mathbf{r}_i,\omega_i ,...)$$
$$= \det | \Phi_j(\mathbf{r}_1,\omega_1) \Phi_k(\mathbf{r}_2,\omega_2) ...\Phi_m(\mathbf{r}_i,\omega_i)...|$$

<u>Hartree-Fock equation:</u> $h(\mathbf{r}) \phi_i(\mathbf{r}) = \varepsilon_i \phi_i(\mathbf{r})$

<u>One-electron Hartree-Fock operator:</u> $h(\mathbf{r}) = -\dfrac{1}{2} \nabla^2$

$$- \sum_A \frac{Z_A}{|\mathbf{r}-\mathbf{R}_A|}$$

$$+ \sum_j^{occ} \{2J_j(\mathbf{r}) - K_j(\mathbf{r})\}$$

<u>Coulomb repulsion operator:</u> $J_j(\mathbf{r}) \phi_i(\mathbf{r}) =\{ \int d\mathbf{r}' \phi_j(\mathbf{r}') \phi_j(\mathbf{r}') |\mathbf{r}-\mathbf{r}'|^{-1}\} \phi_i(\mathbf{r})$

<u>Exchange interaction operator:</u> $K_j(\mathbf{r}) \phi_i(\mathbf{r}) =\{ \int d\mathbf{r}' \phi_j(\mathbf{r}') \phi_i(\mathbf{r}') |\mathbf{r}-\mathbf{r}'|^{-1}\} \phi_j(\mathbf{r})$

We note the underlying physical meaning of the HF field: one determines the motion of a single electron characterized by the kinetic operator $- (1/2) \nabla^2$ in the electrostatic field of fixed nuclei $\{- \sum_A Z_A \, |\mathbf{r}-\mathbf{R}_A|^{-1}\}$. The electron also moves in the interaction field due to its repulsion Coulomb operator $\{\sum_j^{occ} 2 J_j(\mathbf{r})\}$ with the average electron density $\{\sum_j \phi_j(\mathbf{r}') \phi_j(\mathbf{r}')\}$ and the exchange interaction $\{-\sum_j^{occ} K_j(\mathbf{r})\}$.

The Coulomb operator is said to be local, i.e., it does not depend on the orbital on which it is acting, while the exchange operator is nonlocal: the form of the exchange operator actually depends on the orbital $\phi_i(\mathbf{r})$ on which it operates. A multiplicative operator is local, since in order to find the value of the resulting function, one only needs the value of the original function at the same point: A $f(\mathbf{r}) = A(\mathbf{r}) \, f(\mathbf{r})$. An integral operator, on the other hand, is nonlocal: B $f(\mathbf{r}) = \int b(\mathbf{r},\mathbf{r}') \, f(\mathbf{r}') \, d\mathbf{r}'$, and one needs the kernel $b(\mathbf{r},\mathbf{r}')$ of the nonlocal operator. In order to find the value of the resulting function b $f(\mathbf{r}) = g(\mathbf{r})$, one needs the value of the function $f(\mathbf{r})$ in all points, not just at $\mathbf{r}$. Note that the HF equations are highly implicit and nonlinear. Indeed, in order to get the set of orbitals $\{\phi_j(\mathbf{r})\}$, one has to know the form of all the interaction operators $\{J_j(\mathbf{r})\}$ and $\{K_j(\mathbf{r})\}$ which explicitly depend on the solutions $\{\phi_j(\mathbf{r})\}$. As indicated by Figure 1.2, the solution of the problem is iterative.

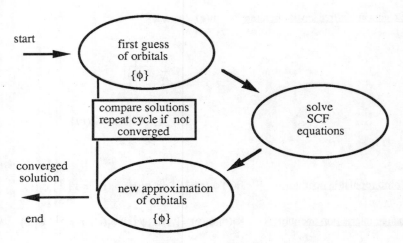

Figure 1.2   Sketch of a Self-Consistent Field (SCF) process

Starting from a guess of the actual solutions $\{\phi_j^{(0)}(\mathbf{r})\}$ (for example, from a strict independent particle solution), a zeroth-order guess of the interaction operators $\{J_i^{(0)}(\mathbf{r})\}$ and $\{K_i^{(0)}(\mathbf{r})\}$ is obtained, leading to a new approximation of the eigenfunctions of the HF equation $\{\phi_i^{(1)}(\mathbf{r})\}$. These solutions are in turn used as "input" for defining better interaction potentials $\{J_i^{(1)}(\mathbf{r})\}$ and $\{K_i^{(1)}(\mathbf{r})\}$ and getting an even more accurate set of solutions $\{\phi_i^{(2)}(\mathbf{r})\}$. The process is repeated until it has converged in the sense that the "input" orbitals $\{\phi_i^{(n-1)}(\mathbf{r})\}$ are identical to the "output" solutions $\{\phi_i^{(n)}(\mathbf{r})\}$ within a predefined threshold. In this scheme, a Self-Consistent Field (SCF) is created. This is the reason why the method is sometimes referred to as the SCF method.

Numerical solutions of the atomic HF equations have been obtained in a systematic way by Hartree's father after his retirement. Hartree's book: "The Calculation of Atomic Structures" [1.5] is actually dedicated to his father with the touching words: "To the memory of my father William Hartree, in recollection of our happy cooperation in work on the calculation of atomic structures". If numerical solutions are rather easily obtained for atomic structures and diatomic molecules, they can hardly be calculated in the case of polyatomic systems. Other methodologies have to be used. The most common of those is the expansion of the atomic and molecular orbitals in terms of basis functions (LCAO approximation), as independently suggested by Roothaan [1.6] and Hall [1.7] in their important 1951 papers. Details about the LCAO approximation are given in Chapter 2.

1.3   PERIODIC MODEL OF A POLYMER CHAIN

We now present some facets of the standard quantum mechanical treatment of polymers. To introduce the concepts, three basic models have been consistently used in the literature. They are represented in Figure 1.3.

The first model is a linear chain of hydrogen atoms. It has appeared as the extension to infinite systems of the first quantum mechanical calculations on an hydrogen atom and of the first calculation on an hydrogen molecule. Historically, a second model is that of an infinite polyene (polyacetylene) and results from the standard π-electron approximation common since the work of Erich Hückel in the early 30's. The third model is that of a polyethylene chain issued from the full machinery of a quantum mechanical treatment of all-electron or all-valence electron systems. Pioneering quantum mechanical calculations

14

on a polyethylene chain were carried out around the end of the 60's (see Section 2.5 and Chart 2.5).

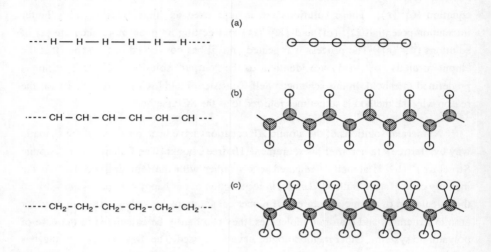

Figure 1.3   Basic models of usual polymeric chains
(a) infinite chain of hydrogen atoms
(b) infinite polyene (polyacetylene)
(c) polyethylene

The linear chain of hydrogen atoms turns out to be a somewhat academic case since its existence is highly speculative. It has nevertheless provided a valuable model. The infinite polyene chain is more attractive for practical purposes. On one side, it is the limit term in the series of linear conjugated polyene molecules: ethylene, butadiene, hexatriene,... a classical set studied in all elementary textbooks of quantum chemistry. On the other side, a renewed and important interest stems from the discovery of highly electrically conducting polyacetylene at the end of the 70's. Polyethylene now plays in polymer quantum chemistry a role similar to those of the helium atom in atomic physics and the hydrogen molecule in molecular physics.

As mentioned in the preface, some limitations are usually introduced in standard quantum polymeric calculations. They are sketched in Figure 1.4 in the case of polyethylene modeled as a zigzag chain built from methylene groups.

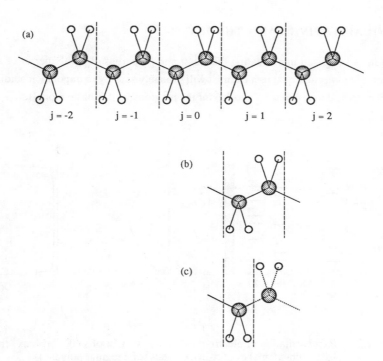

(a)

j = -2   j = -1   j = 0   j = 1   j = 2

(b)

(c)

Figure 1.4   (a) Model of a periodic chain of polyethylene, of
(b) its symmetric unit cell, and of
(c) its asymmetric unit cell

A first limitation is that we usually consider an isolated chain; in practice, the chain is never isolated but exists in a liquid or solid state environment. A second limitation is that the chain is taken to be infinite and perfectly stereoregular. Considering the chain as infinite is not a strong limitation since polymers can have very large molecular weights; in the case of conventional polymers like polyethylene or polypropylene, molecular weights in the range of 10^5-10^6 dalton are achievable. The model also neglects the chain end effects; the importance of such effects has not yet been studied in detail. The crudest assumption of all is that of perfect stereoregularity of the linear chain. However, this allows one to take account of translation symmetry and to apply concepts of condensed matter physics in order to get a rather complete description of the electronic structure of polymers.

1.4   QUALITATIVE BAND THEORY

If he is interested in evaluating the electronic properties of a polymer or of a large oligomer, the molecular quantum chemist will probably start by extrapolation studies like those summarized in Figures 1.5 and 1.6 for the previously cited basic models.

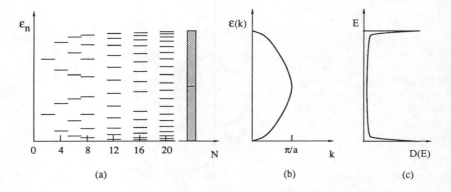

Figure 1.5   Representation of electronic states of a chain of s-like orbitals (chain of
hydrogen atoms) or of π-like orbitals (of a regular polyene)
(a) molecular levels
(b) band structure
(c) density of electronic states

Figure 1.5 represents either a chain of equally spaced s-type orbitals (linear chain of hydrogen atoms) or of $2p_\pi$ orbitals (π orbitals of regular polyacetylene chains; by regular, we mean all the C-C bonds are taken to be of equal lengths). In its leftmost part, Figure 1.5 presents the evolution of the MO levels as a function of the number of hydrogen atoms of the hydrogen chain or of the number of carbon atoms in the oligomeric polyacetylene chains. In the latter case, only the π-orbital levels are conventionally represented. It can be observed that as the number of atoms increases the distance between the energy levels diminishes so as to form, in the infinite limit, occupied energy bands (the core and valence bands) and unoccupied energy bands (the conduction bands). Forbidden energy regions (the forbidden bands or energy gaps) do not exist in the metallic-like situation depicted here where the atoms are equally spaced. The fact that uniformly spaced hydrogen atoms or a non-dimerized polyacetylene have their highest occupied band partially filled will be understood in Section 1.9.3.

In the case of polyacetylene or the hydrogen chain, the polymer is described in its ground state by a fully occupied valence band. One of the unoccupied conduction bands is also represented in Figure 1.5.

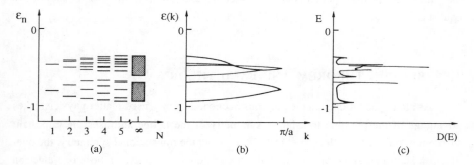

(a) (b) (c)

Figure 1.6 Representation of electronic states of alkanes
    (a) molecular levels of normal alkanes of increasing number N of carbon atoms; in the limit of $N \to \infty$, these levels organize in bands (shaded rectangles) and the neighbouring levels have infinitesimal energy separation
    (b) band structure of polyethylene, an infinitely long alkane chain; the energy levels are expressed in terms of a continuous wavenumber k
    (c) density of electronic states of polyethylene; it measures the number of states per unit energy interval

In the case of polyethylene (Figure 1.6), the unit cell contains two methylene groups (16 electrons). The occupied bands are in this case: (i) two internal core bands deriving from four 1s core electrons of the two carbons per unit cell, (ii) two bands which are usually considered to describe the σ C-C bonds, and (iii) four bands generally associated with the description of the four σ C-H bonds. The 16 electrons in the unit cell are thus fully represented by eight doubly occupied bands. Before developing in detail the theoretical methodology based on lattice periodicity, Figures 1.5 and 1.6 state graphically the links between solid state concepts (which are less familiar to chemists) and molecular theories (which are less familiar to physicists). The interpolation properties between finite oligomers and the infinite polymer have also been emphasized by compact formulas giving the orbital energies (or possibly their structure) as a function of the number of atoms in the chains. Examples of such formulas will be found in Section 1.9.2. dealing with the Hückel methodology.

The density of states (DOS) curves represented in part (c) of Figures 1.5 and 1.6 are most useful for comparisons with experimental data; they correspond to the number of available energy levels per energy unit for infinite systems. They are usually normalized to the number of electrons per unit cell. The three representations in Figures 1.5 and 1.6 are equivalent.

1.5 BLOCH's THEOREM AND BAND THEORY

As already mentioned in the previous Section, the condensed matter physicist takes advantage of the possible translation symmetry of the lattice and uses the important theorem introduced by Bloch [1.8] in 1928. In using the translational symmetry, the one-electron Hamiltonian commutes with the translational operators. In those one-electron Hamiltonians, e.g., Hartree-Fock, where the density is part of their definition, the density must also be symmetric. Therefore, the one-electron states have to be such that the requirements are enforced. In this context, the so-called Bloch functions (molecular orbitals for an infinite one-dimensional chain) are eigenfunctions of the translation operators and, as such, are characterized by a wavenumber, k, defined in the reciprocal space. Bloch's theorem is a direct consequence of the periodicity of the electron density and of the corresponding one-electron potential. For systems having one-dimensional periodicity and a periodic electron density:

$$\rho_n(\mathbf{r}) = |\phi_n(\mathbf{r})|^2 = \rho_n(\mathbf{r}+ja) = |\phi_n(\mathbf{r}+ja)|^2$$

where a is the length of the polymer unit cell in direct space and j is an integer. By taking the square root in the complex mathematical space, Bloch's theorem states the phase relation of the orbitals at periodically related points:

$$\phi_n(\mathbf{r}+ja) = e^{ikja} \phi_n(\mathbf{r})$$

If derived in detail, Bloch's theorem is based on the commutation properties of the one-electron Hamiltonian and the translation operator T(ja) defined such as it translates any function f($\mathbf{r}$) from point $\mathbf{r}$ to point $\mathbf{r}+ja$:

$$T(ja)\ f(\mathbf{r}) = f(\mathbf{r}+ja)$$

We easily note that the one-electron potential function $V(\mathbf{r})$ and the electron density $\rho(\mathbf{r})$ are eigenfunctions of the translation operator with eigenvalues equal to 1 due to the periodic properties of the infinite lattice:

$$T(ja)\ V(\mathbf{r}) = V(\mathbf{r}+ja) = 1.\ V(\mathbf{r})$$

$$T(ja)\ \rho(\mathbf{r}) = \rho(\mathbf{r}+ja) = 1.\ \rho(\mathbf{r})$$

Formally, the eigenfunctions of the translation operator can be represented by the eigenvalue equation:

$$T(ja)\ \phi_n(\mathbf{r}) = \phi_n(\mathbf{r} + ja) = \lambda_j\ \phi_n(\mathbf{r})$$

By using the previous periodic conditions of the electron density of the polymer orbitals:

$$\rho(\mathbf{r}+ja) = T(ja)\ \rho(\mathbf{r}) = T(ja)\ |\phi_n(\mathbf{r})|^2 = |\phi_n(\mathbf{r}+ja)|^2 = \lambda_j^2\ |\phi_n(\mathbf{r})|^2$$

$$= |\phi_n(\mathbf{r})|^2$$

we deduce that the eigenvalue λ_j is a complex number of modulus equal to unity:

$$\lambda_j = e^{ikja}$$

Since the argument of an exponential must be a pure number, a being a length and j a pure number, k must have the dimensions of an inverse length. Thus, the orbitals and their associated energies are functions of k:

$$\phi_n(\mathbf{r}) = \phi_n(k,\mathbf{r})$$

$$\varepsilon_n = \varepsilon_n(k)$$

and can be plotted as functions of k. The representation of the corresponding dispersion curves as a function of k is called an energy band and is given in Figure 1.7 (a).

Standard theorems of solid state physics demonstrate that those energy bands are periodic in the one-dimensional (1D) reciprocal space:

$$\varepsilon_n(k) = \varepsilon_n(k+lg)$$

where g is the reciprocal translation unit length (= $2\pi/a$). Thus, a given orbital $\phi_n(k,\mathbf{r})$ is identical at a k point and at k+lg. As sketched in Figure 1.7 (b), the calculation of the energy bands needs therefore to be performed only for a range of k values equivalent to a single unit cell of the reciprocal lattice. The previous argument is based on the fact that only the values of k within the reciprocal lattice unit cell give nonredundant information. In particular, if comparing two polymer orbitals related by a translation lg in the reciprocal space, we see that:

$$T(ja)\, \phi_n(k,\mathbf{r}) = e^{ikja}\, \phi_n(k,\mathbf{r})$$

and:

$$T(ja)\, \phi_n(k+lg,\mathbf{r}) = e^{i(k+lg)ja}\, \phi_n(k+lg,\mathbf{r})$$

$$= e^{ikja}\, \phi_n(k+lg,\mathbf{r})$$

Therefore, Bloch functions which differ by a reciprocal lattice vector have the same eigenvalue for any translation operator. In other words, we can define a reduced wavenumber k varying from 0 to $2\pi/a$ corresponding to, at most, a translation in the reciprocal space. The length spanned by the maximum interval between all possible reduced vectors is called the first *Brillouin zone* of the polymer.

Another corollary of Bloch's theorem states that the energy bands are symmetric with respect to k=0:

$$\varepsilon_n(k) = \varepsilon_n(-k)$$

The calculation is thus simplified since, using a symmetrized portion of the reciprocal space (the first Brillouin zone ranging from $-\pi/a$ to $+\pi/a$), it needs to be explored only from k=0 to k=+π/a (half the first Brillouin zone) as indicated in part (c) of Figure 1.7.

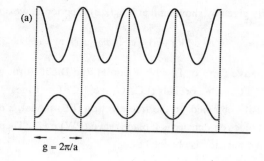

$g = 2\pi/a$

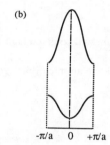

$-\pi/a \quad 0 \quad +\pi/a$

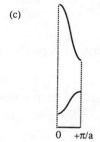

$0 \quad +\pi/a$

Figure 1.7   Representation of energy bands
               (a) in the reciprocal space
               (b) in the first Brillouin zone
               (c) in half the first Brillouin zone

An alternative form of Bloch's theorem: $\phi_n(k,\mathbf{r}+ja) = e^{ikja}\,\phi_n(k,\mathbf{r})$ is obtained when we factorize the Bloch orbitals $\phi_n(k,\mathbf{r})$ into their periodic $u_n(k,\mathbf{r}) = u_n(k,\mathbf{r}+ja)$ and nonperiodic free electron-like contributions e^{ikz}:

$$\phi_n(k,\mathbf{r}) = e^{ikz}\,u_n(k,\mathbf{r})$$

The proof is easily given by showing that if $u_n(k,r)$ is periodic, we obtain:

$$\phi_n(k,r+ja) = e^{ik(z+ja)} u_n(k,r+ja) = e^{ikja} e^{ikz} u_n(k,r) = e^{ikja} \phi_n(k,r)$$

Using this alternative form of Bloch's theorem, the Bloch form of free electron orbitals or of general tight binding orbitals (LCAO-like, LCAO = Linear Combination of Atomic Orbitals) is easily obtained. In the free electron (FE) model, a correct periodic function with constant electron density is a plane wave (PW) e^{-ilgz}. Therefore, a correct Bloch FE orbital in one dimension is obtained as:

$$\phi_n(k,z) = e^{ikz} e^{-ingz} = e^{i(k-ng)z}$$

whose representation in the energy spectrum will be given later (see Section 1.6).

The LCAO Bloch form $\phi(k,r)$ of a single atomic orbital $\chi(r-P)$ centered at point P is thus written as:

$$\phi(k,r) = \sum_j c_j \chi(r-P-ja)$$

$$= N^{-1/2} \sum_j e^{ikja} \chi(r-P-ja)$$

where $N^{-1/2}$ is the normalization factor for a N-cell polymer using an orthonormal atomic orbital basis set.

The LCAO Bloch form which describes the polymer orbital extending over the whole polymer as a Bloch combination of functions centered at the atomic nuclei of polymers is then:

$$\phi_n(k,r) = \sum_j \sum_p c_{npj} \chi_p(r-P-ja)$$

$$= N^{-1/2} \sum_j e^{ikja} \sum_p c_{np}(k) \chi_p(r-P-ja)$$

where n is the band index and k the position vector (wavenumber) in the first Brillouin zone. Note that the polymer orbitals have imaginary components except in two high symmetry points of the first Brillouin zone (k=0, point Γ, and k=π/a, point H) where the Bloch exponential e^{ikja} reduces to real values:

$$\phi_n(\Gamma,\mathbf{r}) = \frac{1}{\sqrt{N}} \sum_j \{ \sum_p c_{np}(0) \chi_p(\mathbf{r}\text{-}\mathbf{P}\text{-ja}) \}$$

$$\phi_n(H,\mathbf{r}) = \frac{1}{\sqrt{N}} \sum_j (-1)^j \{ \sum_p c_{np}(\pi/a) \chi_p(\mathbf{r}\text{-}\mathbf{P}\text{-ja}) \}$$

As will be shown later, the last two relations are useful for getting quick sketches of the band structures of simple polymeric structures.

1.6   QUALITATIVE FREE ELECTRON BAND THEORY

In the free electron (FE) model, the potential of the one-electron equation is considered to be constant over the whole system indicating that the electron has no attractive or repulsive interactions with the framework formed by the nuclei and the other electrons and thus is "free" to move in the polymeric or molecular environment. The free electron model is rather unrealistic from a physical viewpoint. However, it has played such an important role in the theory of metals and is so elegant to apply with the mathematical structure of the band theory that we have decided to introduce it in this part. Furthermore, as we shall see in Chapters 3 and 4, conductivity and nonlinear optical results are still often placed in its conceptual framework. If the constant potential is V_0, it is easily shown that it corresponds to the kinetic motion of a single electron in a zero potential, its energy being shifted by V_0 (as mentioned before, in this Section, we keep the SI unit system to make explicit the dependence with respect to the mass m):

$$\{\frac{-h^2}{8\pi^2 m} \frac{d^2}{dz^2} + V_0\} \phi_n(z) = \varepsilon_n \phi_n(z)$$

or

$$\{\frac{-h^2}{8\pi^2 m} \frac{d^2}{dz^2}\} \phi_n(z) = \{\varepsilon_n\text{-} V_0\} \phi_n(z)$$

$$\left\{\frac{-h^2}{8\pi^2 m}\frac{d^2}{dz^2}\right\}\phi_n(z) = \varepsilon_n^{FE}\,\phi_n(z)$$

with

$$\varepsilon_n^{FE} = \{\varepsilon_n - V_0\}$$

Note that the constant potential must be negative in order to ensure the stability of the molecular or polymeric system and that $-V_0$ is sometimes referred to as the "work function" of the infinite system.

If the system is periodic and infinite, a correct Bloch one-dimensional FE orbital is, in an extended Brillouin zone scheme:

$$\phi(k,z) = e^{ikz}$$

with an associated energy:

$$\varepsilon_n^{FE}(k) = \frac{\dfrac{-h^2}{8\pi^2 m}\dfrac{d^2}{dz^2}\phi(k,z)}{\phi(k,z)}$$

$$= \frac{\dfrac{-h^2}{8\pi^2 m}\dfrac{d^2}{dz^2}e^{ikz}}{e^{ikz}}$$

$$= \frac{h^2 k^2}{8\pi^2 m}$$

We will use later on the identification between the electron mass and the second derivative of the energy bands with respect to k:

$$\frac{1}{m} = \frac{4\pi^2}{h^2}\frac{d^2\varepsilon}{dk^2}$$

In order to take into account the effect of a nonconstant potential, it is common to replace the actual electron mass, m, in the latter expression by an effective mass, m*, which implicitly contains the effect of electron-nucleus and electron-electron interactions.

In a reduced Brillouin zone:

$$k' = k-ng$$

and, as noted above, a correct Bloch one-dimensional FE orbital is, in this case:

$$\phi_n(k,z) = e^{ikz}\, e^{-ingz} = e^{i(k-ng)z}$$

with an associated energy:

$$\varepsilon_n^{FE}(k) = \frac{\dfrac{-h^2}{8\pi^2 m}\dfrac{d^2}{dz^2}\phi_n(k,z)}{\phi_n(k,z)}$$

$$= \frac{\dfrac{-h^2}{8\pi^2 m}\dfrac{d^2}{dz^2}e^{i(k-ng)z}}{e^{i(k-ng)z}}$$

$$= \frac{h^2(k-ng)^2}{8\pi^2 m}$$

When defining the reciprocal lattice, $ng = n2\pi/a$, and the k-position in the first Brillouin zone, $k = \xi 2\pi/a$, the previous equations are rewritten as:

$$\varepsilon_n^{FE}(k) = \frac{h^2(k-ng)^2}{8\pi^2 m} = \varepsilon_0\,(\xi - n)^2$$

with:

$$\varepsilon_0 = \frac{h^2}{2ma^2}$$

The corresponding band structures are plotted in Figure 1.8 and some values of the FE energy bands at high symmetry points are summarized in Chart 1.3.

Chart 1.3   Free electron band structure of a one-dimensional chain

n	$\varepsilon(k) =$	$\varepsilon(\Gamma) =$ $\varepsilon(k=0)$	$\varepsilon(H) =$ $\varepsilon(k=\pi/a)$	$\varepsilon(H') =$ $\varepsilon(k=-\pi/a)$
	$\varepsilon_0 (\xi-n)^2$	$\varepsilon_0 n^2$	$\varepsilon_0 (\frac{1}{2} - n)^2$	$\varepsilon_0 (\frac{1}{2} + n)^2$
0	$\varepsilon_0 \xi^2$	0	$\varepsilon_0 /4$	$\varepsilon_0 /4$
1	$\varepsilon_0 (\xi-1)^2$	ε_0	$\varepsilon_0 /4$	$9 \varepsilon_0 /4$
-1	$\varepsilon_0 (\xi+1)^2$	ε_0	$9 \varepsilon_0 /4$	$\varepsilon_0 /4$
2	$\varepsilon_0 (\xi-2)^2$	$4 \varepsilon_0$	$9 \varepsilon_0 /4$	$25 \varepsilon_0 /4$
-2	$\varepsilon_0 (\xi+2)^2$	$4 \varepsilon_0$	$25 \varepsilon_0 /4$	$9 \varepsilon_0 /4$
3	$\varepsilon_0 (\xi-3)^2$	$9 \varepsilon_0$	$25 \varepsilon_0 /4$	$49 \varepsilon_0 /4$
-3	$\varepsilon_0 (\xi+3)^2$	$9 \varepsilon_0$	$49 \varepsilon_0 /4$	$25 \varepsilon_0 /4$

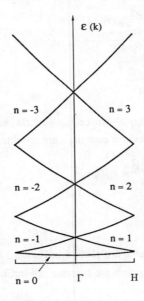

Figure 1.8   Representation of one-electron energy bands of a 1D chain in the free electron model

Since the states due to the N cells of the polymer are unequivocally described by a length equal to $2\pi/a$ of reciprocal space, the relation between the number of states dN and a length dk of the reciprocal space is:

$$dN = \frac{Na}{2\pi} dk$$

If we take into account the fact that each "orbital" state can be occupied by two electrons of opposite spins, the relation between the number of electrons dn and a length dk of the reciprocal space becomes:

$$dn = \frac{Na}{\pi} dk$$

From this expression, we are in a position to compute the Fermi energy ε_F with respect to the number of electrons per unit cell (n/N) where n and N represent the total number of electrons and unit cells, respectively. The Fermi energy is rigourously defined in statistical terms by an occupation probability of a system of indistinguishable particles obeying the Pauli exclusion principle equal to 1/2:

$$n(\varepsilon_F) = \frac{1}{2}$$

if $\varepsilon = \varepsilon_F$ and if the Fermion distribution function is:

$$n(\varepsilon) = \frac{1}{e^{\frac{\varepsilon - \varepsilon_F}{kT}} + 1}$$

Note that, in statistical thermodynamics, the Fermi energy ε_F is a free energy (chemical potential) per particle. In a metallic system, the Fermi level corresponds to the degeneracy point between the top of the valence band and the bottom of the conduction band. In a semiconductor or in an insulator, the Fermi level is strictly speaking located in the middle of the gap, at an equal distance between the "Highest Occupied Molecular Orbital" (HOMO) and the "Lowest Unoccupied Molecular Orbital" (LUMO) at T = 0 K. However, there exists a tradition among molecular quantum chemists to use a simple equivalence of the Fermi energy with the HOMO. Since the total number of electrons, n, is obtained from the integration of the number of states in the Brillouin zone of the polymer:

$$n = \int_{-k_F}^{+k_F} \frac{Na}{\pi} dk = 2 \frac{Na}{\pi} k_F$$

we are now able to calculate the position k_F of the Fermi level with respect to the number of electrons per unit cell:

$$k_F = \frac{n\pi}{2Na} = \frac{\pi}{2a} \frac{n}{N}$$

We deduce the well-known result that if the polymer has one electron per unit cell ($n/N=1$), the first Brillouin zone is only half occupied ($k_F=\pm\pi/2a$) while if the chain has two electrons per unit cell ($n/N=2$), the first Brillouin zone is doubly occupied ($k_F=\pm\pi/a$).

In molecular quantum chemistry, the use of the FE model was independently introduced by Bayliss [1.9], Kuhn [1.10], and Simpson [1.11] for interpreting the visible and UV spectra of conjugated linear chains. These methods form an interesting link with the study of the electronic properties of polyacetylene and its oligomers that will be developed in the next Sections. In the FE model, the basic assumption is that the π electrons move in a constant or almost constant potential field created by the nuclear attractions and the screened Coulomb and exchange interactions of the core and σ-valence electrons of the extended conjugated chains. The FE molecular orbitals ϕ_k are eigenfunctions of the one-electron kinetic operator:

$$\frac{-h^2}{8\pi^2 m} \frac{d^2}{dz^2} \phi_k(z) = \varepsilon_k \phi_k(z)$$

They correspond to a one-dimensional electron gas and have the analytic forms:

$$\phi_k(z) = \sqrt{\frac{2}{L}} \sin \frac{k\pi z}{L}$$

and the energies:

$$\varepsilon_k = \frac{h^2 k^2}{8mL^2}$$

where L is the assumed (delocalization) length of the chain. The one-electron $\pi-\pi^*$ transition energy ΔE of a n π-electron chain results from the excitation of one electron from the HOMO [$k_{HOMO} = n/2$] into the LUMO [$k_{LUMO} = n/2 +1$]:

$$\Delta E = \varepsilon_{LUMO} - \varepsilon_{HOMO}$$

$$= \frac{h^2}{8mL^2} \ [(\frac{n}{2}+1)^2 - (\frac{n}{2})^2]$$

$$= \frac{h^2}{8mL^2} \ [n+1]$$

Note that the chain length L is also an indirect function of the number of electrons n; thus, the final dependence of the FE one-electron $\pi-\pi^*$ transition energy ΔE is inversely proportional to the number of electrons and tends to zero as the number of electrons increases. Bayliss introduced the model for even polyenes, i.e., containing an even number of carbon atoms. Simultaneously, Kuhn applied it to the odd-atom chains of polymethine dyes. It was his pioneering contribution to relate the nature of the first $\pi-\pi^*$ transition to the concept of bond alternation in conjugated chains. Kuhn indeed showed that in the series of polymethine dyes, the bond lengths between all the carbon atoms are equal due to a resonance balance between equivalent extreme forms:

$$>N-(CH=CH)_n-CH=N^+< \quad \leftrightarrow \quad >N^+=CH-(CH=CH)_n-N<$$

All carbon-carbon bonds in the skeleton have 50% double bond character. This fact was later confirmed by X-ray diffraction studies. The calculated FE transition energies of the polymethine series: $>N-(CH=CH)_n-CH=N^+$ (for n=1,2,...5) are in good agreement with the experimental data when taking into account the bond lengths at the chain ends as adjustable parameters. Extrapolation for long chains leads to a zero energy gap, i.e., to a metallic character for an infinite chain. Thus, simple FE model calculations suggest that there is no energy gap between the valence and conduction bands and the limit of the first UV-visible transition for an infinite chain is zero. On the contrary, in the case of polyenes (polyacetylene oligomers), a much poorer agreement with experiment is obtained. In order to reproduce the experimental data, Kuhn had to include a sinusoidal perturbative potential so that the electronic distribution corresponds to alternating single and double bonds. In this case, the extrapolation of the UV spectrum tends to a non-zero energy gap for the infinite chain. Based on the series of oligomers $>(CH=CH)_n<$ (for n=2,3,...6), an extrapolation of the FE transition energies leads to a value of 1.8 eV in close agreement with the experimental energy gap of polymeric polyacetylene.

It will be shown later on that, within the Hückel scheme, bond alternation (i.e., different Hückel "resonance" integrals for single- and double-like bonds) does imply the appearance of an energy gap in the band structure of an infinite polyacetylene chain.

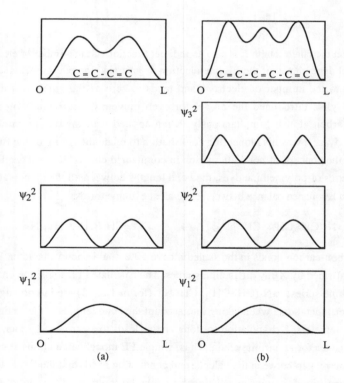

Figure 1.9   π-orbital densities and total densities in one-dimensional free electron models of (a) butadiene and (b) hexatriene

It is striking to note here that the electron density obtained from the FE model (even if it corresponds to a constant potential, i.e., neither attractive nor repulsive locations for the electrons) intrinsically contains the topological elements of bond alternation:

$$\rho(z) = 2\sum_{m}^{occ} \phi_{m}^{2}(z) = \frac{4}{L}\sum_{m}^{occ}\sin^{2}\frac{m\pi z}{L}$$

Due to the increasing number of nodes of orbitals, even polyenes exhibit alternating densities as illustrated by Figure 1.9 in the case of butadiene (2 orbitals, 4 π electrons) and hexatriene (3 orbitals, 6 π electrons); the regions of higher electron densities correspond to the relaxations of bond lengths into stronger, shorter double bonds while the regions of lower densities describe bonds of weaker strength or of longer single bond distances.

Note also that the electron density of the HOMO has roughly the same topology as that of the total electron density, a fact often pointed out in Fukui's theory of frontier orbitals.

1.7 QUALITATIVE LCAO BAND THEORY

If, at first sight, the molecular orbital and solid state representations of Figures 1.5 and 1.6 look rather different, in practice, they are equivalent. This point is more clearly seen when drawing qualitatively the forms of the MO's in a LCAO expansion in terms of atomic orbitals (AO). The number of MO's obtained by the procedure is equal to the number of AO's included in the expansion. Due to first principles, the number of nodes in the orbital (i.e., the number of locations where the orbital vanishes) increases with orbital energy.

Since the π-electron theory is well developed, we choose the case of the polyene series studied in the Hückel approximation. Figure 1.10 is similar to Figure 1.5 complemented by the scheme of the π-molecular orbitals of the carbon backbone to illustrate our point.

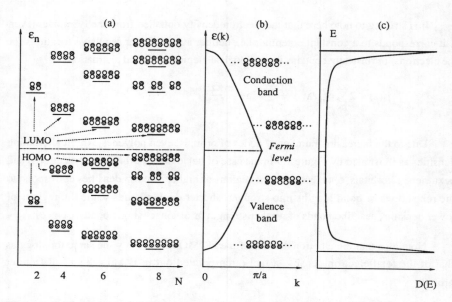

Figure 1.10 Representation of one-electron π energies and molecular or polymeric π orbitals of a regular polyene (polyacetylene) chain
  (a) π molecular levels of alkenes
  (b) π band structure of polyacetylene
  (c) π density of electronic states of polyacetylene

The bonding (no node) and antibonding (one node) π orbitals of ethylene are easily recognized. The four orbitals of butadiene popularized by the work of Woodward and Hoffmann on orbital symmetry are also plotted according to their increasing number of nodes, i.e., their increasing energy. Similar orbital descriptions are found in band theory. For high symmetry points of the first Brillouin zone, the polymer orbitals are real and can be plotted easily. The lowest orbital has no node (symmetric combination of AO's, exp(ikja)=1, if k=0=Γ) while the highest orbital has the maximum number of nodes (most antisymmetric combination, exp(ikja)=(-1)^j if k=π/a=H). Intermediate situations occur near the Fermi energy where, in the case of a regular chain of polyacetylene, degeneracy appears in the structure of the HOMO and the LUMO. In the HOMO, bonding occurs on the double bonds and antibonding on the single bonds; the reverse is true for the LUMO. For other points of the first Brillouin zone, the polymeric orbitals have also imaginary components but still obey the same principles.

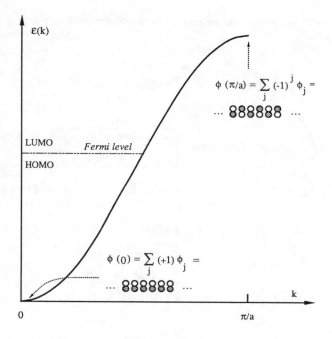

Figure 1.11 Qualitative band structure of a chain with a single π orbital per unit cell

Qualitative molecular orbital techniques are now commonly used to construct π-band structures of conjugated polymers. For example, Figure 1.11 exemplifies the case of an hypothetical chain built from unit cells containing a single π orbital.

The symmetric (bonding) combination of the orbitals is obtained at k=0. The orbital energy increases and the energy band has a positive derivative towards the point k=π/a where the antisymmetric (most antibonding) combination is observed. In Figure 1.12, we consider one unit cell containing two orbitals.

If the orbitals are π-electron atomic orbitals, it corresponds to the case of an all-trans regular polyene where all carbon-carbon bond lengths are set to be equal. The Bloch functions are thus constructed from the well-known π-bonding orbital and π-antibonding orbital of ethylene. With respect to the previous case (a single orbital and a single energy

band), two orbitals are present in the unit cell and two energy bands appear in the first
Brillouin zone.

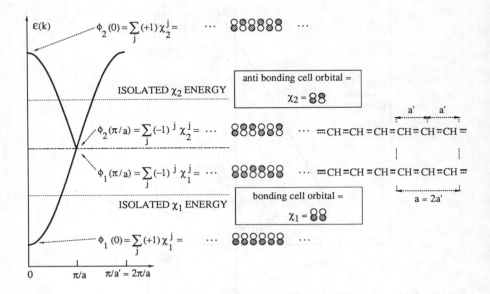

Figure 1.12 Qualitative band structure of a chain with two π orbitals per unit cell

The Brillouin zone is half its original size since the unit cell is twice as large. At k=0,
we observe the positive combination of both basis orbitals while the antisymmetric
combination takes place at the edge of the zone (k=π/a). The lower band has a positive
derivative (increasing number of nodes from the center to the edge of the zone); the higher
band has a negative derivative (decreasing number of nodes from the center to the edge of
the zone). The sign of the derivative of the energy band thus gives an indication on the
bonding and antibonding characters of the various orbitals within the first Brillouin zone.
Figure 1.13 shows the orbital combination of s and p_σ bands.

At k=0, the positive combination of the orbitals means the most bonding situation for
the s orbitals and the most antibonding one for the p_σ. At k=π/a, the opposite situation is
observed, i.e., the most antibonding situation for the s orbitals and the most bonding for
the p_σ ones. The s band has a positive derivative while the p_σ one has a negative one.

Such interpretations are of basic interest when investigating the bonding and antibonding behavior of the orbitals for designing polymers with specific topology as will be mentioned in the applications on highly conducting or optically active macromolecules. Qualitative molecular orbital techniques based on the previous principles are now commonly used to sketch π-band structures of conjugated polymers as illustrated in Section 1.9.5 of the present Chapter.

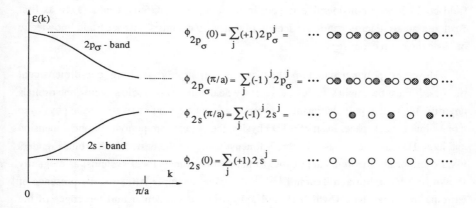

Figure 1.13 Qualitative band structure of a chain with one s orbital and one $2p_\sigma$ orbital per unit cell (reduced zone scheme)

1.8 SPECIFICITY OF POLYMER QUANTUM CHEMISTRY VERSUS SOLID STATE PHYSICS

From a conceptual point of view, it appears that polymer quantum chemistry is an ideal field for cooperation between solid state physicists and molecular quantum chemists. There exists a common interpretation in the discussions concerning orbital energies, orbital symmetry, and gross charges by chemists and solid state physicists who use less familiar terms like first Brillouin zone, dependence of "wave function" (in fact the one-electron wave function is called an orbital by the chemist) with respect to wavenumber k, Fermi surfaces, Fermi contours, and density of states.

Actually, both kinds of scientists use the same quantum mechanical background, the same approximations, and even the same degree of sophistication in order to understand one another. Thus an interesting feature of polymer theoretical chemistry is that its scientific language is partly that of quantum chemistry and partly that of solid state physics. For example, model polymers are described by three-dimensional LCAO orbitals with only one direction of periodicity so that usual concepts of solid state physics such as Bloch functions, Brillouin zones, and energy bands can be applied. Consequently, this field could have a considerable impact in teaching solid state concepts (such as first Brillouin zones, energy bands, or density of states) to chemists with a certain knowledge of basic quantum chemistry.

However, polymer quantum chemistry is not, strictly speaking, one-dimensional physics. It must be emphasized that even if we deal with 1D periodic systems, the orbitals are truly 3D, so that polymer quantum chemistry is not a reduction of solid state physics to a one-dimensional space. In strictly 1D physics, the systems are periodic in one dimension and have 1D wave functions. The three-dimensionality of the basic Bloch orbitals renders invalid important theorems applicable to purely one-dimensional systems, as elegantly shown by McCubbin and Teemull [1.12]. A typical example is that of the presence of extrema in energy bands which should only occur at the center and the edges of the Brillouin zone in a strictly 1D system. For polymers, even in simple cases like the linear zigzag polyethylene chain, some extrema of the energy bands are encountered at arbitrary positions in the first Brillouin zone which are not high symmetry points.

Shockley's theorem (see Peierls book [1.13]) states that the extrema of energy bands in one-dimensional systems may occur only at the center and edges of the reduced Brillouin zone. The term one-dimensional is commonly applied to three types of systems:

(1) a one-dimensional array of one-dimensional potentials,

(2) a one-dimensional chain of three-dimensional "atomic" potentials (such as a linear chain of hydrogen atoms), and

(3) a three-dimensional chain periodic in one dimension only (such as a polyethylene chain).

Shockley's theorem is easily proved in case (1) only. Proofs are given in the classical book on Brillouin zones by Jones [1.14] and in Reference [1.12].

The first demonstration [1.14] is based on the well-known property that a particle in a space-periodic field has average velocity in state k determined by the gradient of the eigenvalue with respect to k [1.15]. In the one-dimensional case, this relation may be written:

$$\frac{dE}{dk} \int_{-a/2}^{a/2} |\phi(k,z)|^2 \, dz = \frac{h^2}{8\pi^2 mi} \int_{-a/2}^{a/2} [\phi^*(k,z) \, \phi'(k,z) - \phi(k,z) \, \phi^{*'}(k,z)] \, dz$$

By writing the general solution $\phi(z)$ as a linear combination of two linearly independent solutions $y_1(z)$ and $y_2(z)$:

$$\phi(z) = A \, y_1(z) + B \, y_2(z)$$

it follows that:

$$[\phi^*(k,z) \, \phi'(k,z) - \phi(k,z) \, \phi^{*'}(k,z)] = (A^*B - B^*A)W$$

where $W = [y_1(z) y_2'(z) - y_1'(z) y_2(z)]$ is the Wronskian of $y_1(z)$ and $y_2(z)$.

According to Bloch's theorem:

$$\phi(a/2) = e^{ika} \phi(-a/2)$$

$$\phi'(a/2) = e^{ika} \phi'(-a/2)$$

and:

$$Ay_1(a/2) = - i \cot(ka/2) \, By_2(a/2)$$

Hence A/B is pure imaginary, $A^*B=-B^*A$ and, therefore:

$$\frac{dE}{dk} = \frac{\dfrac{ah^2}{4\pi^2 mi} \, W \, A^*B}{\displaystyle\int_{-a/2}^{a/2} [|A|^2 y_1^2(z) - |B|^2 y_2^2(z)] \, dz}$$

Since in the range $0<|k|<\pi/a$ neither A nor B can vanish; there can be neither a maximum nor a minimum of $E(k)$ in this range.

Attempts to extend this proof to higher-dimensional chains have not met with any success; it is also equally difficult to establish that the theorem does not hold in these cases. However, it is necessary to find only one example of a linear chain possessing a band extremum within the zone to demonstrate that the theorem is not valid for all types of one-dimensional systems. As mentioned previously, such an example is provided by the band structure of the polyethylene chain exemplified in Figure 1.6. Thus, it is clear that the existence of the minimum is due to a question of "dimensionality" of the chain. The 2D character is introduced by the planar zigzag carbon chain (lying in the x-z plane) and the 3D character is introduced by C-H bonds lying outside this plane. As will be shown later, this energy extremum is mainly due to the interactions of the C-H bonds with the C-C backbone and it will become obvious that the observed minimum corresponds to case (3) mentioned above, i.e., a three-dimensional chain periodic in one dimension only.

1.9   SIMPLE CALCULATIONS OF THE ELECTRONIC STRUCTURE OF CONJUGATED POLYMERS

1.9.1 LCAO Formalism

Accurate solutions of the HF equations have been obtained by numerical procedures only for most of the atoms and some diatomic molecules. It is presently standard to solve the HF equations by expanding the orbitals $\phi_j(\mathbf{r})$ in terms of a set of basis functions $\{\chi_p(\mathbf{r}-\mathbf{P})\}$ centered on the various nuclei of the molecule. The terminology "LCAO" has historically been introduced since, in the early day calculations, Molecular Orbitals (MO) were strictly obtained by Linear Combinations of Atomic Orbitals (LCAO). Nowadays, analytical forms of Atomic Orbitals are also commonly derived by using Linear Combination of Basis Functions (LCBF). These basis functions can in some instances bear little resemblance to the true atomic orbitals. When applying the Self Consistent Field (SCF) scheme, one obtains the so-called SCF-MO-LCAO methodology, illustrated in the left part of Chart 1.4.

Chart 1.4 A comparison of some basic formulas of molecular and polymer quantum chemistry *

Molecular Quantum Chemistry	Polymer Quantum Chemistry

$$\phi_j(\mathbf{r}) = \sum_p c_{jp}\, \chi_p(\mathbf{r}\text{-}\mathbf{P})$$

$$\phi_n(k,\mathbf{r}) = N^{-1/2} \sum_j e^{ikja} \sum_p c_{np}(k)\, \chi_p(\mathbf{r}\text{-}\mathbf{P}\text{-}ja)$$

$$\sum_p c_{jp}\,(h_{pq} - \varepsilon_j\, S_{pq}) = 0$$

$$\sum_p c_{np}(k)\, \{ \sum_j e^{ikja}\, [h_{pq}^j - \varepsilon_n(k)\, S_{pq}^j\,]\}$$

$$\equiv \sum_p c_{np}(k)\, \{\, h_{pq}(k) - \varepsilon_n(k)\, S_{pq}(k)\,\} = 0$$

$$|\,(h_{pq} - \varepsilon\, S_{pq})\,| = 0$$

$$|\{\sum_j e^{ikja}\, [h_{pq}^j - \varepsilon(k)\, S_{pq}^j\,]\}|$$

$$\equiv |\{\, h_{pq}(k) - \varepsilon_n(k)\, S_{pq}(k)\}| = 0$$

$$\rightarrow \quad \varepsilon_1, \varepsilon_2, \varepsilon_3, \ldots$$

$$\rightarrow \quad \varepsilon_1(k),\, \varepsilon_2(k),\, \varepsilon_3(k), \ldots$$

$$\rightarrow \quad D_{pq} = \sum_j c_{jp}\, c_{jq}$$

$$\rightarrow D_{pq}^j = a/2\pi \sum_n \int dk\, c_{np}^*(k) c_{nq}(k)\, e^{ikja}$$

Basic References
J. Ladik, Acta Phys. Hung., **18**, 173 (1965)
J. Ladik, Acta Phys. Hung., **18**, 185 (1965)
J.M. André, L. Gouverneur, and G. Leroy, Int. J. Quantum Chem., **1**, 427 (1967).
J.M. André, L. Gouverneur, and G. Leroy, Int. J. Quantum Chem., **1**, 451 (1967).
G. Del Re, J. Ladik, and G. Biczo, Phys. Rev., **155**, 997 (1967).
J.M. André, J. Chem. Phys., **50**, 1536 (1969).

* reduced version, for extended version, see Chart 2.2

The process would be mathematically exact if it were possible to handle infinite expansions of the molecular orbitals. In practice, the expansion has to be limited to a finite number of basis functions. In the LCAO approach, the search for orbitals in the full three-dimensional space is replaced by the calculation of a finite number of LCAO coefficients.

Turning to applications, it is clear from the previous discussion that the actual procedure must combine equations from molecular quantum chemistry and solid state physics. In molecular quantum chemistry, a molecular orbital is expanded in terms of basis functions; the system of equations and its associated determinant (secular equation) are solved; their eigenvalues are the orbital energies. From the LCAO coefficients, charges and bond orders (projection of the density matrices onto the limited basis used) are calculated.

In polymer quantum chemistry, we take into account the lattice periodicity; the orbitals, the systems of equations and the determinants are no longer real but have imaginary components. This introduces lattice sums which are to be evaluated by adequate procedures. Indeed, from Bloch's theorem, a polymeric orbital has the LCAO form:

$$\phi_n(k,r) = N^{-1/2} \sum_j e^{ikja} \sum_p c_{np}(k)\, \chi_p(r\text{-}P\text{-}ja)$$

By forming the expectation value of a given one-electron operator and by using the variational procedure for getting the best LCAO coefficients $c_{np}(k)$, we obtain for the whole polymer the secular system of equations:

$$\sum_p c_{np}(k)\, \{\sum_j e^{ikja}\, [h_{pq}^{\,j}) - \varepsilon_n(k)\, S_{pq}^{\,j})]\} = 0$$

where the matrix elements $h_{pq}^{\,j}$ and the overlap integrals $S_{pq}^{\,j}$ are respectively defined by:

$$h_{pq}^{\,j} = \int \chi_p(r\text{-}P)\, h(r)\, \chi_q(r\text{-}Q\text{-}ja)\, dr = \int \chi_p(r)\, h(r)\, \chi_q^{\,j}(r)\, dr$$

$$S_{pq}^{\,j} = \int \chi_p(r\text{-}P)\, \chi_q(r\text{-}Q\text{-}ja)\, dr = \int \chi_p(r)\, \chi_q^{\,j}(r)\, dr$$

The compatibility condition of the above system of equations is the vanishing of the secular determinant:

$$|\{\sum_j e^{ikja} [h_{pq}^j - \varepsilon(k) S_{pq}^j]\}| = 0$$

Solving this equation to get the unknown band energies $\varepsilon_n(k)$ produces in the reduced scheme the band structure $\varepsilon(k)$ as a multivalued function of k. We note also that h_{pq}^j is a matrix element of the one-electron operator h between the atomic orbital χ_p centered in the origin unit cell and the atomic orbital χ_q^j centered in cell j. S_{pq}^j has the same meaning for the unit overlap. The key problem is to obtain the exact numerical value or an approximate value of the various matrix elements over the basis functions. Fortunately, it is known from the form of the AO's that these matrix elements should decrease exponentially with the distance between their orbital centers and thus gives rise to a natural convergency of the lattice sums appearing in the secular systems and in the secular determinant. It is to be noted that the dimensions of the matrix equations to be solved are equal to the number of atomic orbitals per unit cell, the effect of the infinite lattice being included in the naturally convergent but formally infinite sums. A comparison of the basic formulas used in molecular and polymer quantum chemistry is given in Chart 1.4.

The previous equations show that in a general way a polymer can be considered as a large molecule. Using this property, it is likely that the usual methods of quantum chemistry could be used to investigate the electronic structure of polymers. As a consequence, it is correct either to solve the equations *ab initio* or to introduce the well-known approximations of molecular quantum chemistry into the formalism of polymeric orbitals.

For instance, we can either consider all the electrons of the system or restrict the number of electrons to be considered, for example, neglecting the core electrons and use approximate expressions for some integrals or some groups of integrals. In the first case, we speak of *ab initio* or nonempirical techniques when considering the whole electronic system and calculating all the necessary integrals in a given model. On the other hand, we use semiempirical or empirical methods when we reduce by physical or mathematical arguments the number of electrons to be considered and the number of integrals to be calculated or evaluated from experimental data. In the next chapter, we analyze the actual methodology and develop the full machinery necessary to compute all valence or all-electron band structures. As it will be shown, the numerical treatment of these techniques requires computer resources ranging from personal computers to supercomputers depending on the type of the approximation introduced. In the next Section, however, we restrict ourselves to a simple empirical model which can be very often solved by hand, does not require important computer resources, and neatly illustrates the basic features of the approach.

1.9.2 Hückel Methodology

The simplest case of an "empirical" model has been independently introduced by Bloch [1.8] in solid state physics where it is known as the "tight binding approximation", and by Hückel [1.16] in quantum chemistry. In the simple Hückel method, only π electrons within purely conjugated organic molecules or chains are taken into account and all interactions except those involving the nearest-neighbours are neglected. This method enjoyed a considerable success due to its simplicity and the validity of its approximations. It is based on the following assumptions.

It is considered that in a conjugated chain the σ electrons form frozen cores with the carbon and hydrogen nuclei. The π electrons are thus attracted by cores of unit effective charge for each unsaturated carbon atom. Due to the orthogonality of the basis functions describing the σ electrons (functions which are symmetric with respect to the molecular plane) and those involving the π electrons (in general, 2p functions which are antisymmetric with respect to the molecular plane), there exists in strictly planar conjugated systems an exact numerical factorization between the matrix elements associated to σ and π electrons. Since the π electrons are less bound to the organic backbones than the σ electrons, it is reasonable to assume that they will be more prone to be involved in physical and chemical processes. To first order, they are responsible for the main electronic properties of the conjugated chain. According to perturbation theory, most molecular properties are determined by energy differences and thus primarily by excitations involving the smallest energy differences. The σ–σ^* energy gaps being large, the corresponding property modifications due to the σ electrons are small compared to the significant perturbations produced by the lower π–π^* excitations. Thus, it is customary to freeze the σ electrons and group them with the ionic backbones of the molecule. However, it is clear that such methods can only be used in calculating those physical properties for which the π electrons are mainly responsible and that the error introduced by the $\sigma-\pi$ separation requires empirical or semiempirical compensation. Implicitly, each π-electron distribution of a conjugated atom should be described by a single $2p_\pi$-atomic orbital. In practice, the Hückel orbitals are not specified. They are not given any precise mathematical form and are moreover abstract quantities. Due to this flexibility, the unspecified basis set is assumed to form an orthonormal set:

$$S_{pq}^{\,j} = \delta_{pq}\,\delta_{0j} \qquad = 1 \text{ for one-center overlap integrals}$$

$$= 0 \text{ for two-center overlap integrals}$$

(where p, q refer to the basis function indices, and 0, j to the unit cell indices).

In the same vein, the Hückel effective one-electron operator h(r) is not specified. Its matrix elements differ from zero only in limited cases and have arbitrary values α and β:

$$h_{pp}^{\,0} = \alpha \qquad \text{for one-center matrix elements; } \alpha \text{ is the site energy often (incorrectly) called Coulomb integral and depends on the type of atom on which the } \pi \text{ orbital is centered;}$$

$$h_{pq}^{\,j} = \beta \qquad \text{for two-center matrix elements associated with chemical bonds; the } \beta\text{'s are usually called resonance integrals or transfer integrals, and depend on the nature of the chemical bond;}$$

$$h_{pq}^{\,j} = 0 \qquad \text{for two-center matrix elements not associated with chemical bonds.}$$

The detailed studies of the 1950's following the development of the Pariser-Parr-Pople methodology have shown that the assumptions used in the Hückel method are coherent. Particularly, the fact that the basis is orthonormal implies that in a similar environment, the one-center matrix element α would have the same numerical value as would also the two-center β integrals.

In a polymeric band structure calculation, the computational scheme is very simple; due to the neglect of nonnearest-neighbour interactions, only the effect of the first translation appears in the secular system and determinant. As a consequence, a very simple k dependence of the π-energy bands is produced. Practical calculations are even easier since the atomic Hückel basis is assumed to be orthogonal; the overlap Bloch sums in those conditions reduce to the δ values (0 or 1):

$$S_{pq}(k) = \sum_j e^{ikja} \, S_{pq}^{\,j} = \sum_j e^{ikja} \, \delta_{pq}\delta_{0j} = \delta_{pq}$$

It is sometimes advocated that there is no longer any need for such a simple theory as the Hückel method with the existence of modern high speed computers and the more sophisticated MO methods now available. However, it must be emphasized that there is still much room for the use of Hückel theory especially as a tool to get some physical insight and as a topological guide into the structure and energy of molecular orbitals and their electron densities. Clear examples will be illustrated in the second part of the book. Its simplicity allows one to develop a powerful "paper and pencil" method and to obtain analytical results easily. In particular, the Hückel method offers the only alternative for obtaining qualitative information on very large systems. In the same way, one of its greatest advantages in the field of polymer chemistry is that analytical interpolation forms for the electronic properties of finite oligomers and infinite polymers can be deduced from closed relationships giving the orbital energies (or possibly their structure) as a function of the number of atoms in the polymeric chains. Examples of such formulas are found in many textbooks [1.17-1.19]. A rather complete analysis in the framework of Hückel methodology is available by the general technique of finite differences [1.20]. For example, the orbital energies of regular chains of N atoms are given in terms of the Hückel parameters α and β by:

$$\varepsilon_j = \alpha + 2\beta \cos(j\pi/N+1) \qquad \text{for linear chains}$$

$$\varepsilon_j = \alpha + 2\beta \cos(2j\pi/N) \qquad \text{for cyclic chains}$$

They correspond to an evolution of the energy gap like:

$$\Delta E = -4\beta \sin(\pi/2(N+1)) \qquad \text{for linear chains}$$

$$\Delta E = -4\beta \sin(\pi/N) \qquad \text{for cyclic chains}$$

For alternant chains of atoms, the mathematical expressions are more elaborate and solutions are, as far as we know, only available for cyclic chains of N atoms and linear chains containing an odd number of N-1 atoms. They read, in both cases, as:

$$\varepsilon_j = \alpha \pm \{ \beta_1{}^2 + \beta_2{}^2 + 2\beta_1\beta_2 \cos (4j\pi/N)\}^{1/2}$$

For the linear case, there also exists one non-bonding orbital at the energy α. It is important to note (that point will be relevant to the discussion developed in the next Section) that in the case of an alternant system, ΔE does not tend to zero as $N\rightarrow\infty$ but instead to a finite value:

$$\lim {}_{N\rightarrow\infty} \Delta E(N) = 2 (\beta_1 - \beta_2)$$

The simplicity of the Hückel method opens up wide possibilities. It is worth noting that the modern interpretation of the high electrical conductivity of doped conjugated polymers in terms of solitons, polarons, or bipolarons was initially formulated in terms of a Hückel model [1.21].

Tables of Hückel parameters [1.22], tables of Hückel molecular orbitals [1.23,1.24], and good books on applications of the Hückel method to organic conjugated systems [1.25-1.29] are widely available. The parameter α of carbon atoms is sometimes taken equal to zero (this merely corresponds to an energy scale shift) while the parameter β is used as the (negative) energy unit. Empirical values of the Hückel parameters are determined by least-square fitting of theoretical and experimental data.

It is striking to note that the value of the parameters depends on the nature of the property considered. Salem [1.27] has taken the example of a series of condensed aromatic hydrocarbons (benzene, naphthalene, anthracene, pentacene,...). From the correlation between the first ionization potential and the negative of the energy of the HOMO (as stated by Koopmans' theorem), a first estimate of $\beta \cong -4$ eV $\cong -90$ kcal/mol is obtained. The same approach yields an empirical value of $\alpha \cong -5.9$ eV $\cong -136$ kcal/mol. On the other hand, the theoretical energy of the first singlet-singlet $\pi-\pi^*$ transition is the difference between the LUMO and HOMO energies. Comparison with experimental data gives a second estimate of $\beta \cong -2.4$ eV $\cong -55$ kcal/mol. Finally, the correlation between the delocalization or resonance energy (i.e., the measure of the molecular extra-stabilization due to electron conjugation with respect to simply additive energy bond scheme) allows for a third estimate of $\beta \cong -0.8$ eV $\cong -18$ kcal/mol. A qualitative reason for the variations in the numerical values of the parameters is that some of the experimental data concern differences of energies while others correspond to absolute (total) energies.

Thus, the large difference between the empirical estimates is due to different combinations of one-electron (kinetic and nuclear attraction) and two-electron (electron repulsion) energy balances in the basic comparison with experiment. A detailed analysis should identify for each type of experiment the terms and the combinations of the SCF-MO-LCAO matrix elements which correspond to α and β in the Hückel theory.

This example is well adapted to call the attention of the user of empirical and semiempirical procedures to the fact that the quality of the selected technique strongly depends on the experimental properties on which they are parameterized and that for given classes of experiments (UV transitions, heats of formation, dipole moments, geometries), different scaling factors or different parameterizations are needed.

1.9.3 A Detailed Example of Hückel Electronic Structure: Polyacetylene

1.9.3.1 Hückel band structure of polyacetylene

The infinite polyene turns out to be a good illustration for the calculation of the band structure of a simple polymeric system. More precisely, we choose the chemical example of an infinite conjugated chain with a single atom in the unit cell or with two atoms per unit cell and, in each case, a lattice vector of length a. Both models are represented in Figure 1.14.

In the first model case, the basis set consists of a single $2p_\pi$-atomic orbital χ^j centered on each of the carbon atoms located in each cell j and serves to form the Bloch orbitals:

$$\phi(k,r) = N^{-1/2} \sum_j e^{ikja} \chi(r-ja)$$

$$= N^{-1/2} \sum_j e^{ikja} \chi^j$$

which are the eigenfunctions of the one-electron equation:

$$h(r)\ \phi(k,r) = \varepsilon(k)\ \phi(k,r)$$

or

$$h(\mathbf{r}) \{N^{-1/2} \sum_j e^{ikja} \chi^j\} = \varepsilon(k) \{N^{-1/2} \sum_j e^{ikja} \chi^j\}$$

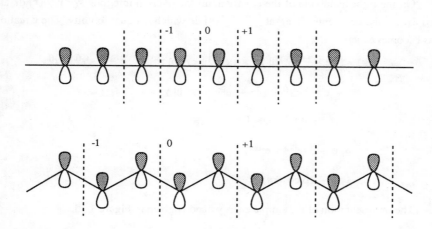

Figure 1.14   π-orbital topology of polyacetylene
      (a) one orbital per unit cell
      (b) two orbitals per unit cell

Multiplying both sides of the equation by χ^0, integrating over d$\mathbf{r}$ and making use of the Hückel approximation, written in this case:

$$\langle \chi^0|h| \chi^0 \rangle = \int \chi^0(\mathbf{r}) \, h(\mathbf{r}) \, \chi^0(\mathbf{r}) \, d\mathbf{r} = \alpha$$
   (one-center integral)

$$\langle \chi^0|h| \chi^1 \rangle = \int \chi^0(\mathbf{r}) \, h(\mathbf{r}) \, \chi^1(\mathbf{r}) \, d\mathbf{r} = \langle \chi^0|h| \chi^{-1} \rangle = \int \chi^0(\mathbf{r}) \, h(\mathbf{r}) \, \chi^{-1}(\mathbf{r}) \, d\mathbf{r} = \beta$$
      (two-center nearest-neighbour integral)

$$\langle \chi^0|h| \chi^j \rangle = \int \chi^0(\mathbf{r}) \, h(\mathbf{r}) \, \chi^j(\mathbf{r}) \, d\mathbf{r} = \langle \chi^0|h| \chi^{-j} \rangle = \int \chi^0(\mathbf{r}) \, h(\mathbf{r}) \, \chi^{-j}(\mathbf{r}) \, d\mathbf{r} = 0$$
   if $j>1$

   (two-center nonnearest-neighbour integral).

we obtain the equation:

$$\sum_j e^{ikja} \int \chi^0(\mathbf{r})\, h(\mathbf{r})\, \chi^j(\mathbf{r})\, d\mathbf{r} = \varepsilon(k) \sum_j e^{ikja} \int \chi^0(\mathbf{r})\chi^j(\mathbf{r})\, d\mathbf{r}$$

On the right-hand side of the equation, all the overlap integrals $\int \chi^0(\mathbf{r})\chi^j(\mathbf{r})\, d\mathbf{r}$ are zero except the one-center integral $\int \chi^0(\mathbf{r})\chi^0(\mathbf{r})\, d\mathbf{r}$ which is equal to unity. The equation thus becomes exactly:

$$\varepsilon(k) = \ldots e^{-ik2a} \langle\chi^0|h|\chi^{-2}\rangle + e^{-ik1a}\langle\chi^0|h|\chi^{-1}\rangle + e^{-ik0a}\langle\chi^0|h|\chi^0\rangle$$

$$+ e^{ik1a}\langle\chi^0|h|\chi^1\rangle + e^{ik2a}\langle\chi^0|h|\chi^2\rangle + \ldots$$

$$= \alpha + e^{-ika}\,\beta + e^{ika}\,\beta$$

$$= \alpha + \beta\,(e^{ika} + e^{-ika})$$

$$= \alpha + 2\beta\,\cos(ka)$$

The representation of the simple energy band is given in Figure 1.15.

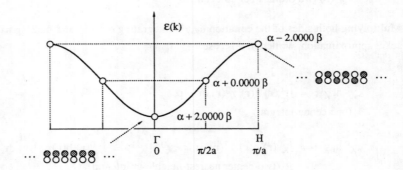

Figure 1.15 Hückel π-band structure of a regular polyacetylene (a single π orbital per unit cell)

It is easy to find the energy for specific points in the first Brillouin zone:

$$\varepsilon(k=0) = \varepsilon(\Gamma) = \alpha + 2\beta$$

$$\varepsilon(k=\pm\pi/2a) = \alpha$$

$$\varepsilon(k=\pm\pi/a) = \varepsilon(H) = \alpha - 2\beta$$

while the Bloch orbitals at the Γ and H points are explicitly:

$$\phi(k=0,r) = \phi(\Gamma,r) = N^{-1/2} \sum_j e^{i0ja} \chi^j \propto \sum_j \chi^j$$

$$\phi(k=\pm\pi/a,r) = \phi(H,r) = N^{-1/2} \sum_j e^{ij\pi} \chi^j \propto \sum_j (-1)^j \chi^j$$

Figure 1.15 shows the structure of those two Bloch orbitals at high symmetry points. It is clearly seen that the first is the maximum bonding orbital while the second is the maximum antibonding orbital as already qualitatively presented in Section 1.7.

In the second case, the basis set consists of two $2p_\pi$-atomic orbitals centered on each of the carbon atoms. For labeling purposes, we call χ_1 the one centered on the lower left carbon atom in Figure 1.14 and χ_2 the other one. From Bloch's theorem, we deduce that a good trial polymeric orbital for the π system has the form:

$$\phi_n(k,r) = N^{-1/2} \sum_j e^{ikja} \sum_p c_{np}(k) \chi_p(r-P-ja)$$

$$= N^{-1/2} \sum_j e^{ikja} \{ c_{n1}(k) \chi_1(r-ja) + c_{n2}(k) \chi_2(r-ja) \}$$

The secular system of equations:

$$\sum_p c_{np}(k) \{ \sum_j e^{ikja} [h_{pq}^j - \varepsilon_n(k) S_{pq}^j] \} = 0$$

explicitly consists of the set of two equations:

$$c_{n1}(k)\{\sum_j e^{ikja} [h_{11}^j - \varepsilon_n(k) S_{11}^j]\} + c_{n2}(k) \{\sum_j e^{ikja} [h_{12}^j - \varepsilon_n(k) S_{12}^j]\} = 0$$

$$c_{n1}(k)\{\sum_j e^{ikja} [h_{21}^j - \varepsilon_n(k) S_{21}^j]\} + c_{n2}(k) \{\sum_j e^{ikja} [h_{22}^j - \varepsilon_n(k) S_{22}^j]\} = 0$$

When using the Hückel approximation (neglect of all but the nearest-neighbour interactions) and defining the matrix elements:

$$h_{11}^j = h_{22}^j = \int \chi_1(\mathbf{r})\, h(\mathbf{r})\, \chi_1^j(\mathbf{r})\, d\mathbf{r} = \int \chi_2(\mathbf{r})\, h(\mathbf{r})\, \chi_2^j(\mathbf{r})\, d\mathbf{r} = \alpha\, \delta_{0j}$$

$$h_{12}^j = \int \chi_1(\mathbf{r})\, h(\mathbf{r})\, \chi_2^j(\mathbf{r})\, d\mathbf{r} = \beta$$

if both the atoms on which χ_1 and χ_2 are centered are nearest-neighbours (h_{12}^0, h_{12}^{-1}, h_{21}^0 and h_{21}^1);

$$h_{12}^j = \int \chi_1(\mathbf{r})\, h(\mathbf{r})\, \chi_2^j(\mathbf{r})\, d\mathbf{r} = 0 \text{ otherwise.}$$

For the evaluation of the overlap matrix elements, the basis is assumed to be orthonormal with respect to each atomic center:

$$S_{pq}^j = \int \chi_p(\mathbf{r})\, \chi_q^j(\mathbf{r})\, d\mathbf{r} = \delta_{pq}\, \delta_{0j}$$

In those conditions, the Hückel secular system of the π electrons for the whole polyacetylene chain is written as:

$$c_{n1}(k)\, \{\alpha - \varepsilon_n(k)\} + c_{n2}(k)\, \beta\, \{1 + e^{-ika}\} = 0$$

$$c_{n1}(k)\, \beta\, \{1 + e^{ika}\} + c_{n2}(k)\, \{\alpha - \varepsilon_n(k)\} = 0$$

and its associated secular determinant:

$$\begin{vmatrix} \alpha - \varepsilon_n(k) & \beta\, \{1 + e^{-ika}\} \\[2mm] \beta\, \{1 + e^{ika}\} & \alpha - \varepsilon_n(k) \end{vmatrix} = 0$$

yields the two π-energy bands of the chain:

$$\varepsilon_1(k) = \alpha + \beta\, \sqrt{2 + 2\cos(ka)} \quad = \alpha + 2\beta\cos\left(\frac{ka}{2}\right)$$

$$\varepsilon_2(k) = \alpha - \beta\, \sqrt{2 + 2\cos(ka)} \quad = \alpha - 2\beta\cos\left(\frac{ka}{2}\right)$$

Clearly the region $0 \leq k \leq \pi/a$ is the only region of the reciprocal space which is necessary for nonredundant information. The representation of the energy as a function of wavenumber k is shown in Figure 1.16. For the sake of clarity, we compare in Chart 1.5 (see Section 1.9.4) the actual relation between the methods used for molecules and for polymers by summarizing the corresponding equations for ethylene (molecule) and polyacetylene (polymer).

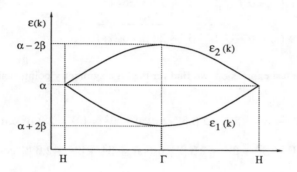

Figure 1.16 Hückel π-band structure of a regular polyacetylene (two π orbitals per unit cell)

Now, in practice, it is found that polyacetylene is formed by a non-regular sequence of carbon-carbon bonds (i.e., there exists an alternation between single-like and double-like bonds) and that a first correction to the simple Hückel scheme would be to distinguish between the nearest-neighbour interactions:

$$h_{12}^0 = h_{21}^0 = \int \chi_1(r) \, h(r) \, \chi_2^0(r) \, dr = \beta'$$

$$h_{12}^1 = h_{21}^1 = \int \chi_1(r) \, h(r) \, \chi_2^1(r) \, dr = \beta$$

In this case, the compatibility condition becomes:

$$\begin{vmatrix} \alpha - \varepsilon_n(k) & \{\beta' + \beta \, e^{-ika}\} \\ \{\beta' + \beta \, e^{ika}\} & \alpha - \varepsilon_n(k) \end{vmatrix} = 0$$

and yields the two π-energy bands of the chain as:

$$\varepsilon_1(k) = \alpha + \sqrt{\beta^2 + \beta'^2 + 2\,\beta\beta' \, \cos(ka)}$$

$$\varepsilon_2(k) = \alpha - \sqrt{\beta^2 + \beta'^2 + 2\,\beta\beta' \, \cos(ka)}$$

From the last expression, we find for the high symmetry points that, at $k=0$ (Γ) and $k=\pi/a$ (H):

$$\Gamma \rightarrow \beta^2 + \beta'^2 + 2\,\beta\,\beta' \, \cos(ka) = \beta^2 + \beta'^2 + 2\,\beta\,\beta' \, \cos 0 = |\beta + \beta'|^2$$

$$H \rightarrow \beta^2 + \beta'^2 + 2\,\beta\,\beta' \, \cos(ka) = \beta^2 + \beta'^2 + 2\,\beta\,\beta' \, \cos \pi = |\beta - \beta'|^2$$

and, consequently:

$$\varepsilon(\Gamma) = \alpha \pm |\beta + \beta'|$$

$$= \alpha \pm 2\beta \qquad\qquad\qquad \text{if } \beta = \beta'$$

and:

$$\varepsilon(H) = \alpha \pm |\beta - \beta'|$$
$$= \alpha \qquad\qquad\qquad\qquad \text{if } \beta = \beta'$$

Figure 1.17 represents the differences between both situations: a polyene with single and double bond alternation ($\beta \neq \beta'$) and a regular polyene ($\beta = \beta'$).

In the latter case, there exists conduction states at infinitesimal distance in energy from the top of the occupied levels. There is no energy gap. This is the typical case of a metal. Should an energy gap exist, a finite excitation energy is required to carry the electrons over that gap. If the energy gap is larger than a limit conventionally set at about 2 eV, the polymer is an insulator. If the energy gap is lower than 2 eV, then a small but non negligible density of electrons can be thermally excited into the upper (conduction) band; the material then exhibits observable electrical conductivity which increases with temperature. The polymer is then a semiconductor.

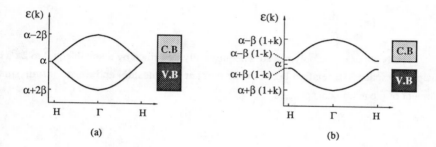

Figure 1.17 Hückel π-band structure (valence band = V.B., conduction band = C.B.) of (a) a regular polyacetylene, and (b) a bond alternant polyacetylene

Note also in Figure 1.17 the analogy which exists between the valence band of an infinite chain and the occupied molecular orbitals of oligomers and between the conduction band and the unoccupied MO's. Furthermore, it is well known from Koopmans' theorem that the energy of the HOMO corresponds, to a good approximation, to the first ionization potential (IP) of the molecule and is the top of the valence band for the polymer. Similarly, the energy of the LUMO is the electron affinity (EA) of the molecule which relates to the bottom of the conduction band of the polymer. Since the simple Hückel method neglects the modifications in the repulsion and exchange terms for the electrons when passing from the ground state to an excited state, the energy difference between the highest occupied and the lowest unoccupied levels is a measure of the first UV transition; in polymers, the same difference (i.e., the energy gap) can be estimated from the difference between the ionization potential and the electron affinity:

$$\Delta E = IP - EA$$

1.9.3.2 Hückel density of states of polyacetylene

Besides the representation of molecular levels and its contraction into energy band schemes, the right-hand side of Figures 1.5 and 1.6 shows a third representation which is the so-called density of states function D(E) of the band. The density function D(E) indicates the variation of the number of allowed energy levels with energy, i.e., the number of allowed electron states per unit energy range:

$$D(E) = \frac{dN}{dE}$$

The N states of the polymer are unequivocally described by a length equal to $2\pi/a$ of reciprocal space, the relation between the number of states dN and dk, an infinitesimal element of the reciprocal space, is:

$$dN = \frac{Na}{2\pi} dk$$

It follows that the density of states is calculated from:

$$D(E) = \frac{dN}{dE} = \frac{(\frac{dN}{dk})}{(\frac{dE}{dk})} = \frac{Na}{2\pi} \frac{1}{(\frac{dE}{dk})} = \frac{Na}{2\pi} (\frac{dk}{dE})$$

From the previous calculation on a regular polyacetylene chain, we have obtained the Hückel expression of the valence π band:

$$\varepsilon(k) = \varepsilon_1(k) = \alpha + 2\,\beta \cos(ka/2)$$

and:

$$\frac{dE}{dk} = -a\beta \sin \frac{ka}{2} = -a\beta \sin\{\arccos \frac{E-\alpha}{2\beta}\}$$

The density of states is directly computed from the previous expressions:

$$D(E) = -\frac{N}{2\pi\beta} \frac{1}{\sin\{\arccos \dfrac{E-\alpha}{2\beta}\}}$$

As expected N(E) is larger than zero since β is a negative quantity and is proportional to the size N of the polymer, i.e., to the number of unit cells. This one-dimensional density of states is represented on the right-hand side of Figure 1.5. In Figure 1.18, we represent $\varepsilon(k)$, dE/dk and D(E). Note that we observe a singularity at $\varepsilon(0) = \alpha + 2\beta$.

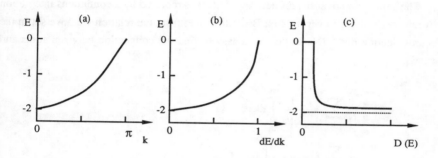

Figure 1.18 Regular polyacetylene:
        (a) Hückel π-valence band $\varepsilon(k)$
        (b) Hückel π-valence band first derivative $dE(k)/dk$
        (c) Hückel π-valence density of states D(E)

This is a consequence of the energy band properties: $\varepsilon(k) = \varepsilon(-k)$ and $d\varepsilon(k)/dk = 0$, leading to the presence of characteristic discontinuities in D(E) for one-dimensional systems. However, the area under the D(E) curve remains finite and is equal to the number of electronic states in the band, namely N; each unit cell containing 2 πelectrons, the N states are thus occupied by two electrons of opposite spins:

$$N = \int_{\alpha+2\beta}^{\alpha} D(E)\, dE = -\frac{N}{2\pi\beta} \int_{\alpha+2\beta}^{\alpha} \frac{1}{\sin(ka/2)}\, dE = \frac{Na}{2\pi} \int_{-\pi/a}^{+\pi/a} dk = N$$

1.9.3.3 Hückel population analysis of polyacetylene

Traditional Hückel calculations yield the shapes of the molecular orbitals as well as their energies. It is thus feasible to compute all data related to the nature of the polymeric orbitals. In molecules, the MO energy levels are discrete and constitute a finite set. If we gradually go to the infinite system, the number of elementary cells tends to ∞. The wavenumber k becomes a continuous variable within the interval $[-\pi/a, +\pi/a]$.

The summation over the discrete set of MO's is replaced by a continuous integration between $-\pi/a$ and $+\pi/a$ over the first Brillouin zone; from the relation we have set in the previous Section $dN = (Na/2\pi) \, dk$, it is easy to find the connection between sums and integrals:

$$\frac{1}{N} \sum_k = \frac{1}{2\pi/a} \int_{-\pi/a}^{+\pi/a} dk$$

$$\frac{1}{N} \sum_k f_k = \frac{1}{2\pi/a} \int_{-\pi/a}^{+\pi/a} f(k) \, dk$$

Let us exemplify in the case of regular polyacetylene the way of calculating the LCAO coefficients of the polymeric orbitals. By putting:

$$\omega = \beta \{1 + e^{ika}\}$$

$$\omega^* = \beta \{1 + e^{-ika}\}$$

the secular system of equations is rewritten as:

$$c_{n1}(k) \{\alpha - \varepsilon_n(k)\} + c_{n2}(k) \, \omega^* = 0$$

$$c_{n1}(k) \, \omega + c_{n2}(k) \{\alpha - \varepsilon_n(k)\} = 0$$

with:

$$|\omega| = \sqrt{2 + 2 \cos ka} \; \beta = 2 \cos \frac{ka}{2} \beta$$

and the two π-energy bands of the chain:

$$\varepsilon_1(k) = \alpha + \beta \sqrt{2 + 2 \cos(ka)} \;\;=\; \alpha + 2\,\beta \cos \frac{ka}{2} = \alpha + |\omega|$$

$$\varepsilon_2(k) = \alpha - \beta \sqrt{2 + 2 \cos(ka)} \;\;=\; \alpha - 2\,\beta \cos \frac{ka}{2} = \alpha - |\omega|$$

The normalization condition of the polymeric orbitals:

$$\int \phi_n^*(k,r)\, \phi_n(k,r)\, dr = \int |\phi_n(k,r)|^2\, dr = 1$$

combined with the secular system of equations gives:

$$c_{k11} = c_{k12} \frac{\omega^*}{|\omega|} = \frac{1}{\sqrt{2}}$$

and:

$$\phi_{kl} = \frac{1}{\sqrt{N}} \left\{ \ldots\, e^{-ika} \left[\frac{1}{\sqrt{2}} \chi_1^{-1} + \frac{1}{\sqrt{2}} \chi_2^{-1} \frac{\omega}{|\omega|} \right] \right.$$

$$+ \left[\frac{1}{\sqrt{2}} \chi_1^{0} + \frac{1}{\sqrt{2}} \chi_2^{0} \frac{\omega}{|\omega|} \right]$$

$$\left. + e^{+ika} \left[\frac{1}{\sqrt{2}} \chi_1^{1} + \frac{1}{\sqrt{2}} \chi_2^{0} \frac{\omega}{|\omega|} \right] + \ldots \right.$$

It is easy to deduce the charges which, in molecules are the sum over the occupied MO's of the square of the LCAO coefficients:

$$q_1 = 2 \sum_k \frac{1}{N} \left(\frac{1}{\sqrt{2}} \right)^2 = \frac{2}{2\pi/a} \int_{-\pi/a}^{+\pi/a} \frac{1}{2}\, dk = 1$$

$$q_2 = 2 \sum_k \frac{1}{N} \left(\frac{\omega^*}{\sqrt{2}\,|\omega|} \right)\left(\frac{\omega}{\sqrt{2}\,|\omega|} \right) = \frac{2}{2\pi/a} \int_{-\pi/a}^{+\pi/a} \frac{1}{2}\, dk = 1$$

and the bond orders which are, in molecules, the sum over the occupied MO's of products of LCAO coefficients. In regular polyacetylene, they are written between adjacent atoms as:

$$l_{12} = 2 \sum_k \frac{1}{N} (\frac{1}{\sqrt{2}})^2 \frac{\omega}{|\omega|} = \frac{2}{2\pi/a} \int_{-\pi/a}^{+\pi/a} \frac{1}{2} \cos \frac{ka}{2} \, dk = \frac{2}{\pi} = 0.6366$$

Note that the Hückel charges are unity, a result well known in all alternant hydrocarbons. The Hückel bond orders between two neighbour atoms are 0.6366. This value is significantly smaller than in ethylene (l_{12}= 1.0000) and slightly smaller than in benzene (l_{12}= 2/3 = 0.6666). On that basis, regular polyacetylene would have, in a first approximation, bond lengths of 1.395 Å if we assume the linear relationship:

$$r_{pq}(Å) = 1.49 - 0.15 \, l_{pq}$$

between bond lengths and bond order (in ethylene, r(Å)=1.34; in benzene, r(Å)= 1.39)

1.9.3.4 Hückel effective electronic mass of polyacetylene

It has been shown in Section 1.6 that the free electron energy (FE) in a constant potential is given by the simple harmonic equation:

$$\varepsilon_n^{FE}(k) = \frac{h^2 k^2}{8\pi^2 m}$$

It is often useful in more complex methodologies to approximate the electronic energy by the same parabolic dependence of the energy with respect to k , even for fairly complex one-electron potentials and consequently to allow the effective electron mass, m*, to take values different from the real electron mass. The effect of the non constant part of the electron-nucleus attraction potential and of the electron-electron Coulomb and exchange interactions is to modify the relationship between the electron band energy, $\varepsilon_n(k)$, and the k location in the first Brillouin zone:

$$\varepsilon_n(k) = \frac{h^2 k^2}{8\pi^2 m^*(k)}$$

In Newton's law, mass is defined as the constant of proportionality between force and acceleration. The correspondence principle in quantum mechanics states that the wave-packet solution of the Schrödinger equation follows the same trajectory as a classical particle which obeys the equation of motion of the corresponding classical Hamiltonian. From those first principles, the effective mass m* is simply related to the curvature of the energy band:

$$\frac{1}{m^*(k)} = \frac{4\pi^2}{h^2} \frac{d^2\varepsilon(k)}{dk^2}$$

In the FE model, by substituting values from the free electron, we deduce immediately that $m^*(k)$ is constant and equal to the actual electron mass:

$$m^{*FE}(k) = m$$

It is easy to compute the electronic effective mass for the various models of polyacetylene introduced in Section 9.3.1.

In the model of an infinite conjugated chain with a single atom in the unit cell, the curvature of the energy band is given by:

$$\frac{d^2\varepsilon}{dk^2} = -2\beta a^2 \cos(ka)$$

and the effective mass is then:

$$m^*(k) = -\frac{h^2}{8\pi^2\beta a^2 \cos(ka)} = \frac{h^2}{8\pi^2|\beta|a^2} \sec(ka)$$

Note that, since β is a negative quantity, the electronic effective mass is positive if $0 \leq k < \pi/2a$ and negative if $\pi/2a < k \leq \pi/a$. It becomes infinite at the Fermi level, $k_F = \pm \pi/2a$. Its k-dependence is illustrated in Figure 1.19 (a).

An infinite effective mass corresponds to an infinite inertia of the electron within the band which is then unable to move. On the other side, an electron with a negative mass is equivalent to a particle with a positive charge. Since their properties are so unlike those of ordinary particles, it is easier to invert the arguments and discuss the properties of "holes" in the electronic bands. The positive effective mass near the bottom of the valence band can also be obtained more easily by the equivalence between the FE and LCAO formulas expanding the trigonometric function in series near the value k=0:

$$\varepsilon_n(k) = \frac{h^2 k^2}{8\pi^2 m^*(k)} + V_0$$

$$= \alpha + 2\beta \cos(ka)$$

$$= \alpha + 2\beta \left\{ 1 - \frac{k^2 a^2}{2} + \dots \right\}$$

$$= \alpha + 2\beta - \beta k^2 a^2 + \dots$$

with the result:

$$m^*(k \to 0) = - \frac{h^2}{8\pi^2 \beta a^2}$$

The order of magnitude of m^* is easily estimated from the value of the interdistance a (of the order of 1.395 Å as indicated in Section 9.3.3. and from the range of β values (0.8 eV $< \beta < 4$ eV) given in Section 9.2. At the bottom of the valence band, one finds: $4.5 \cdot 10^{-28}$ g $< m^* < 22.0 \cdot 10^{-28}$ g to be compared to the rest mass of the free electron ($m = 9.1 \cdot 10^{-28}$ g). The fact that the effective mass of a "conductive" regular polyene is infinite at the Fermi level is an artefact of the nearest-neighbour approximation used in the Hückel method.

The effective masses of the valence $\{m_1^*(k)\}$ and conduction $\{m_2^*(k)\}$ bands of a model of a regular polyacetylene with two atoms (and two π orbitals) per unit cell gives the same result if we note that the translation unit of the latter system a' is twice that of the previous system (a'=2a):

$$m_1^*(k) = - \frac{h^2}{2\pi^2 \beta a'^2} \sec(ka'/2)$$

$$m_2^*(k) = + \frac{h^2}{2\pi^2 \beta a'^2} \sec(ka'/2)$$

Their plot is given in Figure 1.19 (b). The k-dependence of the effective masses of a bond alternant polyacetylene are more elaborate. After a straightforward derivation, they are written as:

$$m_1{}^*(k) = -\frac{h^2}{4\pi^2 \beta \beta' a^2} \ \frac{\{\beta^2 + \beta'^2 + 2\beta\beta'\cos(ka)\}^{3/2}}{(\beta^2 + \beta'^2)\cos(ka) + \beta\beta'\cos^2(ka) + \beta\beta'}$$

$$m_2{}^*(k) = +\frac{h^2}{4\pi^2 \beta \beta' a^2} \ \frac{\{\beta^2 + \beta'^2 + 2\beta\beta'\cos(ka)\}^{3/2}}{(\beta^2 + \beta'^2)\cos(ka) + \beta\beta'\cos^2(ka) + \beta\beta'}$$

The plot of $m_1{}^*(k)$ is given in Figure 1.19 (c).

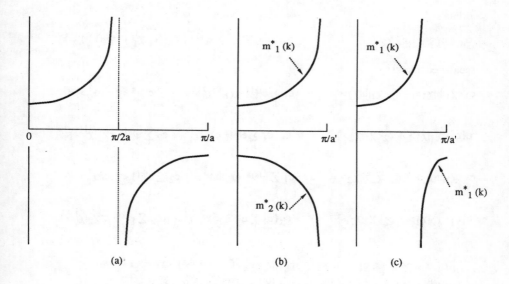

Figure 1.19 Hückel effective electronic mass of the π-band structure
(a) of a regular polyacetylene (a single π orbital per unit cell)
(b) of a regular polyacetylene (two π orbitals per unit cell)
(c) of an alternant trans-polyacetylene, only $m_1{}^*(k)$ is plotted; $m_2{}^*(k)$ is obtained by changing the sign.

1.9.4 Other Examples of Hückel Electronic Structure

Chart 1.5 relates the Hückel equations of ethylene (molecule) to those of polyacetylene (polymer). Its purpose is to exemplify in a simple case the path from molecular quantum chemistry to polymer quantum chemistry and to draw the band structure from the knowledge of the MO energy levels of the monomer.

Chart 1.5 **A comparison of some basic formulas of molecular and polymer quantum chemistry applied to ethylene and polyacetylene in the Hückel methodology**

Molecular Quantum Chemistry
Ethylene: H-(CH=CH)-H

Polymer Quantum Chemistry
Polyacetylene: H-(CH=CH)$_x$-H

---

orbitals:

$$\phi_n = c_1 \chi_1 + c_2 \chi_2$$

$$\phi_n(k) = N^{-1/2} \sum_j e^{ikja} \{ c_{n1}(k) \chi_1^j + c_{n2}(k) \chi_2^j \}$$

secular equations:

$$c_1 <\chi_1|h|\chi_1> + c_2 <\chi_1|h|\chi_2> =$$

$$c_1 \sum_j e^{ikja} <\chi_1|h|\chi_1^j> + c_2 \sum_j e^{ikja} <\chi_1|h|\chi_2^j> =$$

$$\varepsilon\{c_1<\chi_1|\chi_1> + c_2<\chi_1|\chi_2>\}$$

$$\varepsilon_k\{c_1 \sum_j e^{ikja} <\chi_1|\chi_1^j> + c_2 \sum_j e^{ikja} <\chi_1|\chi_2^j>\}$$

$$c_1<\chi_2|h|\chi_1> + c_2<\chi_2|h|\chi_2> =$$

$$c_1 \sum_j e^{ikja} <\chi_2|h|\chi_1^j> + c_2 \sum_j e^{ikja} <\chi_2|h|\chi_2^j> =$$

$$\varepsilon\{c_1<\chi_2|\chi_1> + c_2<\chi_2|\chi_2>\}$$

$$\varepsilon_k\{c_1 \sum_j e^{ikja} < \chi_2 |\chi_1^j> + c_2 \sum_j e^{ikja} < \chi_2|\chi_2^j>\}$$

or

$$c_1 \alpha + c_2 \beta = \varepsilon c_1$$
$$c_1 \beta + c_2 \alpha = \varepsilon c_2$$

$$c_{k1} \alpha + c_{k2} \beta(1 + e^{-ika}) = \varepsilon_k c_{k1}$$
$$c_{k1} \beta(1 + e^{ika}) + c_{k2} \alpha = \varepsilon_k c_{k2}$$

or

$$c_1 (\alpha - \varepsilon) + c_2 \beta = 0$$
$$c_1 \beta + c_2 (\alpha - \varepsilon) = 0$$

$$c_{k1} (\alpha - \varepsilon_k) + c_{k2} \beta(1 + e^{-ika}) = 0$$
$$c_{k1} \beta(1 + e^{ika}) + c_{k2} (\alpha - \varepsilon_k) = 0$$

Chart 1.5 (continued)

secular determinant:

$$\begin{vmatrix} (\alpha - \varepsilon) & \beta \\ \beta & (\alpha - \varepsilon) \end{vmatrix} = 0 \qquad \begin{vmatrix} (\alpha - \varepsilon_k) & \beta(1 + e^{-ika}) \\ \beta(1 + e^{ika}) & (\alpha - \varepsilon_k) \end{vmatrix} = 0$$

energies:

$$\varepsilon_1 = \alpha + \beta \qquad\qquad \varepsilon_{k1} = \alpha + \beta \sqrt{2 + 2\cos(ka)}$$

$$\varepsilon_2 = \alpha - \beta \qquad\qquad \varepsilon_{k2} = \alpha - \beta \sqrt{2 + 2\cos(ka)}$$

The simple methodology we have introduced previously in the case of polyacetylene can also be extended as an exercise to other simple cases of potentially interesting compounds such as polyacenes and polyparaphenylenes (PPP). The solutions are given in Charts 1.6 and 1.7. The qualitative drawing of the energy bands from the MO energy levels of the monomers is detailed in Section 1.9.5.

Chart 1.6 Hückel band structure of polyacenes

---

polymeric scheme and topology:

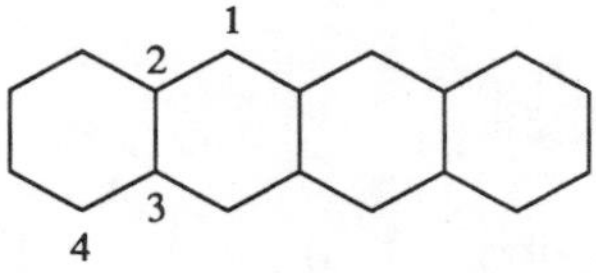

---

trial polymeric orbital:

$$\phi_n(k,r) = N^{-1/2} \sum_j e^{ikja} \sum_p c_{np}(k) \, \chi_p(\mathbf{r}-\mathbf{P}-ja)$$

$$= N^{-1/2} \sum_j e^{ikja} \{ c_{n1}(k) \, \chi_1(\mathbf{r}-ja) + c_{n2}(k) \, \chi_2(\mathbf{r}-ja) + c_{n3}(k) \, \chi_3(\mathbf{r}-ja)$$

$$+ c_{n4}(k) \, \chi_4(\mathbf{r}-ja) \}$$

Chart 1.6 (continued)

non-zero symbolic matrix elements:

$$h_{11}^0 = h_{22}^0 = h_{33}^0 = h_{44}^0 = \alpha$$

$$h_{12}^0 = h_{21}^0 = h_{12}^1 = h_{21}^{-1} = h_{23}^0 = h_{32}^0 = h_{34}^1 = h_{34}^0 = h_{43}^0 = h_{43}^{-1} = \beta$$

---

secular system of equations:

$$\sum_p c_{np}(k) \left\{ \sum_j e^{ikja} [h_{pq}^j - \varepsilon_n(k) S_{pq}^j] \right\} = 0$$

$$c_{n1}(k)\{\alpha - \varepsilon_n(k)\} + c_{n2}(k)\beta\{1+e^{ika}\} \qquad\qquad = 0$$

$$c_{n1}(k)\beta\{1+e^{-ika}\} + c_{n2}(k)\{\alpha - \varepsilon_n(k)\} + c_{n3}(k)\beta \qquad = 0$$

$$c_{n2}(k)\beta \qquad + c_{n3}(k)\{\alpha - \varepsilon_n(k)\} + c_{n4}(k)\beta\{1+e^{ika}\} = 0$$

$$c_{n3}(k)\beta\{1+e^{-ika}\} + c_{n4}(k)\{\alpha - \varepsilon_n(k)\} = 0$$

---

secular determinant:

$$
\begin{vmatrix}
\alpha - \varepsilon_n(k) & \beta\{1 + e^{ika}\} & 0 & 0 \\
\beta\{1 + e^{-ika}\} & \alpha - \varepsilon_n(k) & \beta & 0 \\
0 & \beta & \alpha - \varepsilon_n(k) & \beta\{1 + e^{ika}\} \\
0 & 0 & \beta\{1 + e^{-ika}\} & \alpha - \varepsilon_n(k)
\end{vmatrix} = 0
$$

Chart 1.6 (continued)

π-energy bands:

$$\varepsilon_{kn} = \alpha \pm \beta \sqrt{\frac{1 + 8\cos^2(\frac{ka}{2}) \pm \sqrt{1 + 16\cos^2(\frac{ka}{2})}}{2}}$$

$$= \alpha \pm \beta \sqrt{\frac{5 + 4\cos(ka) \pm \sqrt{9 + 8\cos(ka)}}{2}}$$

or, after some trigonometric manipulations:

$$\varepsilon_{kn} = \alpha \pm \beta \frac{1 \pm \sqrt{1 + 16\cos^2(\frac{ka}{2})}}{2}$$

$$= \alpha \pm \beta \frac{1 \pm \sqrt{9 + 8\cos(ka)}}{2}$$

High symmetry points values:

Γ $\quad\quad\quad\quad\quad\quad \varepsilon(\Gamma) = \alpha \pm 2.5615\,\beta$

$\quad\quad\quad\quad\quad\quad\quad\quad \varepsilon(\Gamma) = \alpha \pm 1.5615\,\beta$

H $\quad\quad\quad\quad\quad\quad \varepsilon(H) = \alpha \pm 1.0000\,\beta$

$\quad\quad\quad\quad\quad\quad\quad\quad \varepsilon(H) = \alpha \pm 0.0000\,\beta$

Note: The π-energy band structure is given in Figure 1.22.

Chart 1.7   Hückel band structure of polyparaphenylenes

polymeric scheme and topology:

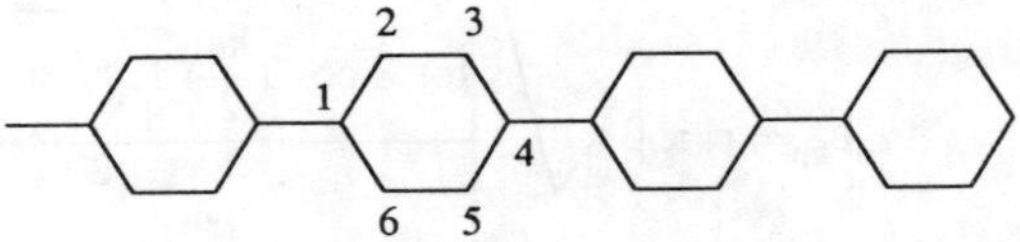

trial polymeric orbital:

$$\phi_n(k,\mathbf{r}) = N^{-1/2} \sum_j e^{ikja} \sum_p c_{np}(k)\, \chi_p(\mathbf{r}\text{-}\mathbf{P}\text{-}ja)$$

$$= N^{-1/2}\sum_j e^{ikja} \{c_{n1}(k)\,\chi_1(\mathbf{r}\text{-}ja) + c_{n2}(k)\,\chi_2(\mathbf{r}\text{-}ja) + c_{n3}(k)\,\chi_3(\mathbf{r}\text{-}ja)$$

$$+ c_{n4}(k)\,\chi_4(\mathbf{r}\text{-}ja) + c_{n5}(k)\,\chi_5(\mathbf{r}\text{-}ja) + c_{n6}(k)\,\chi_6(\mathbf{r}\text{-}ja)\}$$

non-zero symbolic matrix elements:

$$h_{11}^0 = h_{22}^0 = h_{33}^0 = h_{44}^0 = h_{55}^0 = h_{66}^0 = \alpha$$

$$h_{12}^0 = h_{21}^0 = h_{23}^0 = h_{32}^0 = h_{34}^0 = h_{43}^0$$

$$= h_{45}^0 = h_{54}^0 = h_{56}^0 = h_{65}^0 = h_{16}^0 = h_{61}^0 = h_{41}^1 = h_{14}^{-1} = \beta$$

secular system of equations:

$$\sum_p c_{np}(k)\,\{\sum_j e^{ikja}\,[h_{pq}^j - \varepsilon_n(k)\,S_{pq}^j]\} = 0$$

$$
\begin{array}{llll}
c_{n1}(k)\{\alpha\text{-}\varepsilon_n(k)\}+c_{n2}(k)\beta & & +c_{n4}(k)\beta e^{-ika} & +c_{n6}(k)\beta & = 0\\
c_{n1}(k)\beta \quad +c_{n2}(k)\{\alpha\text{-}\varepsilon_n(k)\}+c_{n3}(k)\beta & & & & = 0\\
\quad +c_{n2}(k)\beta \quad +c_{n3}(k)\{\alpha\text{-}\varepsilon_n(k)\}+c_{n4}(k)\beta & & & & = 0\\
c_{n1}(k)\beta e^{ika} \quad +c_{n3}(k)\beta \quad +c_{n4}(k)\{\alpha\text{-}\varepsilon_n(k)\}+c_{n5}(k)\beta & & & & = 0\\
\quad +c_{n4}(k)\beta \quad +c_{n5}(k)\{\alpha\text{-}\varepsilon_n(k)\}+c_{n6}(k)\beta & & & & = 0\\
c_{n1}(k)\beta \quad +c_{n5}(k)\beta \quad +c_{n6}(k)\{\alpha\text{-}\varepsilon_n(k)\} & & & & = 0
\end{array}
$$

Chart 1.7 (continued)

secular determinant:

$$
\begin{vmatrix}
\alpha-\varepsilon_n(k) & \beta & 0 & \beta e^{-ika} & 0 & \beta \\
\beta & \alpha-\varepsilon_n(k) & \beta & 0 & 0 & 0 \\
0 & \beta & \alpha-\varepsilon_n(k) & \beta & 0 & 0 \\
\beta e^{ika} & 0 & \beta & \alpha-\varepsilon_n(k) & \beta & 0 \\
0 & 0 & 0 & \beta & \alpha-\varepsilon_n(k) & \beta \\
\beta & 0 & 0 & 0 & \beta & \alpha-\varepsilon_n(k)
\end{vmatrix} = 0
$$

π-energy bands:

$$\varepsilon_{kn}(A_x) = \alpha \pm \beta$$

$$\varepsilon_{kn}(S_x) = \alpha \pm \beta \sqrt{3 \pm 2\sqrt{1 + \cos(ka)}}$$

$$= \alpha \pm \beta \sqrt{3 \pm 2\sqrt{2}\,\cos(\frac{ka}{2})}$$

High symmetry points values:

Γ

$$\varepsilon(\Gamma) = \alpha \pm 2.4142\,\beta$$

$$\varepsilon(\Gamma) = \alpha \pm 1.0000\,\beta$$

$$\varepsilon(\Gamma) = \alpha \pm 0.4142\,\beta$$

H

$$\varepsilon(H) = \alpha \pm 1.7320\,\beta$$

$$\varepsilon(H) = \alpha \pm 1.7320\,\beta$$

$$\varepsilon(H) = \alpha \pm 1.0000\,\beta$$

Note: the π-energy band structure is given in Figure 1.23

1.9.5 Qualitative Sketch of Hückel Band Structures from MO Calculations

In the previous Sections, we have seen how the π-band structure of a conjugated polymer can be obtained by actually solving Hückel secular systems and determinants. The underlying idea is properly to combine atomic orbitals. It is often very convenient to obtain a qualitative description of those band structures. This can be easily obtained from a knowledge of the molecular orbital (MO) energies of the corresponding monomer. Indeed, the polymer is nothing but a large molecule. Thus, if the MO's of simple molecules are well described, it is very easy to extend the previous concepts to sketch a first guess of the polymeric band structures. The basic idea is that when the MO's of the monomer ψ_n combine to form the polymer orbitals ϕ_n, the nature of the interaction depends on the position in the first Brillouin zone as stated previously [1.19, 1.30-1.32]. The central point Γ has the all-in-phase component:

$$k=0 \qquad \phi_n(\Gamma,r) = N^{-1/2} \sum_j \psi_n^j(r)$$

while the edge of the Brillouin zone corresponds to the maximum out-of-phase description:

$$k=\frac{\pi}{a} \qquad \phi_n(H,r) = N^{-1/2} \sum_j (-1)^j \psi_n^j(r)$$

The principle is that we observe an energy lowering if the molecular orbital interaction results in a bonding interaction and an energy increase otherwise. Sketches of π-band structures of regular and bond alternant trans-polyacetylene, regular and bond alternant cis-polyacetylene, polyacene, and polyparaphenylene are given in Figures 1.20 to 1.23.

The band structure of regular and alternant trans-polyacetylene has been analytically solved in detail in Section 1.9.3. Here, it is shown how their band structures (see Figure 1.20) can be qualitatively constructed from the knowledge of the π MO's and energies of the "monomeric" structure (i.e., -(CH=CH)-). The two π MO's of a double bond are represented on the left part of Figure 1.20. They are respectively bonding and antibonding with respect to the center of the double bond. The "in-phase" combination at $k=\Gamma$ gives respectively the most bonding orbital (no node on the chain) of the polymer and the most antibonding one (one node at each bond). The "out-of-phase" ("plus-minus") combination at k=H gives two identical structures (one node on every other bond), related to each other by a translation of half the unit cell. In regular polyacetylene, they are thus degenerate. The energy of the first band rises with increasing k from k=Γ to k=H; indeed, an antibonding interaction results, so the energy is raised. The second band falls in energy as k increases since the structure of the orbital at k=H is more bonding than the fully antibonding one at k=Γ. The same situation is observed for an alternant trans-polyacetylene except for the fact that the two orbitals at k=H are no longer degenerate. The situation which implies a node at the middle of the bonds which possess a single character is more stable than that where the node is at the middle of the double-like bond.

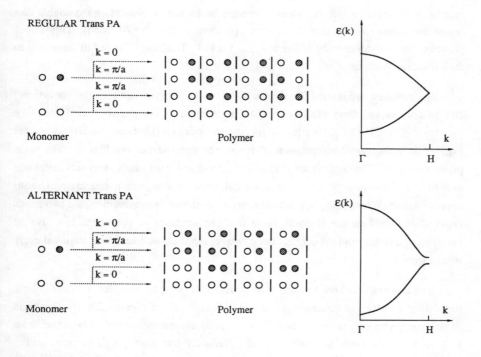

Figure 1.20 Sketch of the π-band structure
(a) of a regular trans-polyacetylene
(b) of an alternant trans-polyacetylene

The π structure of the butadiene skeleton helps in constructing the band structures of regular and alternant cis-polyacetylenes as well as of polyacenes depending on the topology of the selected interactions. The Hückel orbital structure of a butadiene-like monomer -CH=CH-CH=CH- in its cis-conformation (U-shape) is reproduced in the left part of Figure 1.21. The energy rises with the number of nodes. The band structure of regular and bond alternant cis-polyacetylene is plotted in Figure 1.21. At k=Γ, we find the "in-phase" combination of the four orbitals while the "out-of-phase" ones are observed at k=H. In both regular and alternant cis-polyacetylenes, the first and the second band as well as the third and the fourth ones are degenerate at k=H. They have the same orbital bonding structure but translated by half the unit cell. A degeneracy does also exist at k=Γ due to the similarity of the bonding pattern in regular cis-polyacetylene where all interactions between adjacent carbon atoms are assumed to be the same; it is no longer the case in an alternant cis-polyacetylene where the interactions representing the double-like bonds are more intense than those corresponding to the simple bonds. This effect provokes the non-degeneracy of bands 2 and 3 at k=Γ. The bands rise or fall according to their bonding characteristics.

As indicated in Figure 1.22, the band structure of polyacenes ("ladder" polymers) can be constructed from the Hückel orbital structure of the previous butadiene-like monomer -CH=CH-CH=CH- either in its cis-conformation (U-shape, see left part of the Figure) or in its trans-conformation (Z-shape, see right part of the Figure). The same principles apply. The degeneracy at H of the second and third bands is an accidental one due to the same energy of the topological structures within a nearest-neighbour approximation. When going beyond the nearest-neighbour approximation, the two bands cross slightly before the H point. Note that the structure of polyacene can also be considered to result from the connection of two regular trans-polyacetylene chains at every other carbons.

The last example concerns the band structure of polyparaphenylene; it results from translating the MO's of benzene represented in the left part of Figure 1.23. The first band of polyparaphenylene is constructed from the fully symmetric orbital of benzene; from k=Γ to k=H, the fully bonding structure gradually transforms into an antibonding interaction between adjacent phenyl rings. A related situation is sketched for the fully antisymmetric orbital of benzene. The k-dependence of the energy is reversed; the phases on the para carbons are opposite, so that an antibonding interaction results.

71

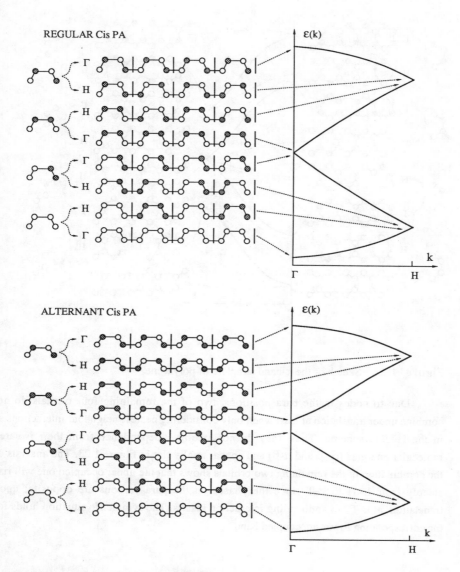

Figure 1.21 Sketch of the π-band structure
        (a) of a regular cis-polyacetylene
        (b) of an alternant cis-polyacetylene

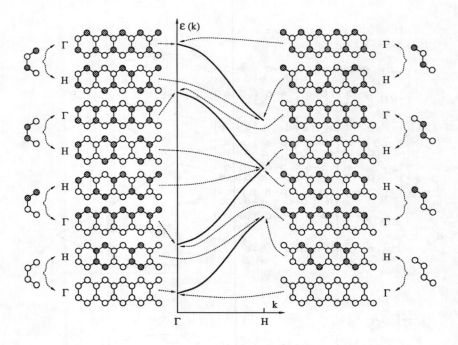

Figure 1.22 Sketch of the π-band structure of polyacene

Due to nodes in the para positions, two of the remaining four orbitals do not combine under translation at least when only considering nearest-neighbour interactions as in the Hückel scheme. Thus, they do not exhibit k-dependence from their isolated molecular energies (α+β and α-β) and appear totally flat in Figure 1.23. The analysis of the combination of the remaining two orbitals shows that the upper occupied one will rise steeply from k=H to k=Γ since the phases on the para carbons are opposite; upon translation, at k=Γ, an antibonding interaction results. The symmetric situation holds for the corresponding lower unoccupied band.

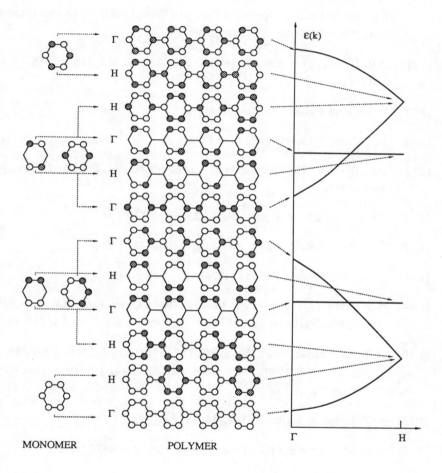

MONOMER   POLYMER

Figure 1.23   Sketch of the π-band structure of polyparaphenylene

References

[1.1]   M. Born and J.R. Oppenheimer, Ann. Physik, **84**, 457 (1927); see also, M. Born and K. Huang, "Dynamical Theory of Crystal Lattices", Oxford University Press, Oxford (1968).

[1.2]   D.R. Hartree, Proc. Cambridge Phil. Soc., **24**, 89, 116, 426 (1928); **25**, 225, 310 (1929).

[1.3]   V. Fock, Z. Physik, **61**, 126 (1930).

[1.4]   J.C. Slater, Phys. Rev., **34**, 1293 (1929); **35**, 509 (1930).

[1.5]   D.R. Hartree, "The Calculation of Atomic Structures", J. Wiley & Sons, New York (1957).

[1.6]   C.C.J. Roothaan, Rev. Mod. Phys., **23**, 69 (1951).

[1.7]   G.G. Hall, Proc. Roy. Soc. (London), **A205**, 541 (1951).

[1.8]   F. Bloch, Z. Physik, **52**, 555 (1928).

[1.9]   N.S. Bayliss, J. Chem. Phys., **16**, 287 (1948); Quart. Revs. (London), **6**, 319 (1952); N.S. Bayliss and J.C. Rivière, Australian J. Sci. Res., **4A**, 344 (1951).

[1.10] H. Kuhn, Chimia Aarau, **2**, 1 (1948); **9**, 237 (1955); J. Chem. Phys., **16**, 840 (1948); **17**, 1198 (1949); Helv. Chim. Acta, **31**, 1441 (1948); Angew. Chem., **71**, 93 (1959).

[1.11] W.T. Simpson, J. Chem. Phys., **16**, 1124 (1948).

[1.12] W.L. McCubbin and F.A. Teemull, Phys. Rev., **A6**, 2478 (1972).

[1.13] R.E. Peierls, "Quantum Theory of Solids", Oxford University Press, Oxford (1955).

[1.14] H. Jones, "The Theory of Brillouin Zones and Electronic States in Crystals", pp.10 et sq., North-Holland Publishing Company, Amsterdam (1962).

[1.15] F. Seitz, "The Modern Theory of Solids", International Series in Pure and Applied Physics, McGraw-Hill Book Company, New York (1940).

[1.16] E. Hückel, Z. Physik, **60**, 423 (1930); **70**, 204 (1931); **76**, 628 (1932).

[1.17] W. Kauzmann, "Quantum Chemistry", p. 682, Academic Press, New York (1957).

[1.18] F.L. Pilar, "Elementary Quantum Chemistry", p. 593, McGraw-Hill, New York (1968).

[1.19] T.A. Albright, J.K. Burdett, and M-H. Whangbo, "Orbital Interactions in Chemistry", p. 299, J. Wiley & Sons, New York (1986).

[1.20] T.K. Rebane, in "Methods of Quantum Chemistry", (M.G. Veselov, Ed.), p. 147, Academic Press, New York (1965).

[1.21] W.P. Su, J.R. Schrieffer, and A.J. Heeger, Phys. Rev., **B22**, 2099 (1980).

[1.22] W.P. Purcell and J.A. Singer, J. Chem. Engineering Data, **12**, 4899, 235 (1967).

[1.23] C.A. Coulson and A. Streitwieser Jr., "Dictionary of π-Electron Calculations", Pergamon, Oxford (1965).

[1.24] E. Heilbronner and H. Bock, "The HMO-Model and its Application, Tables of Hückel Molecular Orbitals", J. Wiley & Sons, London (1976).

[1.25] B. Pullman and A. Pullman, "Les Théories Electroniques de la Chimie Organique", Masson, Paris (1952).

[1.26] A. Streitwieser Jr., "Molecular Orbital Theory for Organic Chemists", J. Wiley & Sons, New York (1961).

[1.27] L. Salem, "Molecular Orbital Theory of Conjugated Systems", Benjamin, New York (1966).

[1.28] E. Heilbronner and H. Bock, "Das HMO-modell und Seine Anwendung", Verlag Chemie (1968).

[1.29]   E. Heilbronner and H. Bock, "The HMO-Model and its Application, Basis and Manipulation", J. Wiley & Sons, London (1976).

[1.30]   J.P. Lowe and S.A. Kafafi, J. Am. Chem. Soc., **106**, 5837 (1984); J.P. Lowe, S.A. Kafafi, and J.P. LaFemina, J. Phys. Chem., **90**, 6602 (1986).

[1.31]   J.M. André, in "Large Finite Systems", (J. Jortner, A. Pullman, and B. Pullman, Eds.), p. 255, D. Reidel Publishing Company, Dordrecht (1987).

[1.32]   R. Hoffmann, "Solids and Surfaces: A Chemist's View of Bonding in Extended Structures", VCH Publisher, New York (1988).

2

THE COMPUTATIONAL MACHINERY OF BAND
THEORY OF POLYMERS

2.1 SUMMARY AND OBJECTIVES

The purpose of this Chapter is

(1) to describe the various analytical expansions of Hartree-Fock polymeric orbitals,

(2) to state the differences between *ab initio* and semiempirical approximations,

(3) to explain the technical difficulties met in *ab initio* calculations of the band structure of polymers, analyze the role of Coulomb and exchange interactions, and write down the expressions for a multipole expansion of the electrostatic interactions,

(4) to state the approximations involved in the semiempirical Zero Differential Overlap and Extended Hückel methods, in the Valence Effective Hamiltonian technique, and in the π-electron semiempirical methods,

(5) to detail the computational tricks used, in practice, to solve specific aspects like integrations in the first Brillouin zone, basis set linear dependences, band indexing difficulties, and density of states calculations,

(6) to discuss in detail the *ab initio* band structures of polyethylene and polysilane and provide a full analysis.

2.2 ANALYTICAL EXPANSION OF HARTREE-FOCK POLYMERIC ORBITALS

In actual calculations, the Linear Combination of Atomic Orbitals (LCAO) scheme implies the definition of a basis set. In principle, the choice of the type of basis functions is arbitrary. It is in practice conditioned by its computational adequacy.

The first idea might be to use, as basis functions, the hydrogenic orbitals, which are the exact eigenfunctions of the hydrogenic Hamiltonian:

$$h(\mathbf{r}) = -\frac{1}{2} \nabla^2 - \frac{Z}{|\mathbf{r}|}$$

Since they are solutions of an Hermitian operator, they form on each atomic site a set of orthogonal functions. Note that they form a complete system only if the functions associated with the continuum are also included. Their radial part is exponentially decreasing [$\propto \exp(-Zr)$] but depends on both the principal quantum number n and the angular quantum number ℓ, which makes their use practically intractable.

Slater-Type Orbitals (STO) proposed in 1930 [2.1] avoid the latter difficulty. Their radial part is also exponentially decreasing [$\propto \exp(-\zeta r)$] but depends only on the principal quantum number n; 2s and 2p orbitals thus have the same radial dependence. The price to be paid is that they are no longer eigenfunctions of an Hermitian operator and thus are no longer orthogonal to each other even on a single site. 1s and 2s STO's thus have a nonorthogonal behaviour. A further difficulty is that, although analytical solutions for all one- and two-center (overlap, kinetic, nuclear attraction, and repulsion) integrals have been published [2.2], tractable expressions for three- and four-center (Coulomb or exchange integrals involving three or four different atoms) integrals are not available in a convenient way. These STO's have been largely used for describing diatomic molecules but not for larger molecules or polymeric systems.

Nowadays, following the suggestion of Boys in 1950 [2.3], use is made of Gaussian-Type Orbitals (GTO) in molecular calculations. Instead of the exponential radial dependence [$\propto \exp(-\zeta r)$], they exhibit a Gaussian-like decay [$\propto \exp(-\alpha r^2)$]. The main advantage resides in the fact that all integrals can now be cast in computationally tractable forms. However, the GTO does not correctly reproduce the STO- or hydrogenic cusp at the nucleus due to the attractive potential discontinuity of $-Z/r$ at $r=0$. Thus, the derivative of a GTO is zero at its origin while, at the same point, the derivative of an STO has a non-zero negative value. One tries to avoid this difficulty by simulating each STO-"atomic-like" orbital by a combination of several "primitive" Gaussians, contracted into a single Gaussian basis function. For instance, in the popular STO-3G basis set, each STO is simulated by a least-squared optimized combination of 3 Gaussians:

$$1s(r) = c_1 e^{-\alpha_1 r^2} + c_2 e^{-\alpha_2 r^2} + c_3 e^{-\alpha_3 r^2}$$

In order to limit adequately the number of basis functions, three main types of basis sets are defined: minimal, extended, and polarized. In a minimal (single-zeta) basis set, only those orbitals which are occupied in the free atoms are considered, i.e., 1s orbitals for hydrogen atoms; 1s, 2s, $2p_x$, $2p_y$, $2p_z$ orbitals for carbon, nitrogen, and oxygen. STO-KG expansions are such minimal basis sets (where K refers to the number of primitive Gaussians used for each STO). They are available for atoms ranging from the first to the fourth row of the periodic table from the original papers issued by Pople's group since 1969 [2.4]. In an extended basis set, each valence orbital is supplemented by a delocalized orbital of the same quantum number: for example, 1s and 1s' orbitals for hydrogen and 1s, 2s, 2s', 3*2p and 3*2p' orbitals for carbon, nitrogen, and oxygen. This allows for more radial flexibility in the spatial representation of the molecular electron density and also partially corrects the deficient description of anisotropic situations by the readjustment of diffuse and localized components through the Self-Consistent Field (SCF) procedure. If one also doubles the core orbitals, the basis set is often referred to as a double-zeta basis. In a polarized basis set, one adds to each valence orbital, basis functions of higher angular quantum number; for example: (2)p orbitals for hydrogen and (3)d orbitals for the carbon, nitrogen, and oxygen atoms. This permits small displacements of the center of the electronic charge distributions away from the atomic positions (charge polarization). List of references for GTO's and tables of atomic functions can be found in the specialized literature [2.5-2.7].

It is easily seen that the more flexible the basis set, the lower the total energy optimized with respect to the variational parameters of the wave function, i.e., the results are closest to the experimental limit. As illustrated in Figure 2.1 by an SCF calculation on the helium atom where a single exponential orbital (Slater-Type Orbital) is used, the total atomic energy of the ground state is -2.847 656 3 a.u. (1 atomic unit of energy is equal to 27.21 eV or 2624.95 kJ/mol). When using two STO's and optimizing their coefficients by the SCF procedure, a lower energy of -2.861 660 is observed without any special optimization of the orbital exponents (ζ_1 and ζ_2 arbitrarily selected to be 1.45 and 2.85). If such an optimization is pursued [2.8], a fully optimized energy of -2.861 670 1 a.u. is obtained for optimized orbital exponents: ζ_1= 1.4461 and ζ_2=2.8622.

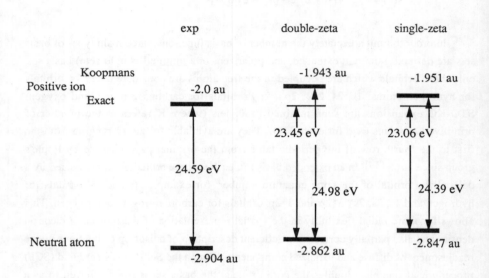

Figure 2.1   Experimental, single-zeta and double-zeta levels of the ground state of He and He$^+$.

With such a double-zeta basis set, the numerical error due to the limitation of the basis set is thus only 0.000 009 9 a.u. = 0.000 26 eV with respect to the Hartree-Fock (HF) limit of -2.861 680 0 a.u. [2.9]. The HF limit is in error compared to the experimental value by 0.042 104 a.u. = 1.145 73 eV. This error is inherent to the HF methodology. The form of the wave function (Slater determinant) does not prevent, in the singlet state, two electrons residing in the same region of space. Such a physical situation is highly disfavored by the strong electrostatic repulsion between particles of the same charge. It is the so-called <u>correlation error</u>. In agreement with the variation theorem, the calculated total energy will be higher than the experimental value, a result easily explained since the HF method gives too much weight to these destabilizing situations found when the electrons are close to one another. The previous results also indicate some interesting trends for the ionization potentials. In the Koopmans' approximation [2.10], the wave function of the ionized species is constructed from the orbitals of the neutral system. In the present case, its energy is the negative of the HF 1s orbital and is overestimated as compared to the exact value (ground state energy of the hydrogenic He^+ = -2.0 a.u.) as imposed by the variation theorem. The relaxation energy is a measure of the importance of the rearrangement effects of the electron density of the neutral system under the ionization process (1.53 eV for a double-zeta basis set, 1.33 eV for the minimal basis set as illustrated by Figure 2.1).

An interesting subminimal basis set is the Floating Spherical Gaussian Orbital (FSGO) set of Frost [2.11]. In this basis set, only 1s Gaussian-type orbitals are used. Each orbital describes an electron pair and some of these orbitals are placed at points of space other than on nuclei, e.g., in bonding regions (see, for example, Figure 2.7 in Section 2.5). Both the orbital exponents and the orbital positions of the Gaussian set are variationally optimized. Due to the fact that the size of the basis set (N) is equal to the number of electron pairs (n/2), the iterative cycles of the usual SCF schemes are completely avoided. Indeed, in that case, the density matrices are unequivocally fixed and computed by an inversion procedure of the overlap matrices. With such an extreme limitation in the size of the basis set (sometimes, as indicated above, referred to as subminimal), the FSGO description of the electronic structure cannot be of full *ab initio* quality. Despite that, this method gives reliable results for given electronic properties: for example, photoelectronic spectra of alkanes [2.12], structure factors of the electronic density [2.13], or conformational stability of polyethylene [2.14]. Quite encouraging results were also obtained in the case of classical bonding situations. On the other hand, by its simplicity, it has permitted to obtain simple insight into the solution of important conceptual problems and the study at moderate cost of the implications of new methodologies and computational techniques on realistic systems. Examples of such studies are an analysis of the role of long-range electrostatic effects in linear chains [2.15] or a simple view of the Hartree-Fock instability problem in connection with the correlation error [2.16]. These features combined with the obvious computational advantages of the method, i.e., no iterative process and a small number of polycenter integrals that do not necessarily need to be stored, has made the FSGO approach very attractive in such instances.

2.3 COMPUTATIONAL TECHNIQUES

2.3.1 *Ab Initio* Versus Semiempirical Schemes

SCF-MO-LCAO methods require a computational effort proportional to the fourth power of the size of the basis set. We call *ab initio* or nonempirical the techniques where all electrons are considered and all the integrals (detailed in Chart 2.2) are calculated. Due to the huge number of integrals to be evaluated in large molecules or polymeric models of chemical significance, very important computing resources are required. In order to get faster and less demanding computational algorithms, it is sometimes imperative to reduce the number of electrons to be considered or of integrals to be computed. For instance, we can reduce the number of electrons by "freezing" n_c (internal or σ) electrons into fixed cores of reduced charge $z_A = Z_A - n_c$ (where Z_A is the nuclear charge) and/or approximate several electron integrals or groups of integrals. In this case, semiempirical methods are generally used to evaluate the rest of the integrals from experimental data and thus to correct the errors introduced by the neglect of some integrals. In general, semiempirical methods are dependent on the square power of the basis set used or of the number of atoms. Chart 2.1 schematizes the transformation from nonempirical (*ab initio*) to semiempirical schemes. The numerical treatment of these non- or semiempirical techniques requires computer resources ranging from PC's to supercomputers depending on the size of the basis sets and on the importance of the approximations introduced.

Chart 2.1 *Ab initio* versus semiempirical schemes

Ab initio   - consider the full one-electron Hartree-Fock operator for a system of n-electrons:

$$h(r) = -\frac{1}{2} \nabla^2 - \sum_A Z_A \, |r-R_A|^{-1} + \sum_j^{occ} \{2 \, J_j(r) - K_j(r)\}$$

- compute from *ab initio* (first principles) all integrals of the basis set in the selected basis set $\{\chi_p\}$

Chart 2.1 (continued)

N^4-dependence — compute full interaction of the averaged electron density:

$$\iint \rho(\mathbf{r}) \; |\mathbf{r}-\mathbf{r}'|^{-1} \rho(\mathbf{r}') \; d\mathbf{r} \; d\mathbf{r}' =$$

$$\sum_p \sum_q \sum_r \sum_s D_{pq} \, D_{rs} \iint \chi_p(\mathbf{r}) \chi_q(\mathbf{r}) |\mathbf{r}-\mathbf{r}'|^{-1} \chi_r(\mathbf{r}') \chi_s(\mathbf{r}') \; d\mathbf{r} \; d\mathbf{r}'$$

<u>Semiempirical</u> — split the n-electrons systems into n_c core electrons (c) and n_v valence electrons (v)

— consider an effective valence one-electron Hartree-Fock operator for a system of n_v valence electrons:

$$h(\mathbf{r}) = -\frac{1}{2}\nabla^2 - \sum_A Z_A \; |\mathbf{r}-\mathbf{R}_A|^{-1} + \sum_j^{n_c} \{2 \, J_j{}^c(\mathbf{r}) - K_j{}^c(\mathbf{r})\}$$

$$+ \sum_j^{n_v} \{2 \, J_j{}^v(\mathbf{r}) - K_j{}^v(\mathbf{r})\}$$

$$= h^c(\mathbf{r}) + \sum_j^{n_v} \{2 \, J_j{}^v(\mathbf{r}) - K_j{}^v(\mathbf{r})\}$$

— evaluate matrix elements of the one-electron Hamiltonian from experimental data (Extended Hückel theory)

or — keep them as parameters (Hückel theory)

or — use the Zero Differential Overlap (ZDO) approximation as in CNDO-like theories

N^2-dependence — keep only the diagonal part of the interaction of the averaged electron density:

$$\iint \rho(\mathbf{r}) \; |\mathbf{r}-\mathbf{r}'|^{-1} \rho(\mathbf{r}') \; d\mathbf{r} \; d\mathbf{r}' =$$

$$\sum_p \sum_r D_{pp} \, D_{rr} \iint \chi_p(\mathbf{r}) \chi_p(\mathbf{r}) |\mathbf{r}-\mathbf{r}'|^{-1} \chi_r(\mathbf{r}') \chi_r(\mathbf{r}') \; d\mathbf{r} \; d\mathbf{r}'$$

In the simple Hückel method [2.17], only the π-electrons of purely conjugated organic molecules are taken into account and all interactions except for nearest-neighbours are neglected. As seen in Chapter 1, this method involves the one-center parameter α and the bond parameter β. It has gained a considerable success by its simplicity and by the soundness of the approximations involved. Around 1950, Mulliken [2.18] and Wolfsberg and Helmholtz [2.19] suggested a very simple type of Hückel parametrization which is easily extendable to σ-bonded systems. In 1963, Hoffmann [2.20] took up this method and applied it to a large variety of saturated and conjugated organic molecules. This technique is by now well-known as the Extended Hückel method. The ZDO-like methods which neglect Zero Differential Overlaps, no longer have their justification in a global evaluation of the matrix elements but more precisely in a close analysis of the energy terms. The LCAO matrix elements are split into their kinetic, nuclear attraction, and repulsion contributions and some integrals are either neglected or approximated by careful procedures. This is the case of the Pariser-Parr-Pople type methods for π-electrons introduced independently by Pariser and Parr [2.21] and by Pople [2.22], or the CNDO (Complete Neglect of Differential Overlap) type methods [2.23-2.25] when considering all-valence electrons. Fundamentally, these procedures attempt either to reproduce by means of one-electron models the results of the many-electron HF theory or to improve the theory-experiment agreement for some selected properties by fitting integrals or groups of integrals to experimental data. The latter approach very often results in an implicit inclusion of part of the correlation effects, although in a very indirect way. In this manner, it is obvious that the computational work is strongly decreased; but, importantly, the incorporation of semiempirical data compensates partly for the errors due to some very crude approximations. The price to be paid is that, since the semiempirical processes do use parameters of experimental origin, the parameterized methods are generally good mostly for the properties to which they have been fitted. It is common to select the adequate semiempirical methods depending on the type of property which is to be investigated. Several textbooks and commented collections of papers cover the field of semiempirical calculations [2.26-2.30].

The techniques which try to simulate the *ab initio* results are also of particular interest. Good results have been recently obtained with simple model potential techniques, for instance the so-called VEH, Valence Effective Hamiltonian, technique. As developed in Section 2.3.3.3, it is supposed in this procedure that the valence Hamiltonian that simulates the Fock operator is the sum of a kinetic term and of various effective atomic potentials for atoms within their characteristic chemical environment. For computational simplicity, these effective potentials are chosen as non-local spherical or nonspherical Gaussian projectors. Parametrization of these potentials is performed by least-square fitting on corresponding valence Fock operators for small model molecules.

A summary of some current *ab initio* and semiempirical procedures is given in Table 2.1. They are classified into two categories:

- self-consistent or iterative procedures where the one-electron operator h(**r**), and hence the matrix elements h_{pq}, explicitly depend on the electron density $\rho(\mathbf{r})$;

- and non-self-consistent or non-iterative techniques where a single evaluation of the matrix elements is performed.

Table 2.1   Empirical, semiempirical, and *ab initio* methods of quantum chemistry

Electrons considered	non-self-consistent methods non-iterative procedures global evaluation of matrix elements h_{pq}	self-consistent methods iterative procedures approximations in the calculation of matrix elements $h_{pq} = h_{pq}\{\rho(\mathbf{r})\}$

all electrons		*ab initio*
		    Clementi→ IBMOL series
		    Pople → GAUSSIAN series
		    Dupuis → HONDO series

valence
$\sigma + \pi$-electrons
	Extended Hückel (EH)
	    Hoffmann
	    VEH
		Zero Differential Overlap (ZDO)
		    CNDO, INDO, MINDO,
		    NDDO, MNDO, AM1, PM3
		    Pople
		    Dewar

π-electrons	Hückel Theory (HMO)
	    Hückel
		Pariser-Parr-Pople (PPP) technique
		    Pariser-Parr
		    Pople

2.3.2 All-Electron *Ab Initio* Methods

2.3.2.1 Principles and difficulties of *ab initio* calculations in polymers

As already mentioned, LCAO-based *ab initio* calculations require the definition of a basis set of size N and imply the calculations of integrals or matrix elements between the N basis functions of the basis set. For atoms, all the basis functions are centered on the same location and the integrals are said to be one-centered. For a molecule or a polymeric cluster of M atoms, the N^2 (or more exactly the $N(N+1)/2$ when taking into account the Hermitian symmetry) kinetic integrals:

$$T_{pq} = \int \chi_p(\mathbf{r}) \{ -\frac{1}{2} \nabla^2 \} \chi_q(\mathbf{r}) \, d\mathbf{r}$$

can be either one- or two-centered while the $M*N(N+1)/2$ nuclear attraction integrals:

$$V_{pq|A} = \int \chi_p(\mathbf{r}) |\mathbf{r} - \mathbf{R}_A|^{-1} \chi_q(\mathbf{r}) \, d\mathbf{r}$$

can be one-, two-, or three-centered.

Finally, one finds $N(N+1)/2 * [\ N(N+1)/2+1]/2$, i.e., approximately $N^4/8$, either one-, two-, three- or four-centered two-electron repulsion integrals:

$$(pq|rs) = \iint \chi_p(\mathbf{r}) \chi_q(\mathbf{r}) |\mathbf{r} - \mathbf{r}'|^{-1} \chi_r(\mathbf{r}') \chi_s(\mathbf{r}') \, d\mathbf{r} \, d\mathbf{r}'$$

For large molecules or large basis sets, a huge number of integrals has to be calculated, which constitutes the bottleneck of first principles (*ab initio*) SCF-MO-LCAO calculations on large oligomeric systems. All electron integrals are then calculated for a specified geometry. If the geometry is not known, guesses of the probable geometries can be obtained through full or partial search of energy minima on the potential surface.

HF *ab initio* calculations provide useful data on ground state properties like bond lengths and angles, molecular conformations and internal rotation barriers, electron densities and dipole moments. Examples of results for small monomers of interest in polymer chemistry are given in Tables 2.2 to 2.6 for several types of basis sets (minimal, extended and polarized).

Table 2.2 Examples of molecular bond lengths (in Å)

Method	CH$_4$ C-H	C$_2$H$_2$ C-C	C$_2$H$_2$ C-H	C$_2$H$_4$ C-C	C$_2$H$_4$ C-H	C$_2$H$_6$ C-C	C$_2$H$_6$ C-H
Semiempirical							
EH	1.02						
CNDO/1	1.11	1.213		1.342		1.520	
CNDO/2	1.116	1.198	1.093	1.320	1.110	1.476	1.117
INDO		1.20	1.10	1.31	1.11	1.46	1.12
MINDO/2	1.100	1.19	1.069	1.337	1.093	1.524	1.109
MINDO/3		1.191	1.076	1.31	1.10	1.486	1.108
NDDO		1.225	1.057	1.410	1.080	1.527	1.094
MNDO		1.19	1.051	1.335	1.089	1.521	1.109
AM1	1.112	1.195	1.061	1.325	1.098	1.501	1.117
Ab Initio							
minimal	1.083	1.168	1.065	1.306	1.082	1.538	1.086
extended	1.083	1.188	1.051	1.315	1.074	1.542	1.084
polarized	1.084	1.186	1.057	1.317	1.076	1.527	1.086
Experimental	1.092	1.203	1.061	1.339	1.085	1.531	1.096

Table 2.3 Examples of barriers to internal rotation (in kcal/mol)

		CH$_3$-CH$_3$	CH$_3$-CH=CH$_2$
Semiempirical	EH	3.04	
	CNDO/1	1.47	
	CNDO/2	2.18	
	MINDO/2		1.89
	NDDO	2.39	
	MNDO	1.0	1.20
	AM1	1.25	
Ab Initio	minimal	2.9	1.25
	extended	2.7	
	polarized		3.0
Experimental		2.88	1.99

Table 2.4   Examples of conformational energy differences (in kcal/mol)

Molecule	conformation	⇐	Semiempirical	⇒	AM1	⇐ *ab initio* ⇒		Expt.
		CNDO	MINDO/1	NDDO		minimal	extended	
n-butane	trans/gauche				0.73	0.9	0.8	0.77
1-butene	skew/cis					0.8	0.8	0.2
1,3-butadiene	trans/gauche	0.63	-1.43	-1.42		1.8	3.5	1.7-2.5

Table 2.5   Examples of relative energies of structural isomers (in kcal/mol)

Molecule	⇐ Semiempirical ⇒			⇐	*ab initio*	⇒	Expt.
	EH	MNDO	AM1	minimal	extended	polarized	
C_3H_4							
propyne	0	0	0	0	0	0	0
allene	23.4	2.5	2.7	17.1	3.4	2.0	1.0-1.6
cyclopropene		26.9	31.4	30.0	39.8	26.5	21.6-21.7
C_4H_6							
trans-1,3-butadiene	0	0	0	0	0	0	0
2-butyne	-9.6	-4.1	2.1	-12.8	3.6	6.5	6.9-8.5
cyclobutene		2.0	15.9	-12.5	18.0	13.4	8.8-11.2
1,2-butadiene	11.1	4.6	7.2	8.5	11.3	12.8	12.5
1-butyne		7.2	7.6	-5.3	9.2	13.0	13.2
methylenecyclopropane		8.9	17.8	5.8	25.6	20.5	21.7
bicyclo[1.1.0]butane		35.1	48.2	11.6	45.7	31.4	23.1-25.6
1-methylcyclopropene		24.7	34.8	17.3	43.5	33.0	32.0
C_4H_8							
isobutene	0	0	0	0	0	0	0
trans-2-butene	2.0	-3.1	-2.1	-0.2	0.5	0.2	1.1
cis-2-butene	5.9	-0.2	-1.0	1.4	2.0	1.8	2.2
cis-1-butene	6.5	2.4	1.6	4.4	3.3	3.6	3.9
methylcyclopropane	27.3			-2.2	15.0	3.6	9.7
cyclobutane	62.0	-9.9	0.2	-17.6	10.4	10.0	10.8
C_4H_{10}							
isobutane	0	0	0	0	0	0	0
trans-n-butane	2.3	-2.9	-1.7	0.1	1.3	0.4	1.6-2.1
C_6H_6							
benzene	0	0	0				0
fulvene	25.6	32.4	40.7				27.0-27.7
$C_{10}H_8$							
naphtalene	0						0
azulene	32.3						32.6

Table 2.6   Examples of dipole moments (in Debye)

| Molecule | ⇐ Semiempirical ⇒ | | | | ⇐ *ab initio* ⇒ | | | |
	CNDO/2	MINDO/1	MNDO	AM1	min.	ext.	pol.	Expt.
propyne	0.43		0.12	0.40	0.50	0.69	0.57	0.75 0.78
cyclopropene			0.42	0.36	0.55	0.55	0.57	0.45
propene	0.36		0.04	0.23	0.25	0.30	0.29	0.36-0.37
propane	0.00		0.01	0.01	0.02	0.06	0.06	0.08
but-1-yne-3-ene					0.37	0.45	0.46	0.4
cyclobutene			0.08	0.17	0.05	0.07	0.03	0.13
1,2-butadiene				0.33	0.39	0.35	0.40	
1-butyne					0.54	0.67	0.65	0.80
methylenecyclopropane					0.22	0.35	0.40	0.40
bicyclo[1.1.0]butane			0.41	0.43	0.58	0.75	0.68	0.68
1-methylcyclopropene					0.83	0.90	0.89	0.84
isobutene	0.65				0.43	0.53	0.44	0.50
cis-2-butene				0.14	0.13	0.13	0.26	
cis-1-butene				0.30	0.34	0.34	0.44	
methylcyclopropane					0.11	0.11	0.10	0.14
isobutane	0.00				0.04	0.11	0.09	0.13
toluene	0.22		0.06	0.27	0.25			0.31-0.36
ammonia	2.08	2.13	1.76	1.85	1.79	2.17	1.95	1.47
phenol	1.73	2.72	1.67	1.24	1.22			1.45-1.55
fluorobenzene	1.66				0.93			1.60-1.66
methanol	1.94	2.48	1.48	1.62	1.51	2.21	2.04	1.69-1.70
water	2.08	2.79	1.78	1.86	1.73	2.44	2.22	1.82-1.85
fluoromethane	1.66				1.16	2.25	2.06	1.85
acetaldehyde	2.53	3.64	2.38	2.69				2.68-2.69
formaldehyde	1.98	3.35	2.16	2.32	1.51	2.63	2.75	2.33-2.34
benzaldehyde	2.50							2.72
acetone	2.90	3.63	2.51	2.92				2.88-2.90

HF *ab initio* methods are also useful for calculating ionization potentials, excitation energies, and dipole strengths. In these latter cases, care must be taken to correctly balance relaxation and correlation effects. HF *ab initio* calculations can also give insight into full conformational potential energy surfaces and molecular associations, two fields of great interest in polymer chemistry. In agreement with the "extrapolation" approach qualitatively described in Chapter 1, the molecular quantum chemist will probably start by an extrapolation study of the properties of size-limited molecules for evaluating the electronic properties of a polymer or of a large oligomer. Such studies are in general very slowly convergent to accurate numerical figures as illustrated in Tables 2.7 and 2.8.

Table 2.7 **Cluster energies of some alkane oligomers (minimal basis set, in a.u.)**

Alkane	$E_{cluster} = E$	E per carbon	$\Delta E = E_{N+1} - E_N$
propane	-116.878 617	-38.959 539	
butane	-155.457 098	-38.864 275	-38.578 481
pentane	-194.035 561	-38.807 112	-38.578 463
		extrapolation	-38.578 5..

The example of $(LiH)_n$ is very illustrative. The model is a linear chain of LiH molecules ($r_{Li-H} = 2$ a.u.) distant from each other by 4 a.u., using a standard minimal basis set STO-3G expansion of the s and p orbitals. This model system is a good test case since it has an inherently polar unit cell due to the different electronegativities of H and Li and thus presents electrostatic interactions which are long-ranged. Table 2.8 gives the numerical results as obtained from the KGNMOL *ab initio* program on molecular clusters of $(LiH)_n$, for odd values of n ranging from 1 to 29, compared with the results of an extrapolated limit.

Examination of the next table calls for the following comments. As expected, the relative cluster energies show very poor convergence due to the role of the end effects. The difference of relative energies between n=27 and n=29 is still of the order of 10^{-4} a.u. The energy increment more rapidly converges; it is 2.10^{-6} a.u. between n=27 and n=29. The conclusion is that an extrapolation approach is generally not practicable for realistic cases due to its poor convergence and the numerous intermediate single-point calculations required to make oneself convinced about the stability of the results.

In the infinite regular polymeric chain as in the molecular or oligomeric cases, nonempirical methods are dependent on the fourth power of the number of basis functions considered. If N_C is the number of cells for which all the overlap and Fock elements are greater than a given threshold (*e.g.*, 10^{-6} a.u.) and N is the size of the basis set in a unit cell, we have to calculate N_C $N(N+1)/2$ Fock integrals and overlap terms. The repulsion operator due to its two-electron character gives rise to much more complex matrices. In usual schemes, those matrices (of number N_C^3) contain $N^2(N+1)^2/8$ integrals as a first approximation. As a consequence, the time consuming part of an *ab initio* calculation has a general $N_C^3 N^4/8$ dependence.

Table 2.8 **Total energies and hydrogen net atomic charge for clusters of $(LiH)_n$ (in a.u.)**

n	E_n	E_n/n	$\Delta E = E_{n+1} - E_n$	q_H
1	-7.817 840	-7.817 840		0.0104
3	-23.493 153	-7.831 051	-7.837 657	0.0474
5	-39.174 549	-7.834 910	-7.840 698	0.0543
7	-54.856 869	-7.836 696	-7.841 160	0.0563
9	-70.539 469	-7.837 719	-7.841 300	0.0571
11	-86.222 190	-7.838 381	-7.841 361	0.0574
13	-101.904 974	-7.838 844	-7.841 392	0.0577
15	-117.587 793	-7.839 186	-7.841 410	0.0578
17	-133.270 635	-7.839 449	-7.841 421	0.0579
19	-148.953 491	-7.839 657	-7.841 428	0.0579
21	-164.636 357	-7.839 827	-7.841 433	0.0580
23	-180.319 229	-7.839 966	-7.841 436	0.0580
25	-196.002 107	-7.840 084	-7.841 439	0.0580
27	-211.684 987	-7.840 185	-7.841 440	0.0581
29	-227.367 871	-7.840 271	-7.841 442	0.0581
Extrapolation			-7.841 449	0.05816

In the *ab initio* methods, all the integrals are correctly evaluated up to infinity in terms of lattice sums as soon as the geometry and the basis functions are defined. A correct computation is still a formidable task at the present time. Indeed, it is proven that long range effects are very important as soon as the unit cell contains permanent dipoles.

Multipole expansions must be used in order to get a satisfactory balance of the electrostatics interactions.

When turning to polymer applications, the numerical procedure combines the equations of the methods of molecular quantum chemistry and solid state physics as shown in Chart 2.2.

Chart 2.2   A comparison of some basic formulas of molecular and polymer quantum chemistry

Molecular Quantum Chemistry	Polymer Quantum Chemistry

Orbital:

$$\phi_j(r) = \sum_p c_{jp}\, \chi_p(r\text{-}P)$$

$$\phi_n(k,r) = \frac{1}{\sqrt{N}} \sum_j e^{ikja} \sum_p c_{np}(k)\, \chi_p(r\text{-}P\text{-}ja)$$

Secular system:

$$\sum_p c_{jp}\, (h_{pq} - \varepsilon_j\, S_{pq}) = 0$$

$$\sum_p c_{np}(k)\, \{ \sum_j e^{ikja}\, [h_{pq}^{\,j} - \varepsilon_n(k)\, S_{pq}^{\,j}\,] \}$$

$$\equiv \sum_p c_{np}(k)\, \{ \, h_{pq}(k) - \varepsilon_n(k)\, S_{pq}(k)\, \} = 0$$

Secular determinant:

$$|\, h_{pq} - \varepsilon\, S_{pq}\, | = 0$$

$$|\, \sum_j e^{ikja}\, [h_{pq}^{\,j} - \varepsilon(k)\, S_{pq}^{\,j}\,]\, |$$

$$\equiv |\, h_{pq}(k) - \varepsilon_n(k)\, S_{pq}(k)\, | = 0$$

One-electron Hartree-Fock operator:

$$h(r) = -\frac{1}{2}\nabla^2 - \sum_A \frac{Z_A}{|r\text{-}R_A|}$$

$$h(r) = -\frac{1}{2}\nabla^2 - \sum_{h,A} \frac{Z_A}{|r\text{-}R_A\text{-}ha|}$$

$$+ \sum_j^{occ} \{\, 2J_j(r) - K_j(r)\, \}$$

$$+ \sum_j^{occ} \int dk\, \{\, 2J_j(k,r) - K_j(k,r)\, \}$$

Chart 2.2 (continued)

Coulomb repulsion operator:

$$J_j(r) \, \phi_i(r) = \{ \int dr' \phi_j(r') \, \phi_j(r') \, |r-r'|^{-1} \} \, \phi_i(r)$$

$$J_j(k,r) \, \phi_i(k',r) = \{ \int dr' \phi_j^*(k,r') \, \phi_j(k,r') \, |r-r'|^{-1} \} \, \phi_i(k',r)$$

Exchange interaction operator:

$$K_j(r) \, \phi_i(r) = \{ \int dr' \phi_j(r') \, \phi_i(r') \, |r-r'|^{-1} \} \, \phi_j(r)$$

$$K_j(k,r) \, \phi_i(k',r) = \{ \int dr' \phi_j^*(k,r') \, \phi_i(k',r') \, |r-r'|^{-1} \} \, \phi_j(k,r)$$

Integrals between molecular or polymeric orbitals:

$$J_{ij} = \int dr \, \phi_i(r) \, J_j(r) \, \phi_i(r)$$
$$= \int dr \, \phi_i(r) \int dr' \phi_j(r') \, \phi_j(r') \, |r-r'|^{-1} \, \phi_i(r)$$
$$= \int dr \int dr' \phi_i(r) \, \phi_i(r) \, |r-r'|^{-1} \, \phi_j(r') \, \phi_j(r')$$

$$J_{ij}(k,k') = \int dr \, \phi_i^*(k,r) \, J_j(k',r) \, \phi_i(k,r)$$
$$= \int dr \, \phi_i^*(k,r) \int dr' \phi_j^*(k',r') \, \phi_j(k',r') \, |r-r'|^{-1} \, \phi_i(k,r)$$
$$= \int dr \int dr' \phi_i^*(k,r) \, \phi_i(k,r) \, |r-r'|^{-1} \, \phi_j^*(k',r') \, \phi_j(k',r')$$

$$K_{ij} = \int dr \, \phi_i(r) \, K_j(r) \, \phi_i(r)$$
$$= \int dr \, \phi_i(r) \int dr' \, \phi_j(r') \, \phi_i(r') \, |r-r'|^{-1} \, \phi_j(r)$$
$$= \int dr \int dr' \phi_i(r) \, \phi_j(r) \, |r-r'|^{-1} \, \phi_j(r') \, \phi_i(r')$$

$$K_{ij}(k,k') = \int dr \, \phi_i^*(k,r) \, K_j(k',r) \, \phi_i(k,r)$$
$$= \int dr \, \phi_i^*(k,r) \int dr' \, \phi_j^*(k',r') \, \phi_i(k',r') \, |r-r'|^{-1} \, \phi_j(k',r)$$
$$= \int dr \int dr' \phi_i^*(k,r) \, \phi_j(k,r) \, |r-r'|^{-1} \, \phi_j^*(k',r') \, \phi_i(k,r')$$

Integrals between atomic orbitals:

$$S_{pq} = \int \chi_p(r) \, \chi_q(r) \, dr$$

$$S_{pq}^j = \int \chi_p(r-P-0a) \, \chi_q(r-Q-ja) \, dr$$
$$= \int \chi_p(r) \, \chi_q^j(r) \, dr$$

$$h_{pq} = \int \chi_p(r) \, h(r) \, \chi_q(r) \, dr$$

$$h_{pq}^j = \int \chi_p(r-P-0a) \, h(r) \, \chi_q(r-Q-ja) \, dr$$
$$= \int \chi_p(r) \, h(r) \, \chi_q^j(r) \, dr$$

$$= T_{pq} - \sum_A Z_A V_{pq|A}$$
$$+ \sum_{r,s} D_{rs} \{ 2(pq|rs) - (pr|qs) \}$$

$$= T_{pq}^j - \sum_{h,A} Z_A V_{pq|A}^{j|h}$$
$$+ \sum_h \sum_l \sum_r \sum_s D_{rs}^{hl} \{ 2 \, ({}_{pq}^{\ j} \, | \, {}_{rs}^{hl}) - ({}_{pr}^{\ h} \, | \, {}_{qs}^{jl}) \}$$

Chart 2.2 (continued)

$T_{pq} = \int \chi_p(r) \{-\frac{1}{2}\Delta\} \chi_q(r) \, dr$

$T_{pq}^{j} = \int \chi_p(r\text{-}P\text{-}0a) \{-\frac{1}{2}\Delta\} \chi_q(r\text{-}Q\text{-}ja) \, dr$

$= \int \chi_p(r)\{-\frac{1}{2}\Delta\} \chi_q^{j}(r) \, dr$

$V_{pq|A} = \int \chi_p(r) \, |r\text{-}R_A|^{-1} \chi_q(r) \, dr$

$V_{pq|A}^{j\ h} = \int \chi_p(r\text{-}P\text{-}0a) \, |r\text{-}R_A\text{-}ha|^{-1} \chi_q(r\text{-}Q\text{-}ja) \, dr$

$= \int \chi_p(r) \, |r\text{-}R_A\text{-}ha|^{-1} \chi_q^{j}(r) \, dr$

$(pq|rs) = \iint \chi_p(r)\chi_q(r) \, |r\text{-}r'|^{-1}.$

$(_{pq}^{j} |_{rs}^{hl}) = \iint \chi_p(r\text{-}P\text{-}0a) \, \chi_q(r\text{-}Q\text{-}ja) \, |r\text{-}r'|^{-1}.$

$\chi_r(r')\chi_s(r') \, dr \, dr'$

$\chi_r(r'\text{-}R\text{-}ha) \, \chi_s(r'\text{-}S\text{-}la) \, dr \, dr'$

$= \iint \chi_p(r)\chi_q^{j}(r) \, |r\text{-}r'|^{-1} \chi_r^{h}(r')\chi_s^{l}(r') \, dr \, dr'$

Results:

$\rightarrow$ $\varepsilon_1, \varepsilon_2, \varepsilon_3, \ldots$

$\rightarrow$ $\varepsilon_1(k), \varepsilon_2(k), \varepsilon_3(k), \ldots$

$\rightarrow$ $D_{pq} = \sum_j c_{jp} \, c_{jq}$

$\rightarrow$ $D_{pq}^{j} = a/2\pi \sum_n \int dk \, c_{np}^{*}(k)c_{nq}(k) \, e^{ikja}$

The equations are based on the fact that the Bloch's theorem implies equivalent relations between coefficients (c_{np}^{0}, c_{np}^{h}) of atomic orbitals (χ_p^{0}, χ_q^{h}) respectively centered in the origin unit cell (0) and in another unit cell (h), separated by a translation (ha):

$$c_{np}^{h} = e^{ikha} \, c_{np}^{0}$$

Hereafter, lower indices (p, q, r, or s) refer to the labeling of a given orbital (χ_p, χ_q, χ_r, or χ_s) within a unit cell while upper indices (j, h, l) refer to the position of a given unit cell. When applying HF or SCF concepts, we consider the motion of a single electron in the field of fixed nuclei and in the averaged Coulomb and Pauli (exchange) fields of all the electrons. In Chart 2.2, we drop the 0 upper indices and do not detail the extent of the summations.

In molecular quantum chemistry, a molecular orbital is expanded in terms of basis functions; secular systems of equations and determinants are solved; their eigenvalues are the orbital energies. From the LCAO coefficients, charges and bond orders (projection of the density matrices onto the limited basis used) are calculated. In polymer quantum chemistry, we take into account the lattice periodicity and solve equations similar to the molecules for each wave number k; the orbitals, the systems of equations, and the determinants are no longer real but have imaginary components. This introduces matrix elements between Bloch functions:

$$h_{pq}(k) = \sum_j e^{ikja} \int \chi_p(\mathbf{r}-\mathbf{P}-0a) \, h(\mathbf{r}) \, \chi_q(\mathbf{r}-\mathbf{Q}-ja) \, d\mathbf{r}$$

$$= \sum_j e^{ikja} \int \chi_p^0(\mathbf{r}) \, h(\mathbf{r}) \, \chi_q^j(\mathbf{r}) \, d\mathbf{r}$$

$$= \sum_j e^{ikja} \, h_{pq}^{\,j}$$

Those matrix elements contain lattice sums which are to be evaluated by adequate procedures. The key problem is thus to obtain the matrix elements over the basis functions:

$$h_{pq}^{\,j} = \int \chi_p^0(\mathbf{r}) \, h(\mathbf{r}) \, \chi_q^j(\mathbf{r}) \, d\mathbf{r}$$

Fortunately, these matrix elements decrease exponentially with the distance between the orbital centers and force the natural convergency properties of those formally infinite lattice sums.

2.3.2.2 The electrostatic balance between the nucleus- and electron-electron interactions

As already mentioned, the calculation of the matrix elements of the Fock operator in the selected basis is in itself a difficult problem due to the presence in extended systems of long-range interactions in the electron-nucleus attraction and in the electron-electron repulsion and exchange interactions. To clearly state this crucial point, we rewrite the detailed form of a single matrix element:

$$h_{pq}^{j} = \int \chi_p(\mathbf{r}) \, h(\mathbf{r}) \, \chi_q^{j}(\mathbf{r}) \, d\mathbf{r}$$

$$= T_{pq}^{j} \qquad \leftarrow \text{kinetic}$$

$$- \sum_{h,A} Z_A \, V_{pq|A}^{j\ h} \qquad \leftarrow \text{nuclear attraction}$$

$$+ \sum_{h} \sum_{l} \sum_{r} \sum_{s} 2 \, D_{rs}^{hl} \, (_{pq}^{\ j}|_{rs}^{hl}) \qquad \leftarrow \text{electron repulsion}$$

$$- \sum_{h} \sum_{l} \sum_{r} \sum_{s} D_{rs}^{hl} \, (_{pr}^{h\,j}|_{qs}^{\ l}) \qquad \leftarrow \text{exchange}$$

The elements of the density matrices are defined by the summation over all the doubly occupied electronic bands and their integration over the first Brillouin zone of the polymer:

$$D_{pq}^{h} = a/2\pi \sum_{n}^{occ} \int dk \, c_{np}{}^*(k) \, c_{nq}(k) e^{ikha}$$

In order to discuss the long-range behaviour of the electrostatic interactions, we rearrange the previous equation into:

$$h_{pq}^{j} = T_{pq}^{j} \qquad\qquad\qquad \leftarrow \text{kinetic}$$

$$+ \sum_{h} \sum_{l} \sum_{r} \sum_{s} 2 \, D_{rs}^{hl} \, \{ (_{pq}^{\ j}|_{rs}^{hl}) \qquad \text{electron-electron}$$

$$\qquad\qquad\qquad\qquad\qquad\qquad\qquad \leftarrow \text{Coulomb}$$

$$- \frac{1}{n_0} S_{rs}^{hl} \sum_{h,A} Z_A V_{pq|A}^{j\ h} \} \qquad \text{electron-nuclei}$$

$$- \sum_{h} \sum_{l} \sum_{r} \sum_{s} D_{rs}^{hl} \, (_{pr}^{h\,j}|_{qs}^{l}) \qquad \leftarrow \text{exchange}$$

where n_0 is the number of electrons per unit cell. In the last equation, we have rescaled the electron-nucleus attraction term to introduce the same double summation over h and l as in electron-electron Coulomb terms by making use of the normalization-like condition:

$$\frac{1}{N} \sum_h \sum_l \sum_r \sum_s 2 D_{rs}^{hl} S_{rs}^{hl} = \sum_l \sum_r \sum_s 2 D_{rs}^{0l} S_{rs}^{0l} = \sum_l \sum_r \sum_s 2 D_{rs}^{l} S_{rs}^{l} = n_0$$

In the programs developed earlier, not enough attention had been paid to the convergency properties of the contribution which is controlled by the balance between the electron density and the sum of the nuclear charges:

$$\sum_h \sum_l \sum_r \sum_s 2 D_{rs}^{hl} \{ (_{pq}^{j}|_{rs}^{hl}) - \frac{1}{n_0} S_{rs}^{hl} \sum_A Z_A V_{pq|A}^{j \; h} \}$$

$$= \int dr \, \chi_p(r) \{ \int dr' \, \rho(r')/|r-r'|^{-1} - \sum_{h,A} Z_A |r-R_A|^{-1} \} \chi_q^j(r)$$

The electron density is defined for polymers by:

$$\rho(r) = \sum_h \sum_l \sum_r \sum_s 2 D_{rs}^{hl} \chi_r^h(r) \chi_s^l(r)$$

$$= \sum_h \sum_l \sum_r \sum_s 2 D_{rs}^{0,l-h} \chi_r^h(r) \chi_s^l(r)$$

where $D_{rs}^{0,l-h}$ is the density matrix element between atomic orbital χ_r in the origin cell (0) and atomic orbital χ_s in cell (l-h). The summations over h and l extend over all the unit cells of the polymer chain, those over r and s over the atomic orbitals present in one cell.

No serious difficulties were anticipated other than those due to the nature and the size of the atomic basis set such as encountered in molecular quantum chemistry. A detailed analysis shows that the formula for the electron density already points to interesting translational symmetry relations. It must be realized that in this formula, the translational symmetry of the periodic polymer is only implicitly introduced by the identities between translationally equivalent elements of the density matrix:

$$D_{rs}^{hl} = D_{rs}^{0,l-h}$$

Two lattice summations acting on all terms are, in principle, to be performed. However, an explicit use of the translational periodicity can be made by a straightforward manipulation [2.31] when using a new index m = l - h:

$$\rho(\mathbf{r}) = \sum_h \sum_l \sum_r \sum_s 2\, D_{rs}^{0,l\text{-}h}\, \chi_r^h(\mathbf{r})\, \chi_s^l(\mathbf{r})$$

$$= \sum_h \sum_m \sum_r \sum_s 2\, D_{rs}^{0m}\, \chi_r^h(\mathbf{r})\, \chi_s^{h+m}(\mathbf{r})$$

$$= \sum_m \sum_r \sum_s 2\, D_{rs}^{0m} \sum_h \chi_r^h(\mathbf{r})\, \chi_s^{h+m}(\mathbf{r})$$

This leads to an important simplification of the formula by letting the h index act only on the atomic orbitals. The use of the same idea for the two-electron integrals also results in a significant simplification of the integral formulas and a decrease in the number of integrals to be stored. This speeds up considerably both the integral and SCF parts of a polymer program with respect to the traditional approach, thereby making feasible routine calculations on polymers of a "chemically" interesting size. It is striking that little attention has been paid to formulations such as this one although it was implicit already in a paper by Pisani and Dovesi in 1980 [2.32] but without direct reference to the electronic density.

From straight chemical intuition, it was usually believed that the range of effectively interacting cells which should be explicitly considered in the actual calculations are very small though no consistent rules for the truncation of j, h, and l summations were actually supplied.

The electrostatic attraction and repulsion terms:

$$+ \sum_h \sum_l \sum_r \sum_s 2\, D_{rs}^{hl} \left({}_{pq}^j \Big|_{rs}^{hl} \right) - \sum_{h,A} Z_A\, V_{pq|A}^{j\,h}$$

are simply the Coulomb energy of a negative charge distribution:

$$P_{pq}^j(\mathbf{r}) = \chi_p(\mathbf{r})\, \chi_q^j(\mathbf{r})$$

in the field of the positive framework of all the nuclei and the negative electron distribution:

$$P_{rs}^{l\text{-}h}(\mathbf{r}) = \chi_r(\mathbf{r})\, \chi_s^{l\text{-}h}(\mathbf{r})$$

Due to cell neutrality, the total charge per cell is zero so that the electrostatic contributions of cells far away from the center of the charge distribution $P_{pq}^j(r)$ go to zero. In this sense, even if individual nuclear attraction and electron repulsion integrals decrease very slowly as a function of the distance (r^{-1} dependence), there is a strict cancellation at medium and large distances between the diverging nuclear attraction and electron repulsion terms since positive nuclear charges exactly compensate the negative electron density of an equal number of electrons.

This is easily illustrated by investigating the contributions of electrostatic interactions from cell (h) with the density $P_{pq}^j(r)$:

$$\int P_{pq}^j(r_1) [-\sum_A Z_A |r_1 - R_A - ja|^{-1} + \sum_l \sum_r \sum_s 2 D_{rs}^{hl} \int P_{rs}^{l-h}(r_2) |r_1 - r_2|^{-1} dr_2] dr_1$$

If the center of charge density $P_{pq}^j(r_1)$ is G_{pq}^j and that of $P_{rs}^{l-h}(r_2)$ is G_{rs}^{l-h}, when G_{pq}^j and G_{rs}^{l-h} are sufficiently far apart, then $r_{12} \approx |G_{pq}^j - G_{rs}^{l-h}| \approx r_{1A} \approx |R_A - G_{pq}^j|$. As a consequence, we can reasonably approximate the former expression by:

$$\approx S_{pq}^j |R_A - G_{pq}^j|^{-1} [\sum_A Z_A + \sum_l \sum_r \sum_s 2 D_{rs}^{01} S_{rs}^{01}]$$

This leads to an absolute cancellation since the two terms within the brackets are, from left to right, the total nuclear charge and the total number of electrons (normalization condition) per unit cell.

This type of argumentation which is in principle correct [2.33] has much delayed a formal, but necessary mathematical analysis of the exact convergence properties of these series, to define better the numerical applicability of the LCAO equations for polymer chains. However, it must be pointed out that as early as 1956, Löwdin [2.34] had already stressed the formal difficulties which would arise in the numerical solutions of this type of equation. Furthermore, starting with the work of O'Shea and Santry [2.35] in 1974 and Ukrainskii [2.36] in 1975, it became gradually apparent that model chains embody also size-related difficulties, mathematically expressed as conditionally convergent lattice sums. This is due to the fact that the centers of positive (nuclear) charges and negative (electron) distributions do not always coincide and that, in almost every case, small but nonnegligible local dipoles do exist. In such a case, the mathematical analysis shows that the convergence of nucleus and electron-electron interactions goes as:

$$\sum_{n=1}^{N} \left\{ \frac{1}{n+\delta} - \frac{1}{n} \right\}$$

which can be very slow if the separation of charges δ is significant as indicated by Table 2.9.

Table 2.9 **Numerical convergence properties of** $\sum_{n=1}^{N} \left\{ \frac{1}{n+\delta} - \frac{1}{n} \right\}$ **with respect to δ (separation of centers of positive and negative charges) and to N (number of terms in the numerical expansion)**

N	$\delta = 0.1$	$\delta = 0.01$	$\delta = 0.001$
10	0.143 989 1	0.015 379 0	0.001 548 6
50	0.151 482 6	0.016 132 2	0.001 623 9
100	0.152 466 2	0.016 230 7	0.001 633 8
1000	0.153 360 8	0.016 320 2	0.001 642 7
10000	0.153 450 7	0.016 329 7	0.001 643 7
10^5	0.153 459 7		
10^6	0.153 460 6		
10^7	0.153 460 7		

Several solutions to this problem have been investigated by the Namur group. A first technique [2.15] makes use of the approximation of the F_0 function which is involved in the calculations of polycenter integrals over Gaussian functions [2.33]. It is thus only valid for the formalism where those functions are involved, i.e., mainly for Gaussian-type integrals. If this approximation is used in the infinite summations of the LCAO formalism, it is possible to single out quantities which are immediately recognized as traditional Madelung expressions. Those Madelung summations, characteristic of the system and of the basis set used, can be evaluated by a direct application of the Laplace transformation [2.37] or the Fourier representation method [2.38].

This has led to another elegant way when using the Fourier transformation technique for evaluating the exact cancellation [2.39] of the direct space matrix elements. This procedure has, however, not been extended since it intrinsically induces some incoherence between the direct space representation of the matrix elements and the reciprocal working space of the Fourier transform.

A more easily implemented solution is based on the multipole expansion which also handles correctly those long-range interactions. A complete formulation has been given for stereoregular linear polymers [2.40] and for periodic helices [2.41] and is summarized below. It only implies the accurate evaluation of the multipole moments of the charge distribution in the unit cell (or in the reduced unit cell for helical polymers). The technique rests on the multipole expansion of the interaction operator r_{ij}^{-1} within the Fock matrix element. This procedure is valid as soon as there is no significant overlap between the charge distributions whose interactions are to be calculated. This technique has the advantage of taking into account the long-range effects in the Fock matrix elements directly in the framework of the exact LCAO formalism. It is thus possible to perform the iterations to relax the charge distribution of the system in its own field. In this way, the induced multipole interactions are automatically adjusted. Exact summations up to infinity are carried out in such a way as to minimize the round-off errors. Selective control on the convergence and more specifically, on the quantities one can afford to compute (*e.g.*, electric moments) is possible and suitable. The method is straightforward to implement into computer programs since the necessary quantities are one-electron integrals which are easily calculated. The number of such integrals grows as the second power of the basis set size N and is proportional to the number of interacting cells N_C explicitly considered in the calculation, i.e., $N^2 N_C$. The computing time is thus negligible with respect to conventional techniques where, as previously cited, a number of $N^4 N_C^3$ terms has to be computed. The method has the additional advantage of being physically appealing through the multipole moments which constitute a traditional framework for interpreting electrostatic interactions.

Table 2.10 provides an example of the better convergency properties of such an approach in the case of an FSGO calculation on a polyethylene chain. The FSGO basis set consists here of two $1s_C$ orbitals, two C-C bond orbitals, and four C-H bond orbitals. In the infinite chain, both carbons are equivalent. In fact, owing to the FSGO basis set used, an asymmetry is artificially introduced within the unit cell which has the same effect as a dipole moment (difference between the barycenters of the orbital distribution and of the nuclei). The effects of this asymmetry become smaller as the number of cells explicitly taken into account gets larger.

Table 2.10 FSGO selected electronic properties of polyethylene (in a.u.)

		Without Multipole Expansion	With Multipole Expansion
Total energy			
	$N_C = 1$	-32.904242	-33.002797
	2	-32.968653	-33.002275
	3	-32.985369	-33.002273
	4	-32.992108	-33.002273
C-C Band width			
	$N_C = 1$	0.26092	0.25194
	2	0.25197	0.25062
	3	0.25094	0.25050
	4	0.25070	0.25050
C-C Band H degeneracy			
	$N_C = 1$	0.05513	0.00289
	2	0.02313	0.00021
	3	0.01198	0.00001
	4	0.00733	0.00000
C-C $\leftrightarrow$ C-H gap			
	$N_C = 1$	0.22930	0.27989
	2	0.26737	0.28419
	3	0.27534	0.28442
	4	0.27885	0.28444

2.3.2.3 Multipole expansion for long-range Coulomb interactions

As noted in the previous Section, in current polymer calculations, it is obviously impossible to deal with very large values of the number of unit cells since the two-electron part of an LCAO polymer calculation involves an enormous number of integrals. There is a need for limiting that number of unit cells to some amenable value N (practically ranging from 1 in the nearest-neighbour approximation to about 5) much less than the value N_C previously introduced. Long-range interactions behave like conditionally and slowly convergent series and actually contribute significantly far beyond this number N. A detailed analysis has solved this problem by the use of a multipole expansion. The form of the matrix elements is cast as:

$$h_{pq}^{j} = T_{pq}^{j} + \sum_{l=-N}^{N} \sum_{r,s} 2 D_{rs}^{l} \sum_{h=-N}^{N} [(_{pq}^{j}|_{rs}^{hh+1}) - \frac{1}{n_0} S_{rs}^{l} \sum_{A} Z_A V_{pq|A}^{j h}]$$

$$+ L_{pq}^{j}(N) - \sum_{h} \sum_{l} \sum_{r} \sum_{s} D_{rs}^{hl} (_{pr}^{h j}|_{qs}^{l})$$

with:

$$L_{pq}^{j}(N) = \{ \sum_{h=-\infty}^{\infty} - \sum_{h=-N}^{N} \} [- \sum_{A} Z_A V_{pq|A}^{j h} + \sum_{l=-N}^{N} \sum_{r,s} 2 D_{rs}^{l} (_{pq}^{j}|_{rs}^{hh+1})]$$

The precise deduction of the multipole expansion gives:

$$L_{pq}^{j}(N) = \sum_{k=0}^{\infty} \sum_{\ell=0}^{\infty} U_{pq}^{j(k,\ell)} a^{-(k+\ell+1)} \Delta_N^{(k+\ell+1)}$$

where:

$$\Delta_N^{(n)} = \sum_{h=1}^{\infty} h^{-n} - \sum_{h=1}^{N} h^{-n} = \zeta(n) - \sum_{h=1}^{N} h^{-n}$$

$\zeta(n)$ is the Riemann zeta function, $U_{pq}^{j(k,\ell)}$ is the interaction of the 2k-pole and 2ℓ-pole moments:

$$U_{pq}^{j(k,\ell)} = \sum_{m=-s(k,\ell)}^{s(k,\ell)} (k+\ell)!(-1)^m[(-1)^k+(-1)^\ell][(k+|m|)!(\ell+|m|)!]^{-1}M_{pq}^{j(k,m)}M^{(\ell,m)*}$$

In the above expressions, s is equal to the smallest numbers between k and ℓ. Capital letter $M^{(\ell,m)}$ refers to the mth component of the 2^ℓth electric moment expressed in spherical coordinates and related to the charge distributions associated with the orbital product $P_{pq}^j(\mathbf{r})$:

$$M_{pq}^{j(k,m)} = \int \chi_p^0(\mathbf{r}) \mid r^k P_k^{|m|} \cos(\theta) e^{im\phi} \mid \chi_q^j(\mathbf{r}) \, d\mathbf{r}$$

or to the total charge (electrons + nuclei) associated with each translational unit:

$$M^{(k,m)} = -\sum_A Z_A R_A^k P_k^{|m|} \cos(\theta_A) e^{im\phi_A}$$

$$+ \sum_{j=-N}^{N} \sum_{p,q} 2 D_{pq}^j \int \chi_p^0(\mathbf{r}) \mid r^k P_k^{|m|} \cos(\theta) e^{im\phi} \mid \chi_q^j(\mathbf{r}) \, d\mathbf{r}$$

Because of electroneutrality constraints:

$$U_{pq}^{j(k,0)} = 0$$

and to avoid coordinate dependence of the results, truncation is performed to sum only $U_{pq}^j{}^{(k,\ell)}$ terms of identical $k+\ell+1$ values. We obtain the working formula:

$$L_{pq}^j(N) = \sum_{k=3,5,\dots;k=odd} \Delta_N^{(k)} a^{-k} \sum_{\ell=1}^{k-1} U_{pq}^{j(k-\ell-1,\ell)}$$

where a selective control can be made by the N and k parameters.

2.3.2.4 The short- or long-range character of the exchange contribution

Important and recent analyses of the exchange term have been given in the literature starting with important papers by Ukrainskii [2.36] and by Piela, André, Fripiat & Delhalle [2.42], completed by more detailed studies by Calais & Delhalle [2.43], and Monkhorst & Kertesz [2.44]. We summarize here a simple approach to the problem.

The exchange contribution:

$$- \sum_h \sum_l \sum_r \sum_s D_{rs}^{hl} \left({}_{pr}^{h}{}_{qs}^{jl} \right) = - \sum_h \sum_l \sum_r \sum_s D_{rs}^{0,l-h} \left({}_{pr}^{h}{}_{qs}^{jl} \right)$$

corrects for the self-repulsion included in the Coulomb part of the SCF one-electron operator as magisterially demonstrated in solid state calculations by Slater [2.45]. It also includes the effect of the Pauli principle on the independent electron model. Clearly, the magnitude of the integrals $\left({}_{pr}^{h}{}_{qs}^{jl} \right)$ is determined, in the same sense as indicated above, by the distributions:

$$P_{pr}^{h} = \chi_p(\mathbf{r}) \, \chi_r^{h}(\mathbf{r}) \quad \text{and} \quad P_{qs}^{jl} = \chi_q^{j}(\mathbf{r}) \, \chi_s^{l}(\mathbf{r})$$

and by the distance between their center of charges. The terms in h and l summations are significant near the origin cell (0) for the summation over h and near the cell (j) for the summation over l:

$$- \sum_h \sum_l \sum_r \sum_s D_{rs}^{hl} \left({}_{pr}^{h}{}_{qs}^{jl} \right)$$

$$\approx - \sum_r \sum_s D_{rs}^{0j} \left({}_{pr}^{0}{}_{qs}^{jj} \right)$$

$$+ D_{rs}^{0j-1} \left({}_{pr}^{1}{}_{qs}^{jj} \right) + D_{rs}^{0j+1} \left({}_{pr}^{-1}{}_{qs}^{jj} \right) + \dots$$

$$+ D_{rs}^{0j+1} \left({}_{pr}^{0}{}_{qs}^{j,j+1} \right) + D_{rs}^{0j-1} \left({}_{pr}^{0}{}_{qs}^{j,j-1} \right) + \dots$$

$$\approx D_{pq}^{0j} \left({}_{pp}^{0}{}_{qq}^{jj} \right)$$

This points out that exchange contributions can be important to Fock matrix elements between widely separated orbitals since the decrease of both two-center repulsion integrals $(^{0\,j\,j}_{pp'qq})$ and elements of density matrices D^{0j}_{pq} can be very slow with distance. In nonmetallic cases, D^j_{pq} decays exponentially and in metallic cases it decays like $|j|^{-1}$. Up to rather recently, little had been undertaken to characterize the convergence of D^{0j}_{pq} with respect to j. However, troublesome numerical results have triggered such an analysis.

From the few existing studies, mainly based on simplified methods such as the Hückel approach, it was known that D^{0j}_{pq} can decay as slowly as $(-1)^j|j|^{-1}$ leading to a $(-1)^j|j|^{-2}$ asymptotic decay of the exchange terms as first reported by Ukrainskii [2.36] in a semiempirical work on a metallic polyacetylene model. In the previously cited analysis of *ab initio* results on an hydrogen chain model [2.42], we have gained insight into the exchange behaviour and reached the conclusions that poor convergency of the exchange energy is connected to the HOMO-LUMO separation (energy gap). In cases where the Hartree-Fock scheme produces good quality results (saturated polymers, insulators,...), it is observed that the exchange potential is of a short-range nature. By contrast, in metallic situations, essentially characterized by a degeneracy of the HOMO and LUMO levels, the exchange potential is of a very long-range nature. However, this situation precisely corresponds to the case where a single-determinant approximation to the ground state wave function does not even provide a qualitatively correct description of the system.

The understanding of the long-range behaviour of the exchange term was further developed by a study due to Monkhorst and Kertesz [2.44]. Using a two-band model system within the Hückel methodology, they recovered the $(-1)^j|j|^{-2}$ asymptotic decay of exchange in the half-filled band (metallic) situation and established a simple relationship between band gap, band width and a large-j decay of the density matrix elements in the filled band (insulating) situation.

More recently, Delhalle, Fripiat, and Harris [2.46] have provided a general analysis of the convergency of D^{0j}_{pq} in terms of basic theorems on the convergence of Fourier series coefficients. They have shown that the convergency of D^{0j}_{pq} is essentially determined by the analytic properties of $D_{pq}(k)$:

$$D_{pq}(k) = \sum_j D^j_{pq} e^{ikja}$$

throughout the first Brillouin zone. In general, it can be stated that fast decaying matrix elements D_{pq}^j correspond to insulators with large band gaps and, at the other extreme, metallic systems will lead to poorly converging D_{pq}^j.

A typical example has been published recently [2.47]. Minimal basis set calculations on the metallic infinite chain of hydrogen atoms have been performed within the computational framework we just described to illustrate the dependence of the Hartree-Fock energy bands on the summation of exchange contributions. The numerical results show the gradual decay of the density of states at the Fermi level as the number of terms in the exchange lattice sums is increased. This last point is a typical deficiency of the Hartree-Fock techniques as applied to infinite systems, independently of their dimensionality.

2.3.2.5 Use of screw symmetry in polymer calculations

Many polymers exist in helical conformations, which leads to large translational units. For instance, the stable conformation of polypropylene is an helix which includes three asymmetric units, -[CH_2-$CH(CH_3)$]-, per turn; the corresponding translational cell includes 27 atoms (18 hydrogens and 9 carbons). Even in the nearest-neighbour approximation, a reliable *ab initio* calculation would be a formidable task. A straightforward way to reduce this numerical effort is to consider explicitly the helical symmetry as it was proposed by Imamura and Fujita and by Blumen and Merkel [2.48].

The common helix designation L*M/t specifies the number L of skeletal atoms in the asymmetric unit of the chain and the number of such asymmetric units M per t turns of the helix in the translational repeat. The translation period is denoted by a, and $\tau = a/M$ is the length of the reduced (helical) period which we hereafter call reduced unit cell or simply cell. The whole polymer is generated from this unit by applying screw symmetry operations $S^j(r) = D(j\alpha)(r) + j\tau$ where $\alpha = 2\pi t/M$ (or $-2\pi t/M$) for a right- (or left-) handed helix. $D(\alpha)$ is a matrix representing a clockwise rotation around the z axis by the angle α; τ is a vector of component $(0,0,\tau)$ directed along z. Chart 2.3 compares the usual equations of polymer quantum chemistry with and without helical translations. Since the size of the unit cell in the direct space is reduced by considering the helical operations, k can be regarded as a wave vector in the extended Brillouin zone ($-\pi/\tau \leq k \leq +\pi/\tau$).

Other important contributions have been published on the group theory developments and applications of the polymer helical symmetry [2.49] as well as on practical procedures to simplify actual calculations [2.50].

Chart 2.3   A comparison of some basic formulas of polymer quantum chemistry with and without helical rotation

Without helix	With helix

Orbital:

$$\phi_n(k,r) = \frac{1}{\sqrt{N}} \sum_j e^{ikja} \sum_p c_{np}(k)\, \chi_p(r\text{-}P\text{-}ja) \qquad \phi_n(k,r) = \frac{1}{\sqrt{N}} \sum_j e^{ikj\tau} \sum_p c_{np}(k)\, \chi_p\{D(\text{-}ja)[r\text{-}S^j(P)]\}$$

Secular system:

$$\sum_p c_{np}(k) \{ \sum_j e^{ikja}\, [h_{pq}^{\,j} - \varepsilon_n(k)\, S_{pq}^{\,j}] \} \qquad \sum_p c_{np}(k) \{ \sum_j e^{ikj\tau}\, [h_{pq}^{\,j} - \varepsilon_n(k)\, S_{pq}^{\,j}] \}$$

$$\equiv \sum_p c_{np}(k) \{ h_{pq}(k) - \varepsilon_n(k)\, S_{pq}(k) \} = 0 \qquad \equiv \sum_p c_{np}(k) \{ h_{pq}(k) - \varepsilon_n(k)\, S_{pq}(k) \} = 0$$

Secular determinant:

$$|\sum_j e^{ikja}\, [h_{pq}^{\,j} - \varepsilon_n(k)\, S_{pq}^{\,j}]| \qquad |\sum_j e^{ikj\tau}\, [h_{pq}^{\,j} - \varepsilon_n(k)\, S_{pq}^{\,j}]|$$

$$\equiv | h_{pq}(k) - \varepsilon_n(k)\, S_{pq}(k) | = 0 \qquad \equiv | h_{pq}(k) - \varepsilon_n(k)\, S_{pq}(k) | = 0$$

A point worth mentioning is that some electron integrals have to be rotated to take into account the helical symmetry as seen from the expression of the Bloch orbitals in Chart 2.3. If we take z as the direction along the polymer chain axis, the s- and p_z-type orbitals are invariant under the helical rotation. The p_x- and p_y-rotated orbitals are obtained from Cartesian p_x- and p_y-ones by the following relations:

$$p_x^j = p_x^j \cos(j\alpha) + p_y^j \sin(j\alpha)$$

$$p_y^j = -p_x^j \sin(j\alpha) + p_y^j \cos(j\alpha)$$

The form of the matrix elements is cast into the same form as in Section 2.3.2.3:

$$h_{pq}^j = T_{pq}^j + \sum_{l=-N}^{N} \sum_{r,s} 2 D_{rs}^l \sum_{h=-N}^{N} [\binom{j}{pq}\binom{hh+1}{rs} - \frac{1}{n_0} S_{rs}^l \sum_A Z_A V_{pq|A}^{j\,h}]$$

$$+ L_{pq}^j (N) - \sum_h \sum_l \sum_r \sum_s D_{rs}^{hl} \binom{h\,j\,\,l}{pr\,qs}$$

with:

$$L_{pq}^j (N) = \{ \sum_{h=-\infty}^{\infty} - \sum_{h=-N}^{N} \} [- \sum_A Z_A V_{pq|A}^{j\,h} + \sum_{l=-N}^{N} \sum_{r,s} 2 D_{rs}^l \binom{j}{pq}\binom{hh+1}{rs}]$$

The detailed deduction of the multipole expansion now gives:

$$L_{pq}^j(N) = \sum_{k=0}^{\infty} \sum_{\ell=0}^{\infty} U_{pq}^{j(k,\ell)} (N)\, \tau^{-(k+\ell+1)}$$

where:

$$U_{pq}^{j(k,\ell)} (N) = \sum_m 2(-1)^{\ell+m} (k+\ell)!\, [(k+|m|)!\,(\ell+|m|)!]^{-1}\, w(k+\ell+1)$$

$$\Delta_{k+\ell+1}(N,m,\alpha)\, M_{pq}^{*j(k,m)}(0)\, M^{(\ell,m)}(0)$$

In the above expressions, $M_{pq}^{j(k,m)}(h)$ and $M^{(\ell,m)}(h)$ are the mth component of the 2^k-pole moment of the electron distribution and of the charge content of cell h, respectively; $w(k+\ell+1)$ takes the value 1 for $k+\ell+1$ odd; +i for $k+\ell+1$ even and right-handed helix; -i for $k+\ell+1$ even and a left-handed helix. The functions $\Delta_{k+\ell+1}(N,m,\alpha)$ are defined by:

$$\Delta_{k+\ell+1}(N,m,\alpha) = \zeta(k+\ell+1,m\alpha) - \sum_{h=1}^{N} \cos(hm\alpha)\, h^{-(k+\ell+1)}$$

$$\zeta(k+\ell+1,m\alpha) = \sum_{h=1}^{\infty} \cos(hm\alpha)\, h^{-(k+\ell+1)}$$

if $k+\ell+1$ is odd, or:

$$\Delta_{k+\ell+1}(N,m,\alpha) = \zeta(k+\ell+1,m\alpha) - \sum_{h=1}^{N} \sin(hm\alpha)\, h^{-(k+\ell+1)}$$

$$\zeta(k+\ell+1,m\alpha) = \sum_{h=1}^{\infty} \sin(hm\alpha)\, h^{-(k+\ell+1)}$$

if $k+\ell+1$ is even.

The functions $\Delta_{k+\ell+1}(N,m,\alpha)$ have to be evaluated only once since the arguments are only related to the type of helix and to the conditions of the calculations (i.e., the input number of cells N). As in Section 2.3.2.3, capital letter $M(\ell,m)$ refers to the mth component of the 2^{ℓ}th electric moment expressed in spherical coordinates and related to the charge distributions associated with the orbital product $P_{pq}^{j}(r)$ or to the total charge (electrons + nuclei) associated with each translational unit. Explicit working formulas are given in [2.51].

2.3.3 All Valence Electron Methods

2.3.3.1 ZDO-like methods

The Zero Differential Overlap (ZDO) approximation assumes that the differential overlap density of two different atomic orbitals is zero:

$$\rho_{pq}(r) = \chi_p(r)\, \chi_q(r) = \delta_{pq}\, \chi_p^{2}(r)$$

Note that, in an all-valence electron calculation, this assumption is satisfied for the orbitals centered on the same nucleus due to the symmetry requirement between orbitals of different angular and magnetic quantum numbers like 2s, 2px, 2py, and 2pz. It is, in general, not rigorously satisfied for orbitals centered on different nuclei which would imply $\chi_p(\mathbf{r})$ or $\chi_q(\mathbf{r})$ or both to be identically zero for each $\mathbf{r}$ value. Inversely, the ZDO assumption automatically implies that the total integrated overlap S_{pq} is zero as it would have been within an orthogonalized basis set:

$$S_{pq} = \int \rho_{pq}(\mathbf{r})\, d\mathbf{r} = \int \chi_p(\mathbf{r})\, \chi_q(\mathbf{r})\, d\mathbf{r} = \int \delta_{pq}\, \chi_p^2(\mathbf{r})\, d\mathbf{r} = \delta_{pq}$$

The ZDO-basis sets are thus assumed to form orthonormal basis sets. In the same logic, nuclear attraction integrals are neglected if they imply orbitals which fulfill the ZDO-requirement:

$$V_{pq|A} = \int \chi_p(\mathbf{r})\, |\mathbf{r}\text{-}\mathbf{R}_A|^{-1}\, \chi_q(\mathbf{r})\, d\mathbf{r}$$

$$= \int \chi_p(\mathbf{r})\, \chi_q(\mathbf{r})\, |\mathbf{r}\text{-}\mathbf{R}_A|^{-1}\, d\mathbf{r}$$

$$= \int \rho_{pq}(\mathbf{r})\, |\mathbf{r}\text{-}\mathbf{R}_A|^{-1}\, d\mathbf{r} = 0 \qquad \text{if } p \neq q$$

In the case of two-electron integrals (pq|rs), the most attractive feature of the ZDO scheme is that it only retains the N^2 one- and two-center Coulomb integrals (pp|qq) and neglects the N^4-N^2 other integrals (mainly three- and four-center), but also the one- and two-center exchange integrals of the type (pq|pq):

$$(pq|rs) = \iint \chi_p(\mathbf{r})\chi_q(\mathbf{r})\, |\mathbf{r}\text{-}\mathbf{r}'|^{-1}\, \chi_r(\mathbf{r}')\chi_s(\mathbf{r}')\, d\mathbf{r}\, d\mathbf{r}'$$

$$= \delta_{pq}\, \delta_{rs} \iint \chi_p(\mathbf{r})\chi_p(\mathbf{r})\, |\mathbf{r}\text{-}\mathbf{r}'|^{-1}\, \chi_r(\mathbf{r}')\chi_r(\mathbf{r}')\, d\mathbf{r}\, d\mathbf{r}'$$

$$= \delta_{pq}\, \delta_{rs}\, (pp|rr)$$

This indeed reduces the computational effort from N^4 to N^2. The main equations of ZDO techniques are summarized in Chart 2.4. Note that z_A is the nucleus charge Z_A partially compensated by core-electron charges and that Q_p and l_{pq} are charges and bond orders in an orthogonalized basis set:

$$Q_p = \sum_j^{occ} 2\, c_{jp}^2 \qquad\qquad\qquad l_{pq} = \sum_j^{occ} 2\, c_{jp}\, c_{jq}$$

Chart 2.4 Principles of Zero Differential Overlap (ZDO) techniques

---

Core matrix elements:
$$h^c_{pq} = \int \chi_p(\mathbf{r})\, h^c(\mathbf{r})\, \chi_q(\mathbf{r})\, d\mathbf{r}$$
$$= \int \chi_p(\mathbf{r})\, [-\tfrac{1}{2}\nabla^2 - \sum_A z_A\, |\mathbf{r} - \mathbf{R}_A|^{-1}]\, \chi_q(\mathbf{r})\, d\mathbf{r}$$

Diagonal:
$$h^c_{pp} = T_{pp} - z_A \int \chi_p(\mathbf{r})\, |\mathbf{r} - \mathbf{R}_A|^{-1}\, \chi_p(\mathbf{r})\, d\mathbf{r} \qquad \Rightarrow U^A_{pp}$$

$$- \sum_{B \neq A} z_B \int \chi_p(\mathbf{r})\, |\mathbf{r} - \mathbf{R}_B|^{-1}\, \chi_p(\mathbf{r})\, d\mathbf{r} \qquad \Rightarrow \sum_{B \neq A} V^B_{pp}$$

Nondiagonal:
$$h^c_{pq} = T_{pq} \cong \beta_{AB}\, S_{pq} \qquad\qquad p \in A,\ q \in B$$

Repulsion and exchange matrix elements:
$$\int \chi_p(\mathbf{r})\, [\sum_j^{occ} 2 J_j(\mathbf{r}) - K_j(\mathbf{r})]\, \chi_q(\mathbf{r})\, d\mathbf{r}$$

Diagonal:
$$\int \chi_p(\mathbf{r})\, [\sum_j^{occ} 2 J_j(\mathbf{r}) - K_j(\mathbf{r})]\, \chi_p(\mathbf{r})\, d\mathbf{r}$$

$$= \sum_j^{occ} \sum_r \sum_s c_{jr}\, c_{js}\, \{2(pp|rs) - (pr|ps)\}$$

$$= \sum_j^{occ} \{2 \sum_r c_{jr}^2\, (pp|rr) - c_{jp}^2\, (pp|pp)\}$$

$$= \sum_r Q_r\, (pp|rr) - \frac{Q_p}{2}\, (pp|pp)$$

Nondiagonal:
$$\int \chi_p(\mathbf{r})\, [\sum_j^{occ} 2 J_j(\mathbf{r}) - K_j(\mathbf{r})]\, \chi_q(\mathbf{r})\, d\mathbf{r}$$

$$= \sum_j^{occ} \sum_r \sum_s c_{jr}\, c_{js}\, \{2(pq|rs) - (pr|qs)\}$$

$$= -\sum_j^{occ} c_{jp}\, c_{jq}\, (pp|qq)$$

$$= -\frac{1_{pq}}{2}\, (pp|qq)$$

When used as such, the previous approximation suffers from an important deficiency since it misses the requirement that the results should be rotationally invariant, i.e., should not depend on the way the molecule is localized in a Cartesian system. Figure 2.2 (a) illustrates this point by the example of a given repulsion integral (2px'2py'|2s2s) calculated in two Cartesian systems of coordinates rotated by 45° in the xy plane.

(a)

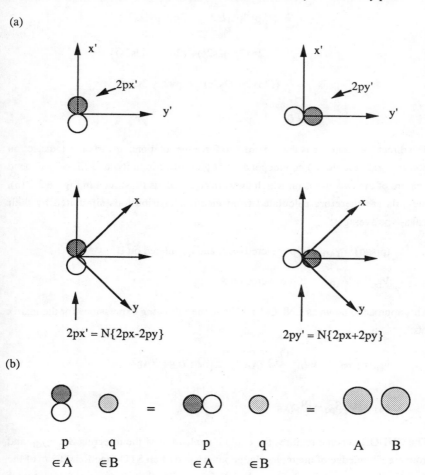

2px' = N{2px-2py}   2py' = N{2px+2py}

(b)

Figure 2.2   (a) Rotational invariance in ZDO scheme
(b) Invariance of electrostatic integrals with respect to orbital type

By expanding the orbital 2px' in one system of coordinate as 2px - 2py in the other system, and 2py' as 2px + 2py, we find that this integral would vanish in the first system of axes while being non-zero in the other:

$$(2px'2py'|2s2s) = 0 \qquad \qquad \text{by the ZDO-approximation}$$

$$= \frac{1}{2} \ \{(2px2px|2s2s) - (2py2py|2s2s)$$

$$+ \quad (2px2py|2s2s) - (2py2px|2s2s)\}$$

$$= \frac{1}{2} \ \{(2px2px|2s2s) - (2py2py|2s2s)\}$$

$$\neq 0$$

The direct consequence is that, in order to force the rotational invariance, the electron repulsion integrals should: (i) be independent of their directionality, and (ii) only depend on the nature of the two atoms on which they are centered, as illustrated in Figure 2.2 (b). The integrals are therefore calculated from electron distributions simulated by their spherical s-type averages:

$$(pp|qq) = \gamma_{AB} \qquad \text{if p centered on A and q centered on B}$$

$$V_{pp}{}^B = V_{AB} \qquad \text{if p centered on A}$$

This approach known as CNDO/1 results in the following expressions for the matrix elements:

$$h_{pp} = U_{pp}{}^A + (Q_A - \frac{Q_p}{2}) \ \gamma_{AA} + \sum_{B \neq A} (Q_B \ \gamma_{AB} - V_{AB})$$

$$h_{pq} = \beta_{pq} \ S_{pq} - \frac{1_{pq}}{2} \ \gamma_{AB}$$

The CNDO/1 scheme requires the explicit evaluation of the integrals S_{pq}, γ_{AB}, and V_{AB} from the knowledge of the molecular geometry and of an STO functional form of the basis functions. These integrals are calculated from Roothaan's one- and two-center integral tables [2.2] using the real (radial and angular) orbital dependence for the overlap integrals S_{pq} and s-like orbitals for the Coulomb integrals γ_{AB} and V_{AB} with Slater exponents obtained from Slater's screening rules [2.1]. The one-center parameters U_{pp}^A are estimated from the valence ionization states of the corresponding atom; this approximation agrees with Koopmans' theorem [2.10] which states that if the electronic relaxation is neglected, the orbital energy is the negative of the corresponding ionization potential. Using general interaction HF rules for open-shell atomic systems, one gets:

$$- IP_p = \varepsilon_p = \int \chi_p(\mathbf{r}) \{h^c(\mathbf{r}) + \sum_{q \neq p} [J_q(\mathbf{r}) - K_q(\mathbf{r})] \}\chi_p(\mathbf{r}) \, d\mathbf{r}$$

with:

$$\int \chi_p(\mathbf{r}) \, h^c(\mathbf{r}) \, \chi_p(\mathbf{r}) \, d\mathbf{r} = U_{pp}^A$$

$$\int \chi_p(\mathbf{r}) \, J_q(\mathbf{r}) \, \chi_p(\mathbf{r}) \, d\mathbf{r} = (pp|qq) = \gamma_{AA}$$

$$\int \chi_p(\mathbf{r}) \, K_q(\mathbf{r}) \, \chi_p(\mathbf{r}) \, d\mathbf{r} = (pq|pq) = 0 \qquad \text{by ZDO approximation}$$

Hence, a valence state ionization energy can be expressed in terms of the core-effective charges z_A and the one-center integrals U_{pp}^A and γ_{AA}:

$$- IP_p = U_{pp}^A + \sum_{q \neq p} \gamma_{AA} = U_{pp}^A + (z_A - 1) \gamma_{AA}$$

which yields the CNDO/1 semiempirical estimate of U_{pp}^A:

$$U_{pp}^A = - IP_p - (z_A - 1) \gamma_{AA}$$

The β_{pq} term is evaluated as a sum of constant characteristics of atoms A and B which are determined in order to fit minimal basis set *ab initio* results on diatomic molecules:

$$\beta_{pq} = \beta_{AB} = \frac{\beta_A^0 + \beta_B^0}{2}$$

The CNDO/2 technique introduces two important modifications in the calculation of the diagonal CNDO matrix elements in order to better describe bond lengths and dipole moments. The first modification concerns the neglect of the so-called "penetration integrals" $(z_B \gamma_{AB} - V_{AB})$ to correct the CNDO/1 description which for instance leads to an erroneous slight binding of the $^3\Sigma_u$ dissociative state of the hydrogen molecule. In an elegant discussion, Pople [2.23] has shown that this incorrect binding is due to a poor balance between electron-electron repulsions and electron-nucleus attractions in the penetration integrals. The HF potential energy curve of this dissociative state is, according to standard principles:

$$E(^3\Sigma_u) = \varepsilon_{1\sigma g} + \varepsilon_{1\sigma u} - J_{1\sigma g\text{-}1\sigma u} + K_{1\sigma g\text{-}1\sigma u} + |R_A\text{-}R_B|^{-1}$$

The CNDO/1 approximations of these various terms are:

$$\varepsilon_{1\sigma g} = \frac{\alpha + \beta}{1 + S} = \alpha + \beta$$

$$\varepsilon_{1\sigma u} = \frac{\alpha - \beta}{1 - S} = \alpha - \beta$$

$$\varepsilon_{1\sigma g} + \varepsilon_{1\sigma u} = 2\alpha = 2h_{1s\text{-}1s} = U_{1s\text{-}1s}{}^A + (1 - \tfrac{1}{2})\,\gamma_{AA} + (\gamma_{AB} - V_{AB})$$

$$J_{1\sigma g\text{-}1\sigma u} = \tfrac{1}{4} \iint \{1s_a(\mathbf{r}) + 1s_b(\mathbf{r})\}^2 \, |\mathbf{r}\text{-}\mathbf{r'}|^{-1} \{1s_a(\mathbf{r'}) - 1s_b(\mathbf{r'})\}^2 \, d\mathbf{r}\, d\mathbf{r'}$$

$$= \frac{\gamma_{AA} + \gamma_{AB}}{2}$$

$$K_{1\sigma g\text{-}1\sigma u} = \tfrac{1}{4} \iint \{1s_a{}^2(\mathbf{r}) - 1s_b{}^2(\mathbf{r})\} \, |\mathbf{r}\text{-}\mathbf{r'}|^{-1} \{1s_a{}^2(\mathbf{r'}) - 1s_b{}^2(\mathbf{r'})\} \, d\mathbf{r}\, d\mathbf{r'}$$

$$= \frac{\gamma_{AA} - \gamma_{AB}}{2}$$

which leads to the specific CNDO/1 expression:

$$E(^3\Sigma_u) =$$

$$2\,U_{1s\text{-}1s}{}^A + \gamma_{AA} \qquad\qquad \Rightarrow \text{independent of the internuclear distance}$$

$$+ \gamma_{AB} - 2\,V_{AB} + |R_A\text{-}R_B|^{-1} \qquad \Rightarrow \text{dependent on the internuclear distance}$$

The poor balance of the term depending on the internuclear distance:

$$\gamma_{AB} - 2\,V_{AB} + |R_A\text{-}R_B|^{-1}$$

results in a net attraction between the two atoms due to the interpenetration of their electron clouds, even though the bond order between them is zero in the state considered. The

CNDO/2 approximation corrects these effects by neglecting the penetration integral and assumes $z_B \gamma_{AB} = V_{AB}$:

$$h_{pp} = U_{pp}{}^A + (Q_A - \frac{Q_p}{2}) \gamma_{AA}$$

$$+ \sum_{B \neq A} (Q_B - z_B) \gamma_{AB} + \sum_{B \neq A} (z_B \gamma_{AB} - V_{AB})$$

$$= U_{pp}{}^A + (Q_A - \frac{Q_p}{2}) \gamma_{AA} + \sum_{B \neq A} (Q_B - z_B) \gamma_{AB}$$

The second modification concerns the one-center parameter $U_{pp}{}^A$ which is now estimated not only from the ionization potential IP_p but also from the electron affinities EA_p of the corresponding orbital atomic valence states:

$$- IP_p = U_{pp}{}^A + (z_A - 1) \gamma_{AA}$$

$$- EA_p = U_{pp}{}^A + z_A \gamma_{AA}$$

$$U_{pp}{}^A = - \frac{1}{2}(IP_p + EA_p) - (z_A - \frac{1}{2}) \gamma_{AA}$$

The CNDO/2 nondiagonal matrix elements h_{pq} are calculated in the same way as in CNDO/1. Note that if all the atoms in the molecule are neutral (so that all $Q_B = z_B$ for all atoms B and $Q_q = 1$ for all orbitals q), the matrix element:

$$h_{pp} = - \frac{1}{2} (IP_p + EA_p) + \{(Q_A - z_A) - (\frac{Q_p}{2} - \frac{1}{2})\} \gamma_{AA}$$

$$+ \sum_{B \neq A} (Q_B - z_B) \gamma_{AB}$$

reduces to the simple form used in the Extended Hückel method (see next Section):

$$h_{pp} = - \frac{IP_p + EA_p}{2}$$

Interestingly, the latter formula also constitutes the theoretical basis for using a single parameter α for all one-center π-electron parameters in the simple Hückel method.

It is not the scope of this Chapter to give a detailed analysis of all the computational techniques generated by Pople's CNDO approach. In the Intermediate Neglect of Differential Overlap (INDO) scheme [2.52], differential overlap is retained for one-center integrals. The method is superior for properties where electron spin distribution is important like singlet-triplet splitting which depends strongly on electron exchange effects. Dewar's Modified Intermediate Neglect of Differential Overlap #3 (MINDO/3) [2.53] has had a great success to predict accurate heats of formation. Pople's first attempt of Neglect of Differential Diatomic Overlap (NDDO) [2.23] assumes ZDO only between AO's on distinct atoms. The parameters are fitted to *ab initio* orbital energies. Integrals like $(s_A p_{iA} | s_B p_{iB})$ or $(p_{jA} p_{iA} | p_{jB} p_{iB})$ are not neglected so that repulsions between different atoms depend on their relative orientations without violating the rotational invariance. MNDO (Modified Neglect of Diatomic Overlap) [2.54] is a carefully parameterized semiempirical version of NDDO. It yields accurate heats of formation and many other useful properties like geometries. It has been very successful in getting optimized geometrical structures of monomers, oligomers, and polymeric chains [2.55]. New parametric molecular models are continuously introduced based on this NDDO approximation. Austin Model #1 (AM1) [2.56] corrects for the major weaknesses of MNDO to reproduce hydrogen bonds. Examples of results obtained by the various ZDO techniques are also illustrated in Tables 2.2 to 2.6.

To summarize, we may say that the aim of Pople and his coworkers was to reproduce the results of *ab initio* SCF-MO-LCAO calculations with less computational effort. At the beginning, the results have suffered from the good and bad features of HF theories: good geometries and charge distributions but poor binding energies. In contrast, the aim of Dewar and coworkers has been to parameterize adequately in order to have a model that would give binding energies within the range of chemical accuracy with the same computational effort as CNDO or INDO.

In a polymeric band structure calculation, practical calculations are usually easy since the valence basis set is considered to be orthogonal; as in the Hückel method, the overlap Bloch sums in those conditions reduce to the δ values (0 or 1):

$$S_{pq}(k) = \sum_j e^{ikja} S_{pq}^j = \sum_j e^{ikja} \delta_{pq} \delta_{0j} = \delta_{pq}$$

The computational scheme is however not as simple as the Hückel scheme since non-nearest interactions are taken into account and are only limited by the convergence of the Bloch sums: $h_{pq}(k) = \sum_j e^{ikja} h_{pq}^j$ (in general about 10-15 unit cells depending on the size of the unit cell). Furthermore, since the method explicitly introduces the electron-electron repulsions, the computational scheme is iterative in order to generate a self-consistent field of Coulomb and exchange interactions. This approach is relatively cheap in terms of computer time since it avoids the full calculation of a huge number of two-electron integrals and plays an interesting role. However, it must be stressed again that it suffers from crude approximations and somewhat arbitrary simplifications. Therefore, the use of such semiempirical techniques should be carefully restricted to the cases for which they have been parametrized. Maybe more so than for any other computational framework, a "black box" type of approach should never be enforced since it could lead to potentially disastrous results.

2.3.3.2 Extended Hückel Method

The Extended Hückel (EH) method uses a nonorthogonal valence minimal basis set:

$$S_{pq}^j = \int \chi_p(r)\chi_q^j(r)\, dv \ne \delta_{pq}\, \delta_{0j}$$

The diagonal matrix elements of the one-electron operator are used as empirical parameters. As in CNDO schemes, they are evaluated from the valence ionization states of the corresponding atom. The Extended Hückel approximation also neglects the effects of the molecular field in the diagonal elements, just as if the electrons had the same motion in the molecule as in the isolated atom. Thus:

$$h_{pp}^0 = \int \chi_p(r)\, h(r)\, \chi_p(r)\, dr = \alpha_p = -IP_p$$

The nondiagonal matrix elements of the one-electron operator are obtained by using Mulliken's approximation [2.18] where differential overlap densities are calculated from one-center orbital densities:

$$\chi_p(r)\, \chi_q^j(r) = \frac{1}{2}\, S_{pq}^j\, [\, (\chi_p)^2(r) + (\chi_q^j)^2(r)\,]$$

and:

$$h_{pq}^j = \int \chi_p(r) \, h(r) \, \chi_q^j(r) \, dr = \beta_{pq}^j = \frac{1}{2} S_{pq}^j [\alpha_p + \alpha_q] = \frac{1}{2} S_{pq}^j [-IP_p - IP_q]$$

It is straightforward to prove that such a scheme does not provide any bonding in an homonuclear diatomic molecule like the hydrogen molecule since its bonding orbital $1\sigma g$ has an energy equal to that of the isolated 1s orbital:

$$\varepsilon_{1\sigma g} = \frac{\alpha + \beta}{1 + S} = -IP_{1s}$$

when considering that:

$$\alpha = \alpha_{1s} = -IP_{1s}$$

$$\beta = \beta_{1s-1s} = \frac{1}{2} S [2\alpha] = -S*IP_{1s}$$

Wolfsberg and Helmholtz [2.19] have proposed to modify Mulliken's approximation by introducing a constant factor K such as:

$$\chi_p(r) \, \chi_q^j(r) = \frac{1}{2} K S_{pq}^j [(\chi_p)^2(r) + (\chi_q^j)^2(r)]$$

This "bonding" correction has been set equal to 1.75 by Hoffmann [2.20] in order to reproduce plausible C-H bond lengths in methane. Consequently, the nondiagonal matrix elements are thus calculated in the EH polymeric method as:

$$h_{pq}^j = \int \chi_p(r) \, h(r) \, \chi_q^j(r) \, dr = \beta_{pq}^j = \frac{1}{2} K S_{pq}^j [\alpha_p + \alpha_q] = \frac{1}{2} K S_{pq}^j [-IP_p - IP_q]$$

The EH scheme requires the explicit evaluation of the overlap integrals S_{pq}^j from the knowledge of the molecular geometry and of the functional form of the basis functions (usually STO's). Since the method does not separate electron-electron interactions from electron-nucleus attractions due to its global evaluation of matrix elements, the total molecular energy is approximated as the sum of the orbital energies:

$$E = \sum_j^{occ} 2 \, \varepsilon_j$$

This way of calculating total energies is equivalent to the assumption that, if they had been explicitly calculated, the total electron-electron Coulomb and exchange interaction energies would exactly balance nuclear-nuclear repulsion energies:

$$\sum_i^{occ} \sum_j^{occ} 2 \, J_{ij} - K_{ij} = \sum_i^{occ} \sum_j^{occ} \{ 2 \int dr \, \phi_i(r) \, J_j(r) \, \phi_i(r) - \int dr \, \phi_i(r) \, K_j(r) \, \phi_i(r) \}$$

$$= \sum_A \sum_{B<A} z_A z_B \, |R_A - R_B|^{-1}$$

In the polymer case, the last two equations are modified to:

$$E = 2 \sum_n^{occ} \int dk \, \varepsilon_n(k)$$

and:

$$\sum_i^{occ} \sum_j^{occ} \iint dk \, dk' \, [2 \, J_{ij}(k,k') - K_{ij}(k,k')]$$

$$= \sum_i^{occ} \sum_j^{occ} \{ 2 \iint dk \, dk' \int dr \, \phi_i(k,r) \, J_j(k',r) \, \phi_i(k,r)$$

$$- \iint dk \, dk' \int dr \, \phi_i(k,r) \, K_j(k',r) \, \phi_i(k,r) \}$$

$$= \sum_A \sum_{B<A} \frac{z_A z_B}{|R_A - R_B|} + \sum_h \sum_A \sum_B \frac{z_A z_B}{|R_A - R_B - ha|}$$

They indicate that the EH scheme in fact neglects the difference in electron-electron, core-core repulsions, and electron-core attractions. These quantities are obviously of opposite signs and should exactly compensate at the infinite distance limit. By its explicit introduction of basis set overlap, Hoffmann's EH method is very powerful for getting a correct description of the orbital topology. It is also an effective preliminary technique for evaluating equilibrium geometries and energy differences between conformers. Binding energies are generally not correctly described. Due to its non-iterative character, charge densities are usually overestimated and are usually poorly described in polar systems. Examples of results obtained by the EH methodology on some oligomers of interest are given in Tables 2.2 to 2.6.

In polymeric calculations, the computational scheme is also interesting since the matrix elements decrease exponentially with distance so that Bloch sums $h_{pq}(k) = \sum_j e^{ikja} h_{pq}^j$ are generally fully converged after 10 to 15 unit cells. Due to the nonorthogonality, diagonalization procedures are more involved and sophisticated band structures can be generated. Since the procedure does not take into account the explicit electron-electron interactions, the method is noniterative and directly produces a stable band structure. A large variety of polymers have been studied by the standard EH method. Typical examples are given in Section 2.5 of this Chapter.

2.3.3.3 A Simulated *Ab Initio* Technique: The Valence Effective Hamiltonian (VEH)

Even *ab initio* minimal basis sets calculations are very time consuming when applied to large systems of practical interest owing to the large number of two-electron integrals to be calculated and stored. Thus, there is also an imperative need for simplified methods that could more economically produce results of *ab initio* quality. The Valence Effective Hamiltonian, the so-called VEH method, originally developed for molecules by Nicolas and Durand [2.57] and extended to polymers by André et al. [2.58], has turned out to be a versatile tool for describing valence electronic structures of hydrocarbons with the same type of accuracy as double-zeta *ab initio* SCF calculations.

Generally speaking, an *effective* or *model* Hamiltonian is intended to simulate a given "exact" Hamiltonian H in some subspace of the Hilbert space spanned by the eigenvectors of H. Given a scalar product in the operator space, one can set up a variational principle to determine the "best" effective Hamiltonian of a given functional form. In the particular case of the VEH method, one simulates the Fock Hamiltonian of an *ab initio* valence pseudopotential calculation. The trial one-electron Hamiltonian is taken as the sum of the kinetic energy and a sum of effective atomic potentials of the atoms within their specific chemical environment and is written in the polymer case as:

$$h(\mathbf{r}) = -\frac{1}{2}\Delta + \sum_h \sum_A V_A^h(\mathbf{r})$$

The summations over h and A are, respectively, extended over cells and atoms within cells. For computational reasons, the atomic effective potentials are chosen as Gaussian projectors:

$$V_A^h(\mathbf{r}) = \sum_l \sum_m \sum_{tu} C_{lmtu}^A \, |_{lmt}^{Ah}> \, <_{lmu}^{Ah}|$$

The summations over l and m define the angular dependence of the projector (l=0 for s-type; l=1 for p-type). For computational facility, the projectors are normalized Gaussians:

$$|_{lmt}^{Ah}> \, = \kappa_{lmt}^{Ah}(\mathbf{r}) = N(x-x_A)^n(y-y_A)^{n'}(z-z_A-ha)^{n''} \exp[-\alpha_{lmt}^A(\mathbf{r}-\mathbf{R}_A-ha)^2]$$

In such a relation, the direction of periodicity is the z direction, ha is a translation in the direct space, $\mathbf{R}_A$ (with components x_A, y_A, z_A) is the center of the Gaussian. Evidently, l=m=0 means a s projector (n=n'=n''=0) while l=1 represents a p projector (for example, p_x is obtained by putting n=1, n'=n''=0). The parameters of the method are therefore the Gaussian coefficients C_{lmtu}^A and exponents α_{lmt}^A. They are determined by a least-square fitting on a series of well-chosen pattern molecules (ethane, butadiene,...). The potentials can be of single-zeta quality ($\sum$ limited to a single term) or of double-zeta quality ($\sum_{tu}$ extended to four terms, two diagonal and two identical cross terms). Furthermore, they can be spherical (isotropic) if the coefficients C_{lmtu}^A for a given atom and a given l value are independent of m. Alternatively, they are nonspherical (anisotropic) if the various angular dependencies on the same atom (p_x, p_y, p_z) have differents weights C_{lmtu}^A and different exponents α_{lmt}^A. In those conditions, the matrix elements of the VEH methodology become:

$$h_{pq}^j = T_{pq}^j - \sum_{h,A} \sum_{l\,m} \sum_{t,u} C_{lmtu}^A \, <_{p|lmt}^{0\;Ah}> \, <_{lmu}^{Ah}|_q^j>$$

where the integrals $<_{p|lmt}^{0\;Ah}>$ and $<_{lmu}^{Ah}|_q^j>$ are mixed integrals between LCAO atomic basis orbitals and the effective potential projectors:

$$\langle^{0}_{p}|^{Ah}_{lmt}\rangle = \int \chi_p(\mathbf{r}\text{-}\mathbf{P}\text{-}0a) \; \kappa^{Ah}_{lmt}(\mathbf{r}\text{-}\mathbf{R}_A\text{-}ha) \; d\mathbf{r}$$

$$= \int \chi_p(\mathbf{r}) \; \kappa^{Ah}_{lmt}(\mathbf{r}) \; d\mathbf{r}$$

$$\langle^{Ah}_{lmu}|^{j}_{q}\rangle = \int \kappa^{Ah}_{lmu}(\mathbf{r}\text{-}\mathbf{R}_A\text{-}ha) \; \chi_q(\mathbf{r}\text{-}\mathbf{Q}\text{-}ja) \; d\mathbf{r}$$

$$= \int \kappa^{Ah}_{lmu}(\mathbf{r}) \; \chi^{j}_{q}(\mathbf{r}) \; d\mathbf{r}$$

The VEH technique reduces the integral evaluation to the calculation of kinetic and overlap integrals between Gaussian functions only (as long as GTO's are chosen as atomic basis). As soon as the integral part over the atomic functions is completed, the integrals between Bloch functions are evaluated. The complex eigenvalue problem producing the band structure is solved for selected k points. Since it further avoids the iterative cycles inherent to all SCF calculations, the method is very fast and economic in computing time. After obtaining the band structures, the standard programs described in Section 2.4 are used to evaluate such electronic properties as the density of states. The method is free from any empirical parameter and, by its very principle, generally gives valence orbital energies comparable to *ab initio* ones depending on the quality of the effective potentials. A list of VEH effective atomic potentials in several molecular environments [2.57-2.61] is given in Table 2.11.

Table 2.11 Tables of VEH parameters

Single-zeta potentials

Isotropic, [2.58]

Atom A	l,m	α_{lm}^{A}	C_{lm}^{A}
C	s	0.60	-1.21
	p	1.12	-1.99
H	s	1.40	-1.30

Table 2.11 (continued)

Double-zeta potentials

1° Anisotropic, [2.58]

Atom A	l,m	$\alpha_{lm1}{}^A$	$\alpha_{lm2}{}^A$	$C_{lm11}{}^A$	$C_{lm22}{}^A$	$C_{lm12}{}^A$ = $C_{lm21}{}^A$
C	s	1.39	0.35	-1.79	-0.47	+0.20
	p_σ	3.25	0.64	-4.66	-0.54	+0.18
	p_π	1.00	0.30	-3.49	-0.84	+1.14
H	s	7.00	0.60	-4.41	-0.43	+0.22

2° Isotropic, [2.59-2.61]

Atom A	l,m	$\alpha_{lm1}{}^A$	$\alpha_{lm2}{}^A$	$C_{lm11}{}^A$	$C_{lm22}{}^A$	$C_{lm12}{}^A$ = $C_{lm21}{}^A$
C	s	3.63	0.33	-10.075	-1.178	+2.077
	p	6.50	0.52	-11.468	-0.443	-0.185
H	s	5.00	0.49	-4.696	-0.500	+0.646
S	s	1.00	0.30	+3.683	+1.521	-3.454
	p	0.90	0.20	-1.899	-0.047	-0.017
C(-S)	s	3.63	0.33	-6.506	-0.514	+0.354
	p	6.50	0.52	-12.380	-0.474	-0.067
N	s	1.50	0.50	+1.520	+0.790	-2.149
	p	3.00	0.35	-4.789	+0.046	-0.353
C(-N)	s	3.63	0.33	-0.486	+0.222	-1.710
	p	6.50	0.52	-1.637	+0.244	-2.834
O	s	5.00	0.50	-8.559	-1.387	+1.107
	p	2.89	0.07	-5.364	-0.066	-0.045
C(-O)	s	3.63	0.33	-5.347	-0.224	-0.212
	p	6.50	0.52	-4.315	+0.063	-2.192

Table 2.11 (continued)

3° Isotropic,

Atom A	l,m	α_{lm1}^A	α_{lm2}^A	C_{lm11}^A	C_{lm22}^A	C_{lm12}^A $= C_{lm21}^A$
C	s	0.60	0.20	-1.721	-0.479	+0.515
	p	6.50	0.52	-9.441	-0.188	-0.927
H	s	5.00	0.49	-3.433	-0.347	+0.178

In polymers, the VEH method has turned out to produce energy gaps of excellent quality. This is a clear indication that part of the electron-hole correlation is included into the virtual one-electron levels by that methodology. The Koopmans transition energies are thus reduced and, accordingly, the calculated values of the polarizabilities are larger (in agreement with the experimental trends). The same behaviour has also been observed in semiempirical polarizability CNDO-type calculations (where the experimental transition energies is a basic ingredient of the parametrization through the correlation decrease of empirically evaluated one-electron integrals).

2.3.4 π-Electron Semiempirical Methods

The separations between kinetic energies, nucleus-electron attractions, and electron-electron repulsions are explicitly taken into account in the semiempirical Pariser-Parr-Pople (PPP) method (which is a ZDO version applicable to π-electrons). Even if historically it was introduced more than ten years before the CNDO, INDO, and NDDO methods, it is not necessary to detail here the form of the matrix elements which is strictly analogous to the forms given in Chart 2.4. The choice of the "complete" neglect of differential overlap, i.e., the original ZDO scheme, is fully justified for $2p_\pi$ atomic orbitals and is not too drastic an approximation. In the case of π-electrons, there is no need for schemes like INDO or NDDO since only diatomic differential overlaps are neglected. An early justification of the ZDO approximation in the PPP method has been given in terms of Löwdin's orbitals [2.62]. It uses Mulliken's [2.18] or Ruedenberg's [2.63] approximations and shows that the three- anf four-center orthogonalized π-repulsion integrals are negligible in this case [2.64]. This justification is fully correct in the case of π-electrons but is sometimes extended to the case of all-valence electrons with less success because of the large overlaps involved between σ orbitals. In order to allow for the explicit calculation of electron integrals, the $2p_\pi$ basis set has to be specified (in general, the analytical form is a STO with the Slater exponent calculated from Slater rules, $\zeta = 1.625 = 3.25/2$). Consequently, the numerical scheme is applicable to strictly planar structures only, a difficulty not met in the simpler heuristic Hückel method where the basis set is unspecified. The main applications of the PPP approach is the interpretation and the prediction of the energies and intensities of π-π^* transitions, as detailed in several textbooks [2.65, 2.66].

2.4 COMPUTATIONAL ASPECTS

2.4.1 Band Structure Calculations

Several *ab initio* molecular SCF-MO-LCAO programs have been written to date and are widely distributed. Among those, let us cite POLYATOM [2.67], and the IBMOL or KGNMOL [2.68], GAUSSIAN [2.69], HONDO [2.70], and TURBOMOLE [2.71] series.

The practice of polymer computations is not so standardized. The Fock and overlap matrices are computed at a given level of approximation (semiempirical or *ab initio*). The calculation is made in direct space. The k-dependent matrices are diagonalized in the reciprocal space and, if necessary, a self-consistent procedure is turned on. Diagonalizations are achieved with the very efficient and self-contained routine CBORIS [2.72] which uses successively the Cholesky decomposition, the Householder tridiagonalization, and the QL algorithm. If difficulties of linear dependence origin are encountered, it is preferable to use the direct transformation into orthogonal Löwdin's orbitals as explained in Section 2.4.3. In all cases, the output consists of the energy bands and of the form of the molecular orbitals of the polymer. An interactive graphical communication can be switched on for:

1. ordering the energy bands,
2. plotting the standard electronic properties, such as band structures, band widths, density of states, simulations of electron spectra, electron densities,

3. calculating electron indices as charges and bond orders, and

4. determining conformations and other properties.

Automatic programs taking into account the effects of long-range interactions are fully implemented.

2.4.2 Summation over Polymeric States and Integrations over First Brillouin Zones

In molecules, the molecular orbital (MO) energy levels are discrete and constitute a finite set. If we gradually go to the infinite system, the number of elementary cells goes up to infinity. The wave number k becomes a continuous variable within the interval $[-\pi/a, +\pi/a]$. The summation over the discrete set of MO's is replaced by a continuous integration between $-\pi/a$ and $+\pi/a$ over the first Brillouin zone as already stated in Sections 1.9.3.2 and 1.9.3.3, according to the relationships:

$$\frac{1}{N} \sum_k = \frac{1}{2\pi/a} \int_{-\pi/a}^{+\pi/a} dk$$

$$\frac{1}{N} \sum_k f_k = \frac{1}{2\pi/a} \int_{-\pi/a}^{+\pi/a} f(k)\, dk$$

The integrals are evaluated by numerical integration over the first Brillouin zone. In practice, the complex eigenvalue problem is solved at 12 nonequidistant k-points in half the Brillouin zone. The abscissas k_i and the weights W_i are listed in Table 2.12. These 12 k points correspond to the abscissas of a 12-point Gauss-Legendre quadrature necessary to numerically compute the density matrix D (the summations are extended over occupied bands):

$$D_{pq}^j = \frac{a}{2\pi} \int_{BZ} dk\, e^{ikja} \sum_{n=1}^{occ} c_{np}^*(k)\, c_{nq}(k)$$

$$= \sum_{i=1}^{12} W_i\, e^{ik_ija} \sum_{n=1}^{occ} c_{np}^*(k_i)\, c_{nq}(k_i)$$

Table 2.12 **Weights and roots of the Legendre polynomial of degree 12 as needed for numerical integrations in half the Brillouin zone**

i	k_i (radians)	A_i
1	0.0289 6448 7993	0.0235 8766 8193
2	0.1506 1226 1496	0.0534 6966 2997
3	0.3614 3603 4181	0.0800 3916 4272
4	0.6482 3944 1541	0.1015 8371 3361
5	0.9930 0795 9289	0.1167 4626 8269
6	1.3740 8014 8713	0.1245 7352 2906
7	1.7675 1250 4877	0.1245 7352 2906
8	2.1485 8469 4301	0.1167 4626 8269
9	2.4933 5321 2049	0.1015 8371 3361
10	2.7801 5661 9408	0.0800 3916 4272
11	2.9909 8039 2093	0.0534 6966 2997
12	3.1126 2816 5596	0.0235 8766 8193

The use of this quadrature scheme is only valid if the cell-index j is relatively small because of the presence of a oscillatory exponential function in the integrand.

2.4.3 Basis Set Linear Dependence

In band structure calculations, a near linear dependence error often occurs when the lowest eigenvalues of the overlap matrix $S(k)$ are of the order of magnitude of 10^{-2} or smaller while linear dependence problems starts in the case of molecules only if the corresponding eigenvalue is as small as 10^{-6}.

The reason for this different behaviour between 1D band structure and molecular calculations partly lies in the truncation of the infinite lattice sums which we have to evaluate. Indeed, the problem can be usually removed if the lattice sums are properly truncated [2.73].

Inherent linear dependences can be also removed by the canonical orthogonalization proposed by Löwdin [2.62, 2.74]. In an orthogonalization procedure, there exist several ways to choose the transformation matrix $T(k)$ which is used to transform the generalized eigenvalue problem:

$$h(k)C(k) = S(k)C(k)E(k)$$

into the classical eigenvalue equations:

$$h'(k)U(k) = U(k)E(k)$$

where:

$$h'(k) = T^{\dagger}(k)h(k)T(k)$$

and:

$$U(k) = T^{-1}(k)\ C(k)$$

The matrix $T(k)$ must obey the relation:

$$T^{\dagger}(k)S(k)T(k) = 1$$

In Löwdin's canonical procedure, the matrix $T(k)$ is defined as:

$$T(k) = W(k)\ s^{-1/2}(k)$$

where $W(k)$ and $s(k)$ are the eigenvectors and the (diagonal) eigenvalues matrices of $S(k)$, respectively. If a given eigenvalue $s_i(k)$ approaches zero, i.e., if a linear dependence exists in the Bloch basis set, the matrix $T(k)$ will have a column with very large values and will not be well behaved. A way to circumvent this problem is to eliminate the columns in the matrix $W(k)$ corresponding to small eigenvalues and, consequently, to work within a reduced transformed orthonormal Bloch basis set. The truncated new functions will span exactly the same region of space as the original one if the eliminated eigenvalues were exactly zero. In eliminating the columns corresponding to small eigenvalues ($< 10^{-4}$), only a small part of the basis set is lost [2.75].

2.4.4 Band Indexing Difficulty

Another interesting point to mention in actual calculations of band structures is the band indexing difficulty. In the reduced zone scheme, the band structure $\varepsilon_n(k)$ is a multivalued function of k and the successive values of n for any given k correspond to levels of increasing energy: first to the inner core electrons, then to valence bands, and finally to unoccupied bands. Due to the prohibitive computing time associated with the complex diagonalization $h(k) C_n(k) = S(k) C_n(k) \varepsilon_n(k)$, the energy bands are calculated for discrete wave number values only, $\varepsilon_n(k_i)$. Bands are usually given an index n, increasing in order of increasing energy, but difficulties arise because different choices can be made at degeneracy points. Two possibilities exist: either band crossing or avoided crossing. These two possibilities correspond to entirely different physical realities; in the first case, there is no band gap, whereas an energy gap does exist in the second case. If point symmetry exists within the unit cell, the problem can be solved by group theory, but when there is a low symmetry and a large number of bands, this becomes practically untractable. Currently, the band indexing is generally carried out by visual inspection or, in a more effective way, by using an interactive graphics interface with the knowledge of the local derivatives $\varepsilon'_n(k_i)$ of the energy eigenvalues. A general technique, easily applicable, for determining locally the k-derivative of all bands in each k-point is obtained by considering the scalar product [2.76]:

$$\varepsilon_n(k) = C^\dagger_n(k)\, h(k)\, C_n(k)$$

with the normalization condition:

$$1 = C^\dagger_n(k)\, S(k)\, C_n(k)$$

Hence:

$$\frac{d\varepsilon_n(k)}{dk} = \frac{dC^\dagger_n(k)}{dk}\, h(k)\, C_n(k) + C^\dagger_n(k)\, \frac{dh(k)}{dk}\, C_n(k) + C^\dagger_n(k)\, h(k)\, \frac{dC_n(k)}{dk}$$

After straightforward matrix manipulation, we get the working formula:

$$\frac{d\varepsilon_n(k)}{dk} = C_n^\dagger(k) \frac{dh(k)}{dk} C_n(k) - \varepsilon_n(k) C_n^\dagger(k) \frac{dS(k)}{dk} C_n(k)$$

which, in an orthogonal atomic or Bloch basis, simplifies to:

$$\frac{d\varepsilon_n(k)}{dk} = C_n^\dagger(k) \frac{dh(k)}{dk} C_n(k)$$

The k dependence of the $h(k)$ and $S(k)$ matrices is very simple:

$$\frac{dh_{pq}(k)}{dk} = \frac{d}{dk} \sum_j e^{ikja} h_{pq}^j = ia \sum_j j\, e^{ikja} h_{pq}^j$$

$$\frac{dS_{pq}(k)}{dk} = \frac{d}{dk} \sum_j e^{ikja} S_{pq}^j = ia \sum_j j\, e^{ikja} S_{pq}^j$$

This procedure has been implemented into an algorithm of general use for polymer consideration. In a practical way, a band structure calculation program is linked to a graphical device and the variation of $\varepsilon_n(k)$ is graphically represented by a rotating fixed length vector centered at $\{\varepsilon_n(k), k\}$ points. Interactive capabilities for zooming various energy zones have been added when the separation between two or more bands is not large enough. Experience has shown that nearly all kinds of bands can be labeled by this procedure and the time and the difficulties of data manipulations are very much reduced.

The following technique is used in practice for the ordering of the band structures. For the first k-point, k_1, the bands are ordered in increasing energy, i.e., the band index m is equal to the energy index n. For the next k-point, k_i, a search is performed for each band to find the energy index n corresponding to that band m. When knowing the energy values ε_n and their derivatives ε'_n at each k_i point, it is possible to extrapolate the energies at k_{i-1}:

$$E_n(k_{i-1}) = \varepsilon_n(k_i) - \varepsilon'_n(k_i)*\Delta k$$

where Δk represents the step between two successive k-points. Similarly, we estimate the energy of the band m at the k_i-point:

$$E_m(k_i) = \varepsilon_m(k_{i-1}) + \varepsilon'_m(k_{i-1}) * \Delta k$$

For each band, we search for the best agreement between the extrapolated energies (E) and the calculated energies (ε) at the k_i-1 and k_i points on the basis of the least-square deviation:

$$\Delta_{i-1} = \varepsilon_m(k_i\text{-}1) - E_n(k_i\text{-}1) \qquad\qquad \Delta_i = E_m(k_i) - \varepsilon_n(k_i)$$

The smallest $R_n = [\Delta_{i-1}]^2 + [\Delta_i]^2$ gives us the energy index n for the band m at the k_i-point. This procedure is repeated for all following k-points.

2.4.5 Density of States Calculations

As soon as the band structure $\varepsilon_n(k)$ has been obtained according to one of the various methods described above, the problem of getting relevant information to be compared to experimental data has to be considered. Indeed, making guesses or interpretations on properties is the most fruitful and rewarding aspect of theoretical studies. Numerous experimental quantities can be correlated with terms deduced from electronic charge distributions and from band structure calculations. Nowadays, it is possible to measure the band structure directly by angle-resolved photoemission. However, this is still the exception because such measurements require high quality samples which is seldom the case with polymers. Thus, in most cases, one can only measure the density of states (DOS) and some transformations to the calculated band structures need to be done to have them in a form readily comparable to experiment. The DOS distribution has its own meaning since in polymeric materials, the number of atoms is very large and the eigenvalues are very dense and bounded; it is therefore more convenient to deal with energy level distribution functions than with individual values. The DOS is defined as the number of allowed energy levels per energy unit:

$$D(E) = \frac{a}{\pi} \sum_{n=1}^{n_0} |\frac{dk}{d\varepsilon_n(k)}|_{\varepsilon_n(k)=E}$$

Such a function has the advantage of providing a synthetic description of one-electron levels and of being closely related to experiment. Though conceptually simple, electronic DOS distributions are not straightforward to evaluate numerically. General computer algorithms for polymers have been proposed [2.77]. Whenever the first derivative of $\varepsilon_n(k)$, $\varepsilon'_n(k)$, goes to zero, there is a large contribution to the DOS, giving rise to singularities in the DOS spectrum. We know in advance there are at least three values of the wave number k for which $\varepsilon'_n(k)$ vanishes identically, i.e., k=0 and k= $\pm\pi/a$. However, other extrema can occur but their number and location cannot be predicted *a priori*. When implementation by computer has to be considered, one has to turn to a new function $D(E_i)$, the DOS histogram. It exhibits the same behaviour as D(E) but gives rise to finite peak height only. It is an average of D(E) over some energy interval ΔE:

$$D(E_i) = \frac{1}{\Delta E} \int_{E_i - \frac{\Delta E}{2}}^{E_i + \frac{\Delta E}{2}} D(E) \, dE \quad ; E_i = E_0 + i \, \Delta E, \ i = 0,1,2,\ldots, \ \Delta E > 0$$

The value of ΔE is chosen according to the particular experiment to be interpreted. $D(E_i)$ is obviously the limiting function of $D(E_i)$ when ΔE goes to zero. $D(E_i)$ is the height of the ith rectangular box centered at E_i; it represents an element of the histogram of the DOS. The histogram method involves generating the eigenvalues $\varepsilon_n(k)$ of the system at a large number of points in the Brillouin zone and then counting the results into boxes that subdivide the energy range. Another approach, more effective in the sense of saving computing time and numerical stability, is currently preferred. Basically, it consists of an analytical or continuous integration over small intervals [$k_i-\Delta k/2$, $k_i+\Delta k/2$] located at wave number values for which a diagonalization has been requested. DOS distributions can be used to test results from energy band structures obtained by various methods. It is known that derivatives enhance variations of a function; therefore, it is no wonder a DOS distribution is a more sensitive test on energy bands since it involves the inverse of the first derivative $\varepsilon'_n(k)$. Examples of the derivation of the DOS distribution from the band structure and comparison with experimental data will be provided in Chapter 3.

2.5 EXAMPLES OF ALL ELECTRON BAND STRUCTURE AND ALL VALENCE ELECTRON SEMIEMPIRICAL BAND STRUCTURE: POLYETHYLENE AND POLYSILANE

Polyethylene provides an excellent example of the calculations of band structures by methodologies ranging from full *ab initio* to the simplest semiempirical ones. The band structure is well described in the literature. The main references are summarized in Chart 2.5.

Chart 2.5 Band structure calculations on polyethylene

Method	Reference
Extended Hückel	- W.L. McCubbin and R. Manne, Chem. Phys. Lett., **2**, 230 (1968). - A. Imamura, J. Chem. Phys., **52**, 3168 (1970). - J.M. André, G.S. Kapsomenos, and G. Leroy, Chem. Phys. Lett., **8**, 195 (1971). - J.M. André and J. Delhalle, Chem. Phys. Lett., **17**, 145 (1972). - J.E. Falk and R.J. Fleming, J. Phys. C: Solid State Phys., **6**, 2954 (1973).
CNDO/2	- K. Morokuma, Chem. Phys. Lett., **6**, 186 (1970). - H. Fujita and A. Imamura, J. Chem. Phys., **53**, 4555 (1970). - J.M. André, G.S. Kapsomenos, and G. Leroy, Chem. Phys. Lett., **8**, 195 (1971). - G. Morosi and M. Simonetta, Chem. Phys. Lett., **8**, 358 (1971). - K. Morokuma, J. Chem. Phys., **54**, 1962 (1971). - J.M. André and J. Delhalle, Chem. Phys. Lett., **17**, 145 (1972). - J.M. André, J. Delhalle, G.S. Kapsomenos, and G. Leroy, Chem. Phys. Lett., **14**, 485 (1972). - B.J. McAloon and P.G. Perkins, Faraday Discuss. II, 1121 (1972). - J.E. Falk and R.J. Fleming, J. Phys. C: Solid State Phys., **6**, 2954 (1973).

Chart 2.5 (continued)

INDO	-	D.L. Beveridge, I. Jano, and J. Ladik, J. Chem. Phys., **56**, 4744 (1972).
MINDO/2	-	D.L. Beveridge, I. Jano, and J. Ladik, J. Chem. Phys., **56**, 4744 (1972).
MINDO/3	-	M.J.S. Dewar, H.S. Sung, and P.K. Weiner, Chem. Phys. Lett., **29**, 220 (1974).
	-	M.J.S. Dewar, Y. Yamaguchi, and S.H. Suck, Chem. Phys. Lett., **51**, 175 (1977).
	-	M.J.S. Dewar, Y. Yamaguchi, and S.H. Suck, Chem. Phys., **43**, 145 (1979).
FSGO	-	J.M. André, J. Delhalle, C. Demanet, and M.E. Lambert-Gérard, Int. J. Quantum Chem., **S10**, 99 (1976).
	-	J.M. André and J.L. Brédas, Chem. Phys., **10**, 367 (1977).
Localized Orbitals	-	B.J. Duke and B. O'Leary, Chem. Phys. Lett., **20**, 459 (1973).
	-	B.J. Duke and B. O'Leary, Chem. Phys. Lett., **32**, 602 (1975).
	-	J. Delhalle, J.M. André, S. Delhalle, C. Pivont-Malherbe, F. Clarisse, and G. Leroy, Theor. Chim. Acta, **43**, 215 (1977).
VEH	-	J.M. André, L.A. Burke, J. Delhalle, G. Nicolas, and Ph. Durand, Int. J. Quantum Chem., **S13**, 283 (1979).
Ab Initio	-	J.M. André, Comput. Phys. Comm., 1, 391 (1970).
	-	J.M. André and G. Leroy, Ann. Soc. Sci. Brux., **84**, 133 (1970).
	-	J.M. André and G. Leroy, Chem. Phys. Lett., **5**, 71 (1970).
	-	E. Clementi, J. Chem. Phys., **54**, 2491 (1971).
	-	J.M. André, J. Delhalle, S. Delhalle, R. Caudano, J.J. Pireaux, and J. Verbist, Chem. Phys. Lett., **23**, 206 (1973).
	-	J.E. Falk and R.J. Fleming, J. Phys. C: Solid State Phys., **6**, 2954 (1973).
	-	J.M. André, J. Delhalle, S. Delhalle, R. Caudano, J.J. Pireaux, and J. Verbist, J. Chem. Phys., **60**, 595 (1974).
	-	W. Richter, Acta Polym., **32**, 657 (1981).

Chart 2.5 (continued)

- R.S. Weidman, K.L. Bedford, and A.B. Kunz,
  Solid State Commun., **39**, 917 (1981).
- A. Karpfen,
  Int. J. Quantum Chem., **19**, 1207 (1981).
- A. Karpfen,
  J. Chem. Phys., **75**, 238 (1981).
- A. Karpfen,
  Phys. Scripta, **T1**, 79 (1982).
- S. Suhai,
  J. Polym. Sci., Polym. Phys. Ed., **21**, 1341 (1983).
- S. Suhai,
  in "Quantum Chemistry of Polymers: Solid State
  Aspects", (J.M. André and J. Ladik, Eds.), p.101,
  D. Reidel Publ. Co., Dordrecht (1984).

Experimental*

- K. Seki, U. Karlsson, R. Engelhardt, and E.E. Koch,
  Chem. Phys. Lett., **103**, 343 (1984).
- N. Ueno, K. Seki, H. Fujimoto, T. Kuramochi,
  K. Sigita, and H. Inokuchi,
  Phys. Rev. B, **41**, 1176 (1990).

* Intramolecular energy band dispersion observed by angle-resolved photoemission with synchroton radiation.

All calculations show a large gap between the valence and conduction bands in agreement with experimental data. The general structure of the valence bands obtained by the various methods appears to be similar. However, CNDO/2 bands in the original parametrization are much too large and the band centers of gravity lie far below the *ab initio* ones. On the other hand, while the band structures show a very similar overall shape, the densities of states are very sensitive to the details of the dispersion curves and thus vary largely according to the approximations used. The comparison between a theoretical DOS and experimental spectrum is excellent for *ab initio* like methodologies. Such a comparison is given in Figure 2.3 in the case of FSGO results. The positions of both theoretical and experimental peaks agree very well and both fine structures are directly comparable.

It should be stressed at this stage that band structure plots do not offer the best representation of valence band properties, particularly in those systems where a large number of bands lies in a narrow energy region. The DOS (usually convoluted for simulating experimental resolution and taking into account specific cross-section effects) histograms appear to give a more significant picture of the distribution of valence levels, when comparing to experimental data in the energy domain.

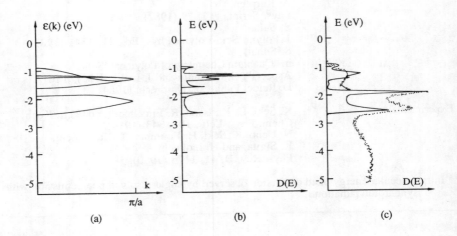

Figure 2.3 FSGO theoretical simulation of the valence photoelectron spectrum of polyethylene
(a) one-electron valence band structure
(b) one-electron valence density of states
(c) simulated (full line) and experimental (dotted lines) XPS valence spectrum

A representation of the *ab initio* STO-3G valence bands of polyethylene -(CH$_2$-CH$_2$)$_n$- is shown in Figure 2.4 (A).

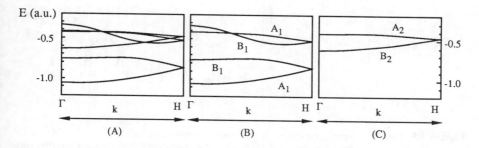

Figure 2.4   (A) *Ab initio* STO-3G structure of the six valence bands of polyethylene
(B) their decoupling into the four bands symmetric with respect to the plane of the zigzag carbon chain and
(C) the two bands antisymmetric with respect to that plane

The symmetry assignments are taken from Mc Cubbin [2.78]. In an LCAO approach (the tight-binding method of solid state physics), the four upper occupied bands are generally assumed to be mainly associated with the C-H bonds. Two bands (B2, A2) form a combination which, in terms of a simple nearest-neighbour LCAO interpretation, are the bonding interactions of the C_{2py} atomic orbital (the axis of the polymer is along the z-direction and the y-direction is into the paper) with the antisymmetric combination of H_{1s} orbitals. In this simple picture, these two bands completely decouple from the two core bands and from the other four valence bands of polyethylene, as seen in Figures 2.4 (B) and 2.4 (C). They obey the usual rules of orbital combination, monotonically increasing or decreasing in energy from Γ to H according to the topology of orbital interactions as explained in Section 1.7. On the other hand and this is a common feature, in all the published calculations on the electronic structure of polyethylene, the other pair of "C-H" valence bands (B1, A1) exhibits a more complicated behaviour involving a clear minimum around k = 0.6 π/a (see Figure 2.4 (B)). This minimum is due to a question of "dimensionality" of the chain. As illustrated in Figure 2.5, the 2D character is introduced by the planar zigzag chain (lying in the x-z plane) while the 3D character is introduced by the C-H bonds lying outside of this plane.

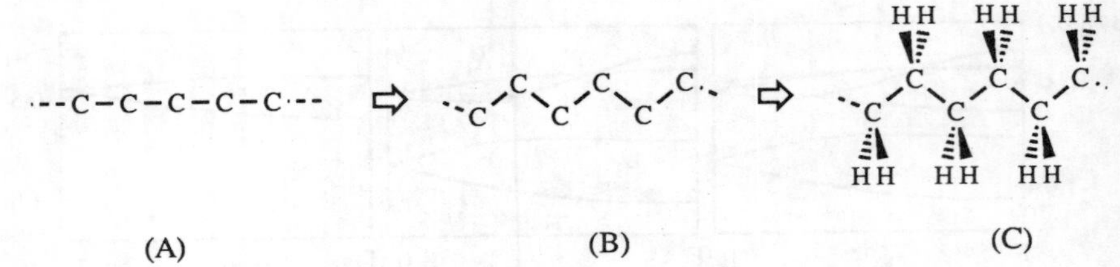

Figure 2.5 Possible dimensionalities of polymeric carbon chains:
(A) a one-dimensional array of one-dimensional potentials; note that this sketch is also applicable to a one-dimensional chain of three-dimensional "atomic" potentials,
(B) a 2D zigzag carbon chain, and
(C) a 3D -$(CH_2\text{-}CH_2)_n$- chain periodic in one dimension only

The questions which can be raised concerning this overall behaviour are:

$1°$ Is this pair of A1-B1 bands associated mainly with combinations of the bonding interactions of the C_{2px} atomic orbital with the symmetric combination of H_{1s} orbitals?

$2°$ Is the sharp dispersion (and the minimum seen in) of the B1 band toward the Γ point due to a strong repulsion between the symmetric combination of H_{1s} orbitals and the C_{2s} orbitals which comes into play in this wave vector region? In short, is this effect due to strong C_{2s}-H_{1s} interaction?

In the following, we analyze some aspects of these questions:

$1°$ We show that the minimum of energy band B1 at a k-point which is not a high symmetry point comes from the 3-dimensionality of the polyethylene skeleton.

$2°$ We take into account the screw axis of order 2 of the chain to simplify the analysis. This reduces the number of valence bands from 6 to 3 and doubles the size of the first Brillouin zone.

$3°$ We compare the orbital pictures issued from the usual minimal basis LCAO scheme (1s for hydrogen, 1s, 2s, 2px, 2py, and 2pz for carbon) and from a simpler (but more intuitive) FSGO-like (Floating Spherical Gaussian Orbital) bond orbital scheme where the electron pairs are represented by bond orbitals not centered on the atoms.

Due to the recent interest in electronic properties of "σ-conjugated" polysilanes [2.79], it is interesting to make a comparative analysis of all-trans polyethylene $-(CH_2-CH_2)_n-$ and polysilane $-(SiH_2-SiH_2)_n-$.

In order further to simplify this analysis, we will make use of the helical symmetry of the polyethylene chain. Polyethylene {or polysilane} can be considered as a zigzag linear polymer with a $-(CH_2-CH_2)_n-$ {or a $-(SiH_2-SiH_2)_n-$} unit cell or as a 1*2/1 helix with an elementary asymmetric unit cell $-(CH_2)_n-$ {or $-(SiH_2)_n-$}. As mentioned previously, Imamura & Fujita and Blumen & Merkel [2.48] have shown that it is computationally practical to reduce the numerical effort by explicitly considering the helical symmetry. The interpretive effort is also greatly simplified and one considers *a priori* the asymmetric unit as the relevant unit for the calculation. The successive displacements of the asymmetric unit involve, at the same time, the translation **and** the rotation of the atomic orbital basis sets.

The *ab initio* STO-3G valence band structures of polyethylene and polysilane are represented in Figure 2.6 for the linear zigzag case and the helical case. The choice of a minimal basis set was compulsory in order to analyze the bonding character of the bands in the simple terms of s and p "chemical" orbitals familiar to organic chemists. It is evident that more elaborate calculations of the band structure of polyethylene do exist in the literature and can be found in the references of Chart 2.5. Recent *ab initio* calculations of the band structure of polysilane can also be found in references [2.80-2.82].

The band structures for the linear and helical cases are in complete agreement. Since in the linear case, the unit cell is twice the size that of the helical case, the linear polyethylene or polysilane have a reciprocal unit cell and hence a first Brillouin zone which is half the size that of the helical case. The agreement is better seen by unfolding, in the linear case, the energy bands according to the symmetry requirements (upper right parts of Figures 2.6 (A) and 2.6 (B)). Conversely, the "folding back" of the dispersion curves for the 6 valence electrons (3 electron pairs) contained in the asymmetric cell $-(CH_2)_n-$ {or $-(SiH_2)_n-$} gives the dispersion curves for the 12 valence electrons (6 electron pairs) of the symmetric unit $-(CH_2-CH_2)_n-$ {or $-(SiH_2-SiH_2)_n-$}.

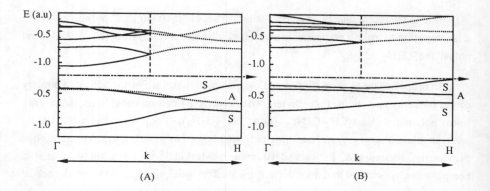

Figure 2.6 Valence band structures of polyethylene (A) and polysilane (B). For both cases, the upper part corresponds to the valence bands of the symmetric unit $-XH_2-XH_2-$ (left) and its partial unfolding (right) while the lower part plots the valence bands of the asymmetric unit $-XH_2-$. S (or A) refers to the symmetry (or antisymmetry) of a given band with respect to the backbone $-X-X-$ plane

Our calculations have been performed on both the simple $-(CH_2)_n-$ {or $-(SiH_2)_n-$} unit cells and the double-sized $-(CH_2-CH_2)_n-$ {or $-(SiH_2-SiH_2)_n-$} unit cells.

The valence energies are summarized in Table 2.13 for three particular points in the first Brillouin zone; the Γ ($k=0$) point, the H ($k=\pi/a$) point, and the middle point ($k=\pi/2a$) in half the first Brillouin zone of the asymmetric unit $-(XH_2)-$ of the polymers; note that the latter point actually corresponds to the H ($k=\pi/a$) point of the symmetric unit $-(XH_2-XH_2)-$.

It is important to stress that the upper occupied symmetric band of both polyethylene and polysilane exhibits a "strange" minimum, even if it is less pronounced in the latter case; it occurs around $k = 0.6\ \pi/a$ for polyethylene and around $k = 0.5\ \pi/a$ for polysilane. The LCAO coefficients of the valence atomic orbitals are summarized in Table 2.14.

Table 2.13   **Energy bands for high symmetry points of polyethylene and polysilane (values in a.u.).**

S (or A) refers to the symmetry (or antisymmetry) of the band with respect to the backbone -X-X- plane

				Polyethylene	Polysilane
$\varepsilon(\Gamma)$	k=0	$\varepsilon_3(\Gamma)$	S	-0.416	-0.329
		$\varepsilon_2(\Gamma)$	A	-0.432	-0.388
		$\varepsilon_1(\Gamma)$	S	-1.059	-0.737
	k=π/2a	ε_3	S	-0.519	-0.357
		ε_2	A	-0.489	-0.410
		ε_1	S	-0.856	-0.643
$\varepsilon(H)$	k=π/a	$\varepsilon_3(H)$	S	-0.326	-0.207
		$\varepsilon_2(H)$	A	-0.640	-0.450
		$\varepsilon_1(H)$	S	-0.760	-0.579

Table 2.14   **LCAO coefficients in minimal basis set calculations on polyethylene and polysilane**

Note that the z direction lies within the plane of the zigzag -X-X- backbone and is the periodicity axis; x lies in the plane of the zigzag -X-X- backbone and is perpendicular to the periodicity axis while y is perpendicular to the plane of the zigzag backbone

		symmetric orbitals		antisymmetric orbitals
		$\varepsilon_1(\Gamma) = -1.059$	$\varepsilon_3(\Gamma) = -0.416$	$\varepsilon_2(\Gamma) = -0.432$
Polyethylene				
C	2s	+ 0.64	- 0.23	--
	2px	- 0.07	+ 0.70	--
	2py	--	--	- 0.54
	2pz	--	--	--
H1	1s	+ 0.14	+ 0.38	- 0.52
H2	1s	+ 0.14	+ 0.38	+ 0.52

TABLE 2.14 (continued)

Polyethylene

		$\varepsilon_1(H) = -0.760$	$\varepsilon_3(H) = -0.326$	$\varepsilon_2(H) = -0.640$
C	2s	- 0.52	--	--
	2px	- 0.33	--	--
	2py	--	--	- 0.53
	2pz	--	- 0.95	--
H1	1s	- 0.30	--	- 0.34
H2	1s	- 0.30	--	+ 0.34

Polysilane

		$\varepsilon_1(\Gamma) = -0.737$	$\varepsilon_3(\Gamma) = -0.329$	$\varepsilon_2(\Gamma) = -0.388$
Si	3s	+ 0.71	- 0.29	--
	3px	- 0.05	+ 0.61	--
	3py	--	--	- 0.43
	3pz	--	--	--
H1	1s	+ 0.19	+ 0.43	- 0.54
H2	1s	+ 0.19	+ 0.43	+ 0.54

		$\varepsilon_1(H) = -0.579$	$\varepsilon_3(H) = -0.207$	$\varepsilon_2(H) = -0.450$
Si	3s	- 0.58	--	--
	3px	- 0.24	--	--
	3py	--	--	- 0.48
	3pz	--	- 0.92	--
H1	1s	- 0.37	--	- 0.45
H2	1s	- 0.37	--	+ 0.45

In a simple picture consisting of chemical bonds, the orbital basis set of the asymmetric unit cell of polyethylene (or polysilane) contains two C-H (or Si-H) bonding orbitals and one C-C (or Si-Si) bonding orbital. In an elementary FSGO (Floating Spherical Gaussian Orbital) scheme, the orbital representation would correspond to that of Figure 2.7.

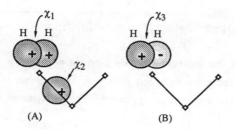

χ_1 χ_3

(A)   (B)

Figure 2.7   Sketch of the basis set bond orbitals:
(A) symmetric combination (with respect to the C-C or Si-Si chain backbone) of the C-H (or of the Si-H) orbital (χ_1) and C-C (or Si-Si) bond orbital (χ_2),
(B) antisymmetric combination (with respect to C-C or Si-Si chain backbone) of the C-H (or of the Si-H) orbital (χ_3)

It is clear from symmetry considerations:

$1°$ that the symmetric combination (with respect to the -C-C- or -Si-Si- backbone) of the C-H (or of the Si-H) orbital (χ_1) will combine with the C-C (or Si-Si) bond orbital (χ_2). In the standard LCAO scheme, only the symmetric orbitals of the backbone atoms, i.e., C_{2s}, C_{2px}, and C_{2pz} (or Si_{3s}, Si_{3px}, and Si_{3pz}) interact with the symmetric combination of H_{1s},

$2°$ while the antisymmetric combination (with respect to the -C-C- or -Si-Si- chain backbone) of the same C-H (or Si-H) orbital (χ_3) will not. This implies that, in this simple model, the antisymmetric band has no contribution from the s, px, and pz backbone orbitals. In the standard LCAO scheme, only C_{2py} (or Si_{3py}) interacts with the antisymmetric combination of H_{1s}. This is confirmed by the inspection of the LCAO coefficients listed in Table 2.14.

The advantage of the bond orbital approach is that the valence band structures of polyethylene and polysilane are now easily modeled by a basis set consisting of only three bond orbitals and that the secular determinant factorizes into a 2*2 (symmetric) block and a 1*1 (antisymmetric) one.

Assuming, in a first approximation, the orthogonality of the bond orbitals, and restricting the interactions to the nearest-neighbours, we obtain the following analytical forms of the symmetric bands:

$$\varepsilon(k) = \frac{1}{2}\{\alpha_{11}+\alpha_{22}+2(\beta_{11}+\beta_{22})\cos(ka) \pm \sqrt{[\alpha_{11}-\alpha_{22}+2(\beta_{11}-\beta_{22})\cos(ka)]^2+4\beta_{12}^2[2+2\cos(ka)]}\}$$

where α_{11} (or α_{22}) is the one-center matrix element of χ_1 (or χ_2); β_{11} (or β_{22}) is the nearest-neighbour interaction between two χ_1's (or between two χ_2's) and β_{12} the nearest-neighbour interaction between χ_1 and χ_2. The antisymmetric band has a simpler energy expression:

$$\varepsilon(k) = \alpha_{33} + 2\beta_{33}\cos(ka)$$

where α_{33} is the one-centered matrix element of χ_3 and β_{33} the nearest-neighbour interaction between two χ_3's.

It is straightforward to fit the various parameters (α_{11}, α_{22}, α_{33}, β_{11}, β_{22}, β_{33}, β_{12}) to the energy values at the points of high symmetry in the first Brillouin zone (Γ and H) and obtain the values listed in Table 2.15.

Table 2.15   Bond orbital parameters for polyethylene and polysilane (in a.u.)

		Polyethylene	Polysilane
one-center matrix element			
XH	α_{11}	-0.6400	-0.5160
XX	α_{22}	-0.6400	-0.4100
XH	α_{33}	-0.5365	-0.4185
two-center matrix element			
XH-XH	$2\beta_{11}$	+0.1190	+0.0630
XX-XX	$2\beta_{22}$	-0.3130	-0.2030
XX-XH	$2\beta_{12}$	-0.2380	-0.1870
XH-XH	$2\beta_{33}$	+0.1045	+0.0305

These sets of parameters allow a nice fit to the STO-3G valence band structures of polyethylene and polysilane. The parametrized band structures also exhibit the "strange" minimum of the band B1. As can be seen from Table 2.16, the parameters reproduce fairly well the values of the energies at the three selected points of the first Brillouin zone for which the explicit analytical forms are:

k = 0

$$\varepsilon_3(\Gamma) = \frac{1}{2} [\, \alpha_{11} + 2\beta_{11} + \alpha_{22} + 2\beta_{22} + \sqrt{(\alpha_{11} + 2\beta_{11} - \alpha_{22} - 2\beta_{22})^2 + 16\beta_{12}^2} \,]$$

$$\varepsilon_2(\Gamma) = \alpha_{33} + 2\beta_{33}$$

$$\varepsilon_1(\Gamma) = \frac{1}{2} [\, \alpha_{11} + 2\beta_{11} + \alpha_{22} + 2\beta_{22} - \sqrt{(\alpha_{11} + 2\beta_{11} - \alpha_{22} - 2\beta_{22})^2 + 16\beta_{12}^2} \,]$$

k = π/2a

$$\varepsilon_3 = \frac{1}{2} [\, \alpha_{11} + \alpha_{22} + \sqrt{(\alpha_{11} - \alpha_{22})^2 + 8\beta_{12}^2} \,]$$

$$\varepsilon_2 = \alpha_{33}$$

$$\varepsilon_1 = \frac{1}{2} [\, \alpha_{11} + \alpha_{22} - \sqrt{(\alpha_{11} - \alpha_{22})^2 + 8\beta_{12}^2} \,]$$

k = π/a

$$\varepsilon_3(H) = \alpha_{22} - 2\beta_{22}$$

$$\varepsilon_2(H) = \alpha_{33} - 2\beta_{33}$$

$$\varepsilon_1(H) = \alpha_{11} - 2\beta_{11}$$

Table 2.16 **Parametrized energy bands and orbital charges for high symmetry points of polyethylene and polysilane (in a.u.)**

		energy parametrized	energy STO-3G	charge χ_1	charge χ_2	charge χ_3
Polyethylene						
k = 0						
$\varepsilon_3(\Gamma)$	S	-0.416	-0.416	0.837	0.163	--
$\varepsilon_2(\Gamma)$	A	-0.432	-0.432	--	--	1.000
$\varepsilon_1(\Gamma)$	S	-1.059	-1.059	0.164	0.836	--
k = π/2a						
ε_3	S	-0.472	-0.519	0.501	0.499	--
ε_2	A	-0.536	-0.489	--	--	1.000
ε_1	S	-0.808	-0.856	0.499	0.501	--
k = π/a						
$\varepsilon_3(H)$	S	-0.326	-0.326	0.000	1.000	--
$\varepsilon_2(H)$	A	-0.640	-0.640	--	--	1.000
$\varepsilon_1(H)$	S	-0.760	-0.760	1.000	0.000	--
Polysilane						
k = 0						
$\varepsilon_3(\Gamma)$	S	-0.329	-0.329	0.694	0.306	--
$\varepsilon_2(\Gamma)$	A	-0.388	-0.388	--	--	1.000
$\varepsilon_1(\Gamma)$	S	-0.737	-0.737	0.301	0.699	--
k = π/2a						
ε_3	S	-0.320	-0.357	0.312	0.688	--
ε_2	A	-0.418	-0.410	--	--	1.000
ε_1	S	-0.605	-0.643	0.683	0.317	--
k = π/a						
$\varepsilon_3(H)$	S	-0.207	-0.207	0.000	1.000	--
$\varepsilon_2(H)$	S	-0.450	-0.450	--	--	1.000
$\varepsilon_1(H)$	S	-0.579	-0.579	1.000	0.000	--

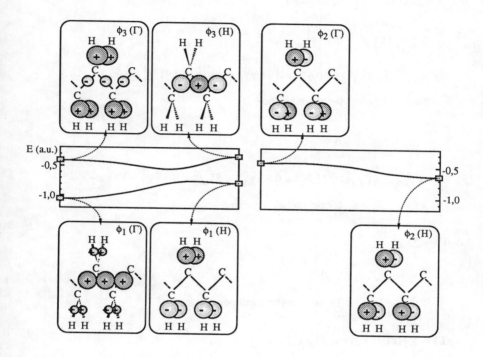

Figure 2.8   Sketch of the polymer orbitals in terms of bond orbitals in the case of polyethylene. Left: symmetric combinations (with respect to the C-C backbone) $\phi_1(\Gamma)$, $\phi_3(\Gamma)$, $\phi_1(H)$, $\phi_3(H)$. Right: antisymmetric combination (with respect to C-C backbone) $\phi_2(\Gamma)$, $\phi_2(H)$. Note that the results for polysilane results are similar but the amplitudes of the orbitals are to be rescaled according to the values of $\phi_i(\Gamma)$ and $\phi_i(H)$ given in the text

Table 2.16 also lists the calculated orbital charges. Sketches of the orbital structure are illustrated for the case of polyethylene in Figure 2.8. The analytical forms of these orbitals are given below:

k = 0

$$\phi_3(\Gamma) \propto \sum_j \{c_{31}(\Gamma)\, \chi_1^j + c_{32}(\Gamma)\, \chi_2^j \}$$

$$= \sum_j \{0.91\, \chi_1^j - 0.40\, \chi_2^j \}$$

$$= \ldots + 0.91\, \chi_1^{-1} - 0.40\, \chi_2^{-1} + 0.91\, \chi_1^0 - 0.40\, \chi_2^0 + 0.91\, \chi_1^1 - 0.40\, \chi_2^1 + \ldots$$

for polyethylene

$$= \sum_j \{0.83\, \chi_1^j - 0.55\, \chi_2^j \}$$

$$= \ldots + 0.83\, \chi_1^{-1} - 0.55\, \chi_2^{-1} + 0.83\, \chi_1^0 - 0.55\, \chi_2^0 + 0.83\, \chi_1^1 - 0.55\, \chi_2^1 + \ldots$$

for polysilane

$$\phi_2(\Gamma) \propto \sum_j \chi_3^j$$

$$= \ldots + \chi_3^{-1} + \chi_3^0 + \chi_3^1 + \ldots$$

for polyethylene and polysilane

$$\phi_1(\Gamma) \propto \sum_j \{c_{11}(\Gamma)\, \chi_1^j + c_{12}(\Gamma)\, \chi_2^j \}$$

$$= \sum_j \{0.41\, \chi_1^j + 0.90\, \chi_2^j \}$$

$$= \ldots + 0.41\, \chi_1^{-1} + 0.90\, \chi_2^{-1} + 0.41\, \chi_1^0 + 0.90\, \chi_2^0 + 0.41\, \chi_1^1 + 0.90\, \chi_2^1 + \ldots$$

for polyethylene

$$= \sum_j \{0.55\, \chi_1^j + 0.83\, \chi_2^j \}$$

$$= \ldots + 0.55\, \chi_1^{-1} + 0.83\, \chi_2^{-1} + 0.55\, \chi_1^0 + 0.83\, \chi_2^0 + 0.55\, \chi_1^1 + 0.83\, \chi_2^1 + \ldots$$

for polysilane

$$k = \pi/a$$

$$\phi_3(H) \propto \sum_j (-1)^j \chi_2^j$$

$$= \dots - \chi_2^{-1} + \chi_2^0 - \chi_2^1 + \dots$$

for polyethylene and polysilane

$$\phi_2(H) \propto \sum_j (-1)^j \chi_3^j$$

$$= \dots - \chi_3^{-1} + \chi_3^0 - \chi_3^1 + \dots$$

for polyethylene and polysilane

$$\phi_1(H) \propto \sum_j (-1)^j \chi_1^j$$

$$= \dots - \chi_1^{-1} + \chi_1^0 - \chi_1^1 + \dots$$

for polyethylene and polysilane

From Figure 2.8, we can easily interpret the energy evolution of the bands through orbital topology. The case of the antisymmetric orbital ϕ_2 is the most evident. Starting from the most antibonding situation at Γ (antisymmetric with respect to the plane of the zigzag chain and antisymmetric with respect to two successive atoms of the chain), energy decreases (the orbital gets slightly stabilized) at the H point where there exist bonding contributions between C-H bonds located on the same side of the main zigzag chain. Note that the dispersion is more important for polyethylene $\{\varepsilon_2(\Gamma)-\varepsilon_2(H)=0.208$ a.u.$=5.66eV\}$ than for polysilane $\{\varepsilon_2(\Gamma)-\varepsilon_2(H)=0.062$ a.u.$=1.69$ eV$\}$. Furthermore, as expected from the difference in C-H and Si-H bond energies, the barycenter of band 2 is higher in energy (less stable) for polysilane ($\alpha_{33}=-0.4185$ a.u.$=-11.38$ eV) than for polyethylene ($\alpha_{33}=-0.5365$ a.u.$=-14.59$ eV). The interaction between two successive antisymmetric combinations (χ_3) of X-H bond orbitals is positive since they are in opposite phase due to the presence of the screw axis. It is less repulsive for the two Si-H bonds ($2\beta_{33}=+0.0305$ a.u.$=0.83$ eV), which are more distant from one another than for the more interacting C-H bonds ($2\beta_{33}=+0.1045$ a.u.$=2.84$ eV). Schematic energy relationships for the antisymmetric band ε_2 are shown in the right part of Figure 2.9 for polyethylene (B) and polysilane (B'). Figure 2.9 is a complete energy sketch of the interactions which are discussed in this Section.

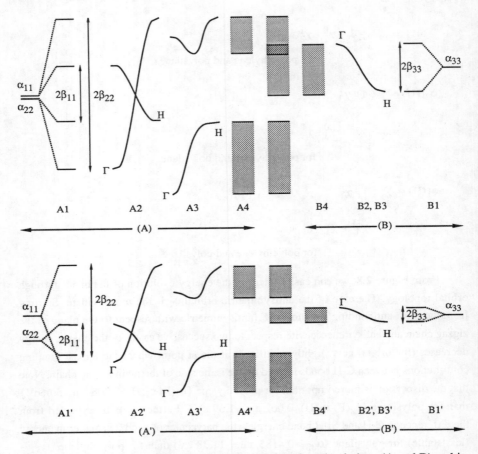

Figure 2.9 Energy sketch of the main interactions in polyethylene (A and B) and in polysilane (A' and B').

Symmetric combinations (with respect to the -X-X- backbone). A1 and A1': sketch of the χ_1 and χ_2 interactions; A2 and A2': sketch of the noninteracting χ_1 and χ_2 Bloch orbital energies; A3 and A3': sketch of the interacting χ_1 and χ_2 Bloch orbital energies, A4 and A4': sketch of the occupied symmetric valence energy region.

Antisymmetric combinations (with respect to the -X-X- backbone). B1 and B1': sketch of the χ_3 interactions; B2, B2' and B3, B3': sketch of the χ_3 Bloch orbital energies; B4 and B4': sketch of the occupied antisymmetric valence energy region.

The interpretation of the behaviour of the symmetric bands 1 and 3 is more complex. The most stable situation is that of $\phi_1(\Gamma)$, i.e., the bonding combination of χ_1 and χ_2 at point Γ. The main bonding contribution in this state is that of the chain X-X χ_2 orbital. It is more dominant in polyethylene $\{c_{11}(\Gamma)=0.41; c_{12}(\Gamma)=0.90\}$ than in polysilane $\{c_{11}(\Gamma)=0.55; c_{12}(\Gamma)=0.83\}$. Its antibonding counterpart $\phi_3(\Gamma)$ is the antibonding combination of χ_1 and χ_2; it is dominated by the X-H χ_1 orbitals: polyethylene $\{c_{31}(\Gamma)=0.91; c_{32}(\Gamma)=-0.40\}$, polysilane $\{c_{31}(\Gamma)=0.83; c_{32}(\Gamma)=-0.55\}$. It is striking to observe that ϕ_1 gradually transforms into the antisymmetric (with respect to the chain atoms) combination of the X-H χ_1 orbitals. The transformation is so complete that $\phi_1(H)$ has no contribution from the bond X-X χ_2 orbitals. On the other hand, ϕ_3 transforms into the antisymmetric (with respect to the chain atoms) combination of the X-X χ_2 orbitals. $\phi_3(H)$ has no contribution from the X-H χ_1 orbitals.

The latter statement is important since it is generally believed that the highest occupied orbital of the tetrahedrally σ-bonded chains mainly originates in a X-H contribution. It is shown here that it is mainly the X-X bond orbital with nodes at each atom of the chain. This result is fully confirmed by inspection of the LCAO coefficients listed in Table 2.13; it is clearly seen there that the $\phi_3(H)$ orbitals have only a C_{2pz} contribution in polyethylene or a Si_{3pz} contribution in polysilane. Turning to Figure 2.9, the sketches (Figure 2.9-A2 and A2') of the non-interacting χ_1 and χ_2 Bloch orbitals [with energy dependence $\varepsilon_{XH}(k) = \alpha_{11} + 2\beta_{11}\cos(ka)$ and $\varepsilon_{XX}(k) = \alpha_{22} + 2\beta_{22}\cos(ka)$] demonstrate that, in both polyethylene and polysilane, the origin of the previous behaviour is due to an avoided crossing of the bands (compare $\varepsilon_{XH}(k)$ and $\varepsilon_{XX}(k)$, in Figure 2.9-A2 and A2', to $\varepsilon_1(k)$ and $\varepsilon_3(k)$ in Figure 2.9-A3 and A3'). Note also the almost parallel behaviour of the XH symmetric χ_1 and antisymmetric χ_2 combination (compare $\varepsilon_{XH}(k) = \alpha_{11} + 2\beta_{11}\cos(ka)$ in Figure 2.9-A2 and A2', to $\varepsilon_2(k) = \alpha_{33} + 2\beta_{33}\cos(ka)$ in Figure 2.9-B2 and B2').

Now, we turn to the initial question of the origin of the minimum in band ε_3. It is clear from the analytical formula of energy bands ε_1 and ε_3 that in the right part of the first Brillouin zone of the asymmetric unit, the interaction term $4\beta_{12}^2(2+2\cos(ka))$ progressively decreases to vanish at the H point. Thus, the dispersion of the 3-th band of the asymmetric -XH$_2$- unit cell toward the H point (or the dispersion of the B1 band of the symmetric -XH$_2$-XH$_2$- unit cell toward the Γ point) **is not** due to the repulsion of the X-H and X-X bonds; it is mainly due to the repulsion of the node structure of the dominating

X-X bond densities (or C_{2pz}, or Si_{3pz} atomic densities) which strongly interact in this region and issue from the avoided crossing of $\varepsilon_{XH}(k)$ and $\varepsilon_{XX}(k)$. In this region where $k \rightarrow \pi/a$, $\cos(ka) \rightarrow -1$ and $2+2\cos(ka) \rightarrow 0$, the energy band is mainly determined by:

$$\varepsilon_3(k) \approx \alpha_{22} + 2\beta_{22}\cos(ka) + \frac{\beta_{12}^2[2 + 2\cos(ka)]}{|(\alpha_{11} - \alpha_{22}) + 2(\beta_{11} - \beta_{22})\cos(ka)|}$$

$$\approx \alpha_{22} + 2\beta_{22}\cos(ka)$$

Since β_{22} and $\cos(ka)$ are negative in this region, the contribution will be increasingly antibonding toward the H point where the repulsion of successive -XX- bonds dominates. Note also that the minimum is obtained when:

$$(\alpha_{11} - \alpha_{22}) + 2(\beta_{11} - \beta_{22})\cos(ka) = 0$$

i.e., when the intensities of the XH-XH and XX-XX interactions exactly compensate:

$$\varepsilon_{XH}(k) = \alpha_{11} + 2\beta_{11}\cos(ka) = \varepsilon_{XX}(k) = \alpha_{22} + 2\beta_{22}\cos(ka)$$

At the energy minimum of band 3, there is thus a non-negligible XX-XH interaction as seen by introducing the former compensation into the $\varepsilon_3(k)$ expression:

$$\varepsilon_3(k)_{min.} = \alpha_{11} + 2\beta_{11}\cos(ka) + \sqrt{\beta_{12}^2[2+2\cos(ka)]}$$

$$= \alpha_{11} + 2\beta_{11}\cos(ka) + 2\beta_{12}\cos(ka/2)$$

Our interpretation on the evolution of band ε_3 is now the following: starting from Γ where the XX-XX contribution is less significant, the XX-XX contribution gradually increases toward H. The dominant stabilization with respect to the k-increase comes from the XH-XH interactions (compare $\varepsilon_3(k)$ in Figure 2.9-A3 and A3' to $\varepsilon_{XH}(k)$ in Figure 2.9-A2 and A2'). As stated previously, the XX-XX and XH-XH interactions equilibrate at the minimum and in the right part of the Brillouin zone, the energy increase is provoked by the strong repulsive character of the XX-XX ($2\beta_{22}\cos(ka)$) in this region (compare $\varepsilon_3(k)$ in Figure 2.9-A3 and A3' to $\varepsilon_{XX}(k)$ in Figure 2.9-A2 and A2').

It is sometimes erroneously stated that the σ-conjugation in polysilane is due to Si-Si bonds of smaller bond energies than the Si-H ones while the opposite situation, i.e., C-C bonds stronger than C-H, exists in the polyethylene chain. Figure 2.9 shows that there are actually no major differences in the behaviour of the X-X and X-H bonds. The energy domain of the Si-Si bonds is, however, more embedded within the Si-H region than it is the case between C-C and C-H bonds. Note that the standard thermodynamical data corroborate this view (bond energies in kJ/mol, C-C: 348; C-H: 412; Si-Si: 176; Si-H: 318 [2.83]).

Along the same idea, the set of parameters is compatible with the chemical role of σ conjugation as proposed by Dewar [2.84]. In his work, Dewar shows that any two-orbital CH_2 unit in a long paraffin plays the same role as a two-orbital =CH-CH= unit in linear conjugated polyenes. The comparison of the one-site X-H α_{11} interactions (-0.6400 a.u. for polyethylene and -0.5160 a.u. for polysilane) as well as the ratio of the - XX-XX- nearest-neighbour to the one-site chain interactions $(2\beta_{22}/\alpha_{22} = 0.489$ for polyethylene, $= 0.495$ for polysilane) can be used in this scheme.

Finally, a rough estimate of the electron effective mass can also be made from the analytical curve of the HOMO level near $\varepsilon_3(H)$; we find that in the vicinity of H, $\varepsilon_3(k) \approx \alpha_{22} + 2\beta_{22}\cos(ka) \approx h^2k^2/8\pi^2m$ and $m^*(k) = h^2/8\pi^2a^2\beta_{22}$. As expected, the electron effective mass is negative according to the bonding (stabilizing) character of β_{22}; not forgetting the unit cell size effect, the numerical values $a^2\beta_{22}$ (the variables which define the denominator of the effective mass, m*) are in the ratio: polysilane / polyethylene = [-0.2030 a.u. * 3.609335^2 a.u.^{2}] / [-0.3130 a.u. * 2.376618^2 a.u.^{2}] = 1.50; the effective mass in polyethylene is thus 1.5 times larger than in polysilane, a result in complete agreement with the calculated values of Teramae and Takeda [2.81].

156

References

[2.1] J.C. Slater, Phys. Rev., **36**, 57 (1930).

[2.2] C.C.J. Roothaan, J. Chem. Phys., **19**, 1445 (1951).

[2.3] S.F. Boys, Proc. Roy. Soc. (London), **A200**, 542 (1950).

[2.4] W.J. Hehre, R.F. Stewart, and J.A. Pople, J. Chem. Phys., **51**, 2657 (1969); W.J. Hehre, R. Ditchfield, R.F. Stewart, and J.A. Pople, J. Chem. Phys., **52**, 2769 (1970); W.J. Pietro, B.A. Levi, W.J. Hehre, and R.F. Stewart, Inorg. Chem., **19**, 2225 (1980); W.J. Pietro, R.F. Hout, Jr., E.S. Blurock, W.J. Hehre, D.J. DeFrees, and R.F. Stewart, Inorg. Chem., **20**, 3650 (1981), W.J. Pietro, and W.J. Hehre, J. Comput. Chem., **4**, 241 (1983).

[2.5] W.J. Hehre, L. Radom, P.v.R. Schleyer, and J.A. Pople, "Ab Initio Molecular Orbital Theory", J. Wiley & Sons, New York (1986).

[2.6] S. Huzinaga, "Physical Sciences Data, 16: Gaussian Basis Sets for Molecular Calculations", Elsevier, Amsterdam (1984).

[2.7] R. Poirier, R. Kari, and I.G. Csizmadia, "Physical Sciences Data, 24: Handbook of Gaussian Basis Sets", Elsevier, Amsterdam (1985).

[2.8] E. Clementi, J. Chem. Phys., **40**, 1944 (1964).

[2.9] L. Allen, E. Clementi, and H. Gladney, Rev. Modern Phys., **35**, 465 (1963).

[2.10] T. Koopmans, Physica, **1**, 104 (1933).

[2.11] A.A. Frost, J. Chem. Phys., **47**, 3707 (1967).

[2.12] J.M. André, M.E. Lambert-Gérard, and C. Lamotte, Bull. Soc. Chim. Belg., **85**, 745 (1976).

[2.13] J.M. André and J.L. Brédas, Chem. Phys., **20**, 367 (1977).

[2.14] J.L. Brédas, J.M. André, and J. Delhalle, Chem. Phys., **45**, 109 (1980).

[2.15]   J. Delhalle, J.M. André, C. Demanet, and J.L. Brédas, Chem. Phys. Lett., **54**, 186 (1978).

[2.16]   J.M. André, J. Delhalle, and C. Lamotte, Bull. Soc. Chim. Belg., **89**, 691 (1980).

[2.17]   E. Hückel, Z. Physik, **60**, 423 (1930); **70**, 204 (1931); **76**, 628 (1932).

[2.18]   R.S. Mulliken, J. Chim. Phys., **46**, 497, 675 (1949).

[2.19]   M. Wolfsberg and L. Helmholtz, J. Chem. Phys., **20**, 837 (1952).

[2.20]   R. Hoffmann, J. Chem. Phys., **39**, 1397 (1963); **40**, 2474, 2480, 2745 (1964).

[2.21]   R.G. Parr, J. Chem. Phys., **20**, 239, 1499 (1952); **21**, 568 (1952); R. Pariser and R.G. Parr, J. Chem. Phys., **21**, 466, 767 (1953).

[2.22]   J.A. Pople, Trans. Far. Soc., **49**, 1375 (1953).

[2.23]   J.A. Pople, D.P. Santry, and G.A. Segal, J. Chem. Phys., **43**, S129 (1965); J.A. Pople and G.A. Segal, J. Chem. Phys., **43**, S136 (1965); **44**, 3289 (1966).

[2.24]   J.A. Pople and D.L. Beveridge, "Approximate Molecular Orbital Theory", McGraw Hill, New York (1970).

[2.25]   J.N. Murrel and A.J. Harget, "Semiempirical Self-Consistent Field Molecular Orbital Theory of Molecules", Wiley-Interscience, London (1972).

[2.26]   O. Sinanoglu and K.B. Wiberg, "Sigma Molecular Orbital Theory", Yale University Press, New Haven and London (1970).

[2.27]   G. Klopman and B. O'Leary, "Introduction to All-Valence Electrons S.C.F. Calculations", Springer-Verlag, Berlin (1970).

[2.28]   R. Daudel and C. Sandorfy, "Semiempirical Wave-Mechanical Calculations on Polyatomic Molecules", Yale University Press, New Haven and London (1971).

[2.29] G.A. Segal (Ed.), "Semiempirical Methods of Electronic Structure Calculations, Part A: Techniques, Part B: Applications", Plenum Press, New York (1977).

[2.30] J. Sadlej, "Semiempirical Methods of Quantum Chemistry", Ellis Horwood Limited, Chichester (1985).

[2.31] J.M. André, D.P. Vercauteren, and J.G. Fripiat, J. Comput. Chem., 5, 349 (1984).

[2.32] C. Pisani and R. Dovesi, Int. J. Quantum Chem., 17, 501 (1980).

[2.33] J.M. André and J. Delhalle, in "Quantum Theory of Polymers", (J.M. André, J. Delhalle, and J. Ladik, Eds.), pp.1 et sq., NATO-ASI Series C39, D. Reidel Publishing Company (1978).

[2.34] P.O. Löwdin, Advan. Phys., 5, 1 (1956).

[2.35] F. O'Shea and D.P. Santry, Chem. Phys. Lett., 25, 164 (1974).

[2.36] I.I. Ukrainskii, Theoret. Chim. Acta, 38, 139 (1975).

[2.37] J. Delhalle, J.G. Fripiat, and L. Piela, Int. J. Quantum Chem., S14, 431 (1980).

[2.38] F.E. Harris and H. Monkhorst, Phys. Rev., B2, 4400 (1970); F.E. Harris, J. Chem. Phys., 56, 4422 (1972); F.E. Harris, in "Theoretical Chemistry: Advances and Perspectives", (H. Eyring and D. Henderson, Eds.),Vol. 1, pp.147 et sq., Academic Press, New York (1975).

[2.39] J.M. André, J.G. Fripiat, C. Demanet, J.L. Brédas, and J. Delhalle, Int. J. Quantum Chem., S12, 233 (1978).

[2.40] J. Delhalle, L. Piela, J.L. Brédas, and J.M. André, Phys. Rev., B22, 6254 (1980).

[2.41] L. Piela, J.M. André, J.L. Brédas, and J. Delhalle, Int. J. Quantum Chem., S14, 405 (1980).

[2.42] L. Piela, J.M. André, J.G. Fripiat, and J. Delhalle, Chem. Phys. Lett., **77**, 143 (1981).

[2.43] J. Delhalle and J.L. Calais, J. Chem. Phys., **85**, 5286 (1986); J.L. Calais and J. Delhalle, in "Understanding Molecular Properties", (J. Avery et al., Eds.), pp. 511 et sq., D. Reidel Publishing Company (1987); J. Delhalle and J.L. Calais, Phys. Rev., **B35**, 9460 (1987); J. Delhalle and J.L. Calais, Int. J. Quantum Chem., **S21**, 115 (1987).

[2.44] H.J. Monkhorst, Phys. Rev., **B20**, 1504 (1979); H.J. Monkhorst and M. Kertesz, Phys. Rev., **B24**, 3015 (1981).

[2.45] J.C. Slater, Phys. Rev., **81**, 385 (1951); **82**, 538 (1952).

[2.46] J. Delhalle, J.G. Fripiat and F.E. Harris, Int. J. Quantum Chem., **S18**, 141 (1984); J. Delhalle and F.E. Harris, Int. J. Quantum Chem., **22**, 219 (1985); J. Delhalle and F.E. Harris, Phys. Rev., **B31**, 6755 (1985).

[2.47] J. Delhalle, M.H. Delvaux, J.G. Fripiat, J.M. André, and J.L. Calais, J. Chem. Phys., **88**, 3141 (1988).

[2.48] A. Imamura and H. Fujita, J. Chem. Phys., **61**, 115 (1974); A. Blumen and C. Merkel, Phys. Status Solidi, **B83**, 425 (1977).

[2.49] I. Bozovic, J. Delhalle, and M. Damnjanoviç, Int. J. Quantum Chem., **20**, 1143 (1981); M. Damnjanovic, I. Bozovic, and N. Bozovic, J. Phys. A, Math. Gen., **16**, 3937 (1983); **17**, 747 (1984); I. Bozovic and J.Delhalle, Phys. Rev. B, **29**, 4733 (1984).

[2.50] M. Springborg and R.O. Jones, J. Chem. Phys., **88**, 2652 (1988).

[2.51] J.M. André, D.P. Vercauteren, V.P. Bodart, and J.G. Fripiat, J. Comput. Chem., **5**, 535 (1984).

[2.52] J.A. Pople, D.L. Beveridge, and P.A. Dobosh, J. Chem. Phys., **47**, 2026 (1967).

[2.53] R.C. Bingham, M.J.S. Dewar, and D.H. Lo, J. Am. Chem. Soc., **97**, 1285, 1294, 1302, 1307 (1975); M.J.S. Dewar, Science, **187**, 1037 (1975).

[2.54] M.J.S. Dewar and W. Thiel, J. Am. Chem. Soc., **99**, 4899, 4907 (1977); M.J.S. Dewar and M.L. McKee, J. Comput. Chem., **4**, 84 (1983).

[2.55] J.L. Brédas, B. Thémans, J.M. André, A.J. Heeger, and F. Wudl, Synth. Met., **11**, 343 (1985); J. Riga, Ph. Snauwaert, A. De Pryck, R. Lazzaroni, J.P. Boutique, J.J. Verbist, J.L. Brédas, J.M. André, and C. Taliani, Synth. Met., **21**, 223 (1987); B. Thémans, J.M. André, and J.L. Brédas, Synth. Met., **21**, 149 (1987); J.M. Toussaint, B. Thémans, J.M. André, and J.L. Brédas, Synth. Met., **28**, C205 (1989).

[2.56] M.J.S. Dewar, E.G. Zoebisch, E.F. Healy, and J.J.P. Stewart, J. Am. Chem. Soc., **107**, 3902 (1985)

[2.57] G. Nicolas and Ph. Durand, J. Chem. Phys., **70**, 2020 (1979).

[2.58] J.M. André, L.A. Burke, J. Delhalle, G. Nicolas, and Ph. Durand, Int. J. Quantum Chem., **S13**, 283 (1979).

[2.59] J.L. Brédas, R.R. Chance, R. Silbey, G. Nicolas, and Ph. Durand, J. Chem. Phys. **75**, 255 (1981).

[2.60] J.L. Brédas, R.R. Chance, R. Silbey, G. Nicolas, and Ph. Durand, J. Chem. Phys. **77**, 371 (1982).

[2.61] J.L. Brédas, B. Thémans, and J.M. André, J. Chem. Phys., **78**, 6137 (1983).

[2.62] P.O. Löwdin, J. Chem. Phys., **18**, 365 (1950).

[2.63] K. Ruedenberg, J. Chem. Phys., **19**, 1433 (1951).

[2.64] J.M. André and G.Leroy, Bull. Soc. Chim. Belges, **78**, 421 (1969).

[2.65] J.N. Murrell, "The Theory of the Electronic Spectra of Organic Molecules", Chapman and Hall, London (1971).

[2.66] H.H. Jaffé and M. Orchin, "Theory and Applications of Ultraviolet Spectroscopy", J. Wiley & Sons, New York (1962).

[2.67] I.G. Csizmadia, M.C. Harrison, J.W. Moscowitz, S. Seung, B.T. Sutcliffe, and M.P. Barnett, POLYATOM, Quantum Chemistry Program Exchange, **11**, 47 (1964); D.B. Newmann, H. Basch, R.L. Korregay, L.C. Snyder, J. Moskowitz, C. Hornback, and P. Liebman, POLYATOM 2, Quantum Chemistry Program Exchange, **11**, 199 (1973).

[2.68] A. Veillard, IBMOL 4, Computer Program, IBM Research Laboratory, San Jose, Calif., (1968); E. Ortoleva, G. Castiglione, and E. Clementi, IBMOL 6, Comput. Phys. Comm., **19**, 337 (1980); R. Gomperts and E. Clementi, KGNMOL, Quantum Chemistry Program Exchange, Bulletin **8**, 538 (1988).

[2.69] W.J. Hehre, W.A. Lathan, M.D. Newton, R. Ditchfield, and J.A. Pople, GAUSSIAN 70, Quantum Chemistry Program Exchange, **11**, 236 (1973); J.S. Binkley, R. Whiteside, P.C. Hariharan, R. Seeger, W.J. Hehre, M.D. Newton, and J.A. Pople, GAUSSIAN 76, Quantum Chemistry Program Exchange, **11**, 368 (1978); J.S. Binkley, R.A. Whiteside, R. Krishnan, R. Seeger, D.J. DeFrees, H.B. Schlegel, S. Topiol, L.R. Kahn, and J.A. Pople, GAUSSIAN 80, Quantum Chemistry Program Exchange, **13**, 406 (1981); J.S. Binkley, M.J. Frisch, K. Rahgavachari, D.J. DeFrees, H.B. Schlegel, R.A. Whiteside, E.M. Fluder, R. Seeger, and J.A. Pople, GAUSSIAN 82, Carnegie Mellon University, Pittsburgh PA; M.J. Frisch, J.S. Binkley, H.B. Schlegel, K. Rahgavachari, R.L. Martin, J.J.P. Stewart, F.W. Bobrowicz, D.J. DeFrees, R. Seeger, R.A. Whiteside, D.J. Fox, E.M. Fluder, and J.A. Pople, GAUSSIAN 86, release C, Carnegie-Mellon University, Pittsburgh PA.

[2.70] M. Dupuis, J. Rys, and H.F. King, HONDO, Quantum Chemistry Program Exchange, **11**, 336, 338 (1977); **13**, 401 (1981); M. Dupuis, J.D. Watts and H.O. Villar, HONDO7, Quantum Chemistry Program Exchange, Bulletin **8**, 544 (1988).

[2.71] R. Ahlrichs, M. Bär, M. Häser, H. Horn, and C. Kölmel, Chem. Phys. Lett. **162**, 165 (1989).

162

[2.72] J. Zupan, Ann. Soc. Sci. Brux., Ser.1, **89**, 337 (1975).

[2.73] S. Suhai, P.S. Bagus, and J. Ladik, Chem. Phys., **68**, 467 (1982).

[2.74] P.O. Löwdin, Rev. Mod. Phys., **39**, 259 (1967).

[2.75] A. Szabo, N.S. Ostlund, "Modern Quantum Chemistry ", Mc Graw-Hill (1989).

[2.76] J.M. André, J. Delhalle, G.S. Kapsomenos, and G. Leroy, Chem. Phys. Lett., **14**, 485 (1972).

[2.77] J. Delhalle and S. Delhalle, Int. J. Quantum Chem., **11**, 349 (1977); J. Delhalle, D. Thelen and J.M. André, Computers and Chemistry, **3**, 1 (1979).

[2.78] W.L. McCubbin, in "Electronic Structure of Polymers and Molecular Crystals", (J.M. André and J. Ladik, Eds.), NATO ASI Series, Vol B 9, p.171, Plenum Press, New York (1975).

[2.79] J.M. Zeigler and F.W. Gordon Fearon (Eds.), "Silicon-Based Polymer Science, A Comprehensive Resource", Advances in Chemistry Series **224** , American Chemical Society, Washington (1990).

[2.80] R.D. Miller, Polym. News, **12**, 326 (1987).

[2.81] H. Teramae and K. Takeda, J. Am. Chem. Soc., **111**, 1281 (1989).

[2.82] J.W. Mintmire and J. Ortiz, in "Silicon-Based Polymer Science, A Comprehensive Resource", (J.M. Zeigler and F.W. Gordon Fearon, Eds.), Advances in Chemistry Series **224**, p. 543, American Chemical Society, Washington (1990).

[2.83] See, for example, P.W. Atkins, "Physical Chemistry", Third edition, Table 4.5, p.822, Oxford University Press, Oxford (1986).

[2.84] M.J.S. Dewar, J. Am. Chem. Soc., **106**, 669 (1984).

3

QUANTUM APPROACH TO ELECTRICAL CONDUCTION PHENOMENA IN CONJUGATED ORGANIC POLYMERS

3.1 SUMMARY AND OBJECTIVES

The purpose of this chapter is

(1) to give an introduction to the field of conducting polymers,

(2) to determine the relationship existing between the geometric structure and the electronic structure in these systems,

(3) to gain a deep insight into the mechanisms leading to the appearance of high electrical conductivity upon redox or protonic acid doping of conjugated polymers,

(4) to offer some predictions in terms of novel macromolecular architectures that could result in low bandgap characteristics,

(5) to exemplify by a set of selected examples the application of quantum chemistry to the molecular design of organic compounds having a high conductivity.

3.2 INTRODUCTION TO THE FIELD OF CONDUCTING POLYMERS

Prior to the 70's, polymeric materials were essentially considered to be electrically insulating and in fact used as such in a number of applications. However, in the past fifteen years, a number of major discoveries has provided the possibility of controling the electrical conductivity of some types of polymers over a wide range, all the way from the insulating regime up to the metallic regime. These discoveries have triggered a revolution in our perception and knowledge of the organic solid state. They have also led to substantial efforts to synthesize electrically conducting polymers for technological applications. These applications are starting to materialize due to the steady improvement in the mechanical characteristics, environmental stability, solubility, and processibility of the materials that can nowadays be prepared.

The genuine interest in conducting plastics stems from the fact that they offer the opportunity to combine in a single material the electrical properties of metals with the plasticity, light weight, (usually) low cost, and synthetic engineering feasibility which are typical of polymers. Many application areas are envisioned. In the first place, we can cite the domain of secondary batteries, which currently constitutes the area where applications appear to be in the most advanced stage. Polyaniline batteries are now being produced in Japan by Bridgestone and Seiko; polypyrrole batteries developed by Varta and BASF in Germany are in the final phases of tests before reaching the market place; new battery concepts involving polyacetylene or polyparaphenylene are proposed by the researchers at Allied-Signal. Other applications involve the development of products for electromagnetic and electrostatic shielding, semiconductor devices, polymer-modified electrodes, electrochromic displays, and anticorrosion covers.

As general introductions to the field of conducting polymers, we may refer the reader to a number of collective works [3.1-3.4] including the Proceedings of the 1986, 1988, and 1990 International Conferences on Synthetic Metals and the Proceedings of the 1989 Advanced Research Workshop on "Conjugated Polymeric Materials: Opportunities in Electronics, Optoelectronics, and Molecular Electronics".

Our goal is here twofold: (i) to illustrate the application to an important area of research, of most of the quantum-chemical techniques which have been described in the first part of this book; and (ii) to show how the theoretical results have helped in the rationalization of the fascinating experimental data obtained in the field of conducting polymers as well as to share with the reader the exciting concepts which have been uncovered in the past decade. After a brief presentation in this Section of the history of conducting polymers, Sections 3.3 to 3.5 will concentrate on the electronic properties of the undoped compounds. Section 3.3 is devoted to a general introduction on the electronic structure of conjugated systems; the consequences of the ground-state degeneracy in polyacetylene, the prototypical and topologically simplest conjugated polymer, are outlined; we then describe the electronic structures of polyparaphenylene, polypyrrole, polythiophene, polyaniline, and their derivatives; the evolution of the electronic properties as a function of chain length is also discussed. In Section 3.4, we illustrate the influence on the electronic properties, of torsions along the backbones of conjugated polymeric chains. This allows us to rationalize the thermochromic and solvatochromic effects which have been recently discovered namely in polythiophene derivatives and to illustrate the interplay between conformation and electronic structure. Section 3.5 focusses on the theoretical efforts which are paid to try and design novel polymeric architectures leading to small bandgap systems, i.e., polymers which would be intrinsically conducting or semiconducting without the need of doping. Sections 3.6 and 3.7 are concerned with doped polymers. In Section 3.6, we present a general discussion on the nature of charged species introduced by the doping process: solitons, polarons, and bipolarons; the peculiarities of degenerate ground state systems (such as trans-polyacetylene or the fully oxidized form of polyaniline, pernigraniline) are pointed out. The evolution of the electronic structure at high doping level is illustrated in Section 3.7 in the case of polyacetylene, polythiophene, and the emeraldine (semioxidized form) of polyaniline. Finally, in Section 3.8, we describe the evolution of the geometrical and electronic structures upon photoexcitation of polyacetylene; this Section can be viewed as an introduction to the fourth part of the book dealing with the nonlinear optical properties of conjugated compounds.

At first, it is therefore informative to recall the main discoveries which have led to the development of conducting polymers. Although a number of works hinted at the possibility of achieving significant electrical conductivities in polymers as early as in the sixties [3.5], it may be reasonably assessed that the first key development was the finding in 1973 that the inorganic polymer polysulfurnitride is a metal [3.6]. The room temperature conductivity of polysulfurnitride is of the order of 10^3 S/cm, to be compared to about 6×10^5 S/cm for copper and 10^{-14} S/cm for polyethylene (Note that S stands for Siemens which is a unit of conductance equivalent to the reciprocal Ohm). Polysulfurnitride becomes superconducting (i.e., offers no resistance to electrical current) below a critical temperature of 0.3 K [3.7]. Although this critical temperature is very low by today's standards, it is worth stressing that polysulfurnitride is still the only polymeric material known to date to possess the property of superconductivity. These discoveries were of particular importance because they provided an "existence theorem" for highly conducting polymers and stimulated the enormous amount of work that proved to be necessary in order to synthesize other polymeric conductors. The metallic character of polysulfurnitride is an intrinsic property of the material related to the presence of one unpaired electron per unit cell. As a result, the highest occupied band is half-occupied and there exists no forbidden energy gap between the highest occupied and lowest unoccupied electronic levels.

Although its other physical properties, the least not being its explosive nature, prevented it from entering any technological application, polysulfurnitride provided the field with further insight that eventually led to the discovery of an entirely new class of conducting polymers. By the mid-70's, it was observed that the room-temperature conductivity of polysulfurnitride can be enhanced by one order of magnitude following exposure to bromine and similar oxidizing agents [3.8]. In this case, the conducting entity is no longer a neutral polymer but rather a polymeric multication, charge neutrality being preserved by incorporating into the material the reduced form of the oxidizing agent (such as Br_3^- when the polymer is exposed to bromine).

The major breakthrough in the area of conducting plastics occurred when the same redox chemistry was applied by the group of MacDiarmid, Heeger, and Shirakawa to an intrinsically insulating organic polymer, polyacetylene. Polyacetylene, which has been known since the early 30's in the form of an ill-defined black powder, was obtained for the first time in the form of a free-standing film in the early 70's: this was actually the consequence of a preparation mistake involving the use of what was up to then believed to be much too large a concentration of the catalyst. In 1977, it was found that polyacetylene, which has an intrinsic conductivity much lower than 10^{-5} S/cm, can be made highly conducting (i.e., electrical conductivity on the order of 10^3 S/cm) by exposing it to oxidizing or reducing agents [3.9].

We may note that this process is often referred to as "doping" by analogy with the doping of inorganic semiconductors such as silicon, germanium, or gallium arsenide. For reasons that will become clear in Section 3.6, this is a rather misleading analogy and the process is best viewed as a redox reaction. The insulating pristine neutral polymer is converted into an ionic complex consisting of a polymeric cation (or anion) and a counterion which is the reduced form of the oxidizing agent (or the oxidized form of the reducing agent). When condensed-matter physics terminology is employed to describe the phenomenon, the use of an oxidizing reagent is said to correspond to a p-type doping (the polymer chain becoming positively charged) and that of a reducing agent, to an n-type doping (the polymer chain becoming negatively charged). From this simple discussion, it is clear that an important criterion in selecting potentially highly electrically conducting polymers is the ease with which the system can be oxidized or reduced. This largely accounts for the choice of π-bonded unsaturated polymers which, like polyacetylene have small ionization potentials and/or large electron affinities. Electrons of π character can be relatively easily removed or added to form a polymeric ion without too much disruption of the σ bonds which are primarily responsible for holding the polymer together. In the case of a saturated polymer such as polyethylene or polypropylene, for example the oxidation process would require higher energy and remove σ electrons, leading to bond breaking and chemical decomposition. Even in the case of conjugated polymers, it is important to try and choose relatively mild oxidizing agents which do not cause any disruptive chemistry.

Following polyacetylene, this basic principle has been applied with success to an increasingly large number of other organic polymers including polyparaphenylene, polypyrrole, polythiophene, polyaniline, and their derivatives [3.2-3.4]. The structures of the principal conducting polymer systems are displayed in Figure 3.1. The maximum dopant (i.e., counterion) concentrations are typically on the order of several mol % per repeat unit of the chain backbone, for instance up to about 15 % in the case of polyacetylene doped by sodium, $([CH]Na_{0.15})x$.

In a number of instances, it was found that oxidation or reduction of the polymer can be achieved electrochemically by subjecting the neutral polymer to the appropriate oxidizing or reducing voltage in an electrochemical cell. The charges appearing on the polymer chains are then neutralized by counterions coming from the electrolyte solution. An interesting group of conducting polymers consists of those prepared by the electrochemical oxidation and simultaneous polymerization of a monomer which reacts at the anode of an electrochemical cell. This group includes polypyrrole, polythiophene, and their numerous derivatives. The possibility of electrochemically cycling the polymer between the neutral and doped states constitutes the basis for the application of conducting polymers in batteries.

More recently, another class of conjugated polymers was uncovered in which it is found that, besides the possibility of increasing the conductivity through redox chemistry, an insulator-to-conductor transition can also be obtained through acid-base chemistry which does not change the number of electrons on the polymer chains. The prototypical example of this type of compounds is polyaniline [3.10], whose fascinating properties will be extensively discussed in the following Sections.

Despite their early promises, conducting polymers took a long time to turn into real conducting "plastics" because of their insolubility and infusibility which prevented them from being treated by conventional polymer processing techniques. The situation, however, dramatically improved in the mid-eighties when studies on polythiophene or polypyrrole compounds showed that derivatization of the aromatic rings (namely on the β-carbon position with alkyl tails containing at least four carbons) allowed the resulting polymer to be soluble in most common organic solvents [3.11]. Derivatization with ethane sulfonate or butane sulfonate side-chains even allows for solubilization in water and led to the concept of self-doped polymer: when the polymer chains are oxidized, the sulfonate groups which are covalently linked to the polymer backbones become the counterions [3.12].

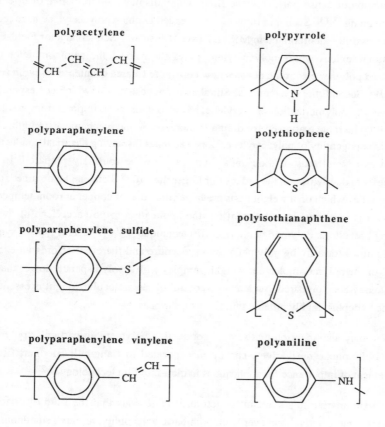

Figure 3.1   Structure of some of the most important conjugated polymers

Major improvements have also been brought to the synthetic procedures of conjugated polymers leading to materials which can be much better characterized, are more highly ordered, and display significantly better mechanical and electrical properties. These improvements can be examplified in the case of polyacetylene. First, in the early 80's, a new synthetic pathway based on the use of a soluble precursor route opened the way to the formation of dense polyacetylene films which display a good degree of orientation upon stretching [3.13]. Such thin films of the so-called Durham polyacetylene have recently been successfully incorporated in MISFET (Metal-Insulator-Semiconductor Field-Effect-Transistor) devices where polyacetylene plays the role of the semiconductor [3.14]; conjugated polymer electronic devices now constitute an area of intense research activity [3.15, 3.16]. Second, the conventional Shirakawa synthesis [3.17], which corresponds to a Ziegler-Natta polymerization of acetylene, has also undergone important improvements [3.18, 3.19] which have allowed for dramatic increases in the mechanical strength and the electrical conductivity of polyacetylene. Doped samples displaying electrical conductivities as high as several tens of thousands S/cm can now be routinely prepared; in some instances, conductivities on the order of or larger than 10^{+5} S/cm can be obtained, that is, conductivities which on a weight basis are larger than that of copper at room temperature. Finally, it is very important to point out the application to polyacetylene of the Ring Opening Metathesis Polymerization (ROMP) technique [3.20, 3.21]. This type of synthesis indeed allows for a living polymerization procedure and therefore affords an excellent control of the polymer molecular weight; samples with a polydispersity lower than 1.07 have already been prepared. Such a good control of the molecular weight is essential for instance to deposit highly ordered thin films on surfaces.

We may also mention that several conducting polymers, in particular polypyrrole and polythiophene derivatives, can be incorporated in Langmuir-Blodgett films, an achievement of importance for applications in ultra-thin film technology [3.22].

To summarize this short introduction, it is obvious that the field of conducting polymers is more active than ever from both basic and applied research standpoints. It is also clear that, as has been so often the case before, further developments require strong interactions and collaboration between chemists, physicists, and materials scientists, experimentalists and theoreticians alike. Conducting polymers typify the example of an area where only a multidisciplinary approach can be of significance. We will try to illustrate such an approach in the following Sections.

3.3 GENERAL CHARACTERISTICS OF THE ELECTRONIC STRUCTURE OF CONJUGATED POLYMERS

In this Section, we present the general features of the electronic structure of conjugated polymers in their pristine state. We first examine polyacetylene, which is the simplest among conjugated polymers from the topological point of view. We then focus on polymers containing aromatic rings such as polyparaphenylene, polypyrrole, and polythiophene. The peculiarities of the electronic structure of the different forms of polyaniline (leucoemeraldine, emeraldine, and pernigraniline) are pointed out. Finally, the evolution of the electronic properties as a function of chain length is briefly described. We stress that all the calculations presented here are carried out on single polymer chains. The possible influence of interchain interaction effects is not explicitly taken into account. The excellent overall agreement between the results of our calculations and the experimental data, however, justifies our approach and indicates that interchain interactions usually do not affect the electronic properties of the polymer in a significant way. On the other hand, it is clear that they are essential to comprehend the macroscopic transport properties, since charge carrier have to hop from chain to chain.

Throughout this Section, some aspects of the electronic structure will be emphasized, for instance the values of the ionization potential, the electron affinity, the width of the highest occupied or lowest unoccupied electronic band, and the bandgap. These parameters are of particular importance in the context of conducting polymers:

(i) The ionization potential value determines whether a given electron acceptor (oxidizing agent) is able to ionize at least partly the polymer chain, which is of importance for a p-type doping. The electron affinity value plays an equivalent role with respect to the n-type doping process.

(ii) The width of the highest occupied valence band, or the lowest unoccupied band, is a measure of the extent of electronic delocalization along the polymer chain. It can be very roughly correlated with the mobility of possible charge carriers in that band. It is, however, far from being sufficient to describe the transport properties.

(iii) The value of the bandgap is indicative of the intrinsic electrical properties of the polymer. Above 2 eV, the polymer can be considered to be an insulator with intrinsic electrical conductivities lower than 10^{-10} S/cm at room temperature; for bandgap values smaller than 2 eV, the polymer is conventionally described as a semiconductor with electrical conductivities depending strongly on the actual value of the bandgap. A zero bandgap value, as in the case of polysulfurnitride, leads to a metallic conductivity, typically larger than 10^{+2} S/cm at room temperature.

3.3.1 Polyacetylene

Three nonequivalent backbone geometries can be envisioned for polyacetylene: the all-trans form (which is thermodynamically the most stable form) and two cis forms, cis-transoid and trans-cisoid. The traditional Shirakawa synthesis leads to the cis form which is converted into the all-trans form upon heating or doping [3.23].

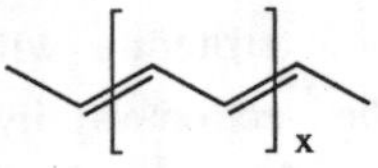

all-trans 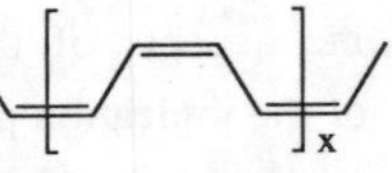   trans-cisoid 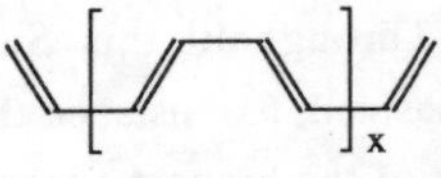   cis-transoid

The Valence Effective Hamiltonian (VEH) valence band structure of all-trans-polyacetylene is displayed in Figure 3.2 together with the (unconvoluted) electronic density of states [3.24]. The five occupied valence bands can be classified into two categories: (i) the bands denoted ε_1 to ε_4 in the Figure are σ bands with carbon 2s and $2p_\sigma$ and hydrogen 1s contributions; (ii) band ε_5 is the π band. It must be stressed that the width of the π band (on the order of 6 eV) is by far the largest among all conjugated polymers that we have investigated to date. Such a huge bandwidth is consistent with the fact that polyacetylene displays the highest electrical conductivities upon doping. The bandgap is calculated to be 1.5 eV, in excellent agreement with the experimental data which show an absorption onset at 1.4 eV.

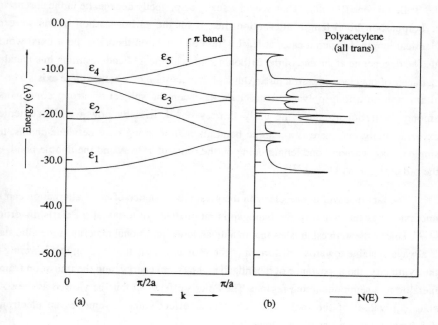

Figure 3.2   VEH band structure (a) and valence electronic density of states (b) of trans-polyacetylene

It is also interesting to note that the configuration, be it all-trans or cis, has rather little influence on the characteristics of the π band [3.24]. As a result, since the electronic properties of interest are dominated by the π band, all three polyacetylene forms have similar values for the ionization potential (about 4.7 eV) and the width of the highest occupied valence band. The bandgap is slightly larger for the cis form. The conformation has thus major effects only on the σ bands, the differences there being large enough to be possibly detected by well-resolved UPS photoemission spectra.

We observe in the electronic band structure of trans-polyacetylene that bands ε_1 and ε_2 on the one hand, and bands ε_3 and ε_4 on the other, are almost degenerate at the end of the Brillouin zone ($k = \pi/a$). They would actually be perfectly degenerate in the absence of carbon-carbon bond-length alternation, i.e., if all the carbon-carbon bonds were of identical length. In such a case, at least within one-electron theories, the π band would also be degenerate at the end of the Brillouin zone with the π^* band, resulting in a metallic character of the electronic structure. Although the question of whether there exists a bond alternation on a long polyacetylene chain has been the subject of strong controversy among theoreticians since the early thirties [3.25], it is by now well documented experimentally that there is a marked bond alternation along polyacetylene chains, the single carbon-carbon bond length being on the order of 1.36 Å and the double bond, on the order of 1.44 Å [3.26, 3.27].

The fact that within one-electron theories, the existence of bond alternation can be understood as the origin of the bandgap is rationalized in terms of a Peierls distortion [3.28]. Peierls theorem establishes that in a (quasi)onedimensional material, a metallic state is always unstable towards a distortion of the geometric structure leading to a lowering of the symmetry, the appearance of a bandgap at the Fermi energy, and the transition from a metallic to a semiconducting regime. The Peierls effect can thus be viewed as the solid state equivalent of the molecular Jahn-Teller effect since it removes an electronic degeneracy through a symmetry lowering.

The presence of bond alternation has very important consequences. If we consider an infinite all-trans-polyacetylene chain, two energetically equivalent bond isomer forms (hereafter referred to as phase A and phase B) can be derived and are obtained from one another simply by exchanging the single and the double bonds:

A   B

The ground state of trans-polyacetylene is therefore said to be doubly degenerate since there are two geometric structures that correspond to the same minimal total energy. This leads to a double-well potential energy curve for trans-polyacetylene, when the total energy is plotted as a function of the degree of bond-length alternation; see Figure 3.3. At the end of the 70's, it was recognized that such a twofold degeneracy opened the way for the possible existence of nonlinear excitations on all-trans-polyacetylene chains, corresponding to the formation of so-called solitons [3.29, 3.30]. Generally speaking, solitons are excitations of a system leading from one minimum of the potential to another minimum *of the same energy*. In trans-polyacetylene, they constitute topological kinks that extend over several bonds and gradually lead from a phase A chain segment to a phase B chain segment, as illustrated on top of Figure 3.4.

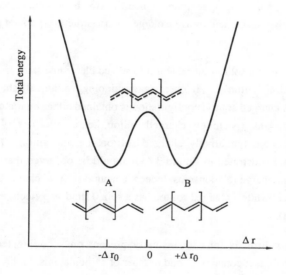

Figure 3.3   Sketch of the total energy curve for an infinite trans-polyacetylene chain as a function of the degree of bond-length alternation Δr

The possibility of obtaining an isolated soliton defect in trans-polyacetylene is thus related to the fact that although the lattice geometry to the left of the defect (here, phase A) is not identical to that to the right of the defect (phase B), both phases A and B have identical total energies. Within the soliton, starting from one side of it, the double bonds gradually elongate and the single bonds gradually shorten in such a way that, arriving at the other side of the defect, the single-double bond characters have been fully exchanged. The number of carbons over which the exchange takes place (i.e., the soliton width) is currently agreed to be on the order of 20 [3.31, 3.32].

At this stage, it is worth pointing out that: (i) solitons are not unlike the defect examined in the early sixties by Pople and Walmsley [3.33], which can be said to correspond to a soliton of zero-width, this being a high energy configuration; (ii) solitons cannot be obtained in cis-polyacetylene since the two bond isomer forms that can be envisioned (the cis-transoid and trans-cisoid geometric structures) do not possess identical total energies.

The presence of a soliton introduces a localized electronic state around mid-gap. In the case of a neutral soliton (i.e., in chemical terminology, a radical), the mid-gap state is half-occupied. A charged soliton configuration is obtained either by removing an electron from the mid-gap state (positively charged soliton, i.e., a cation) or by adding an extra electron to that state (negatively charged soliton, i.e., an anion). These electronic configurations are illustrated in Figure 3.4. It should be observed that solitons possess very peculiar spin-charge relationships. Indeed, a neutral soliton (charge 0) corresponds to a radical (spin 1/2) while a charged soliton has a spin 0 (mid-gap electronic state empty or doubly occupied).

A neutral soliton is thus intrinsically present on a conjugated segment of a polyacetylene chain that contains an odd number of carbon atoms. In this respect, the allyl radical corresponds to the shortest polyacetylene chain carrying a soliton. In the case of chains containing an even number of carbon atoms, solitons, be them neutral or charged, can only be created in pairs, as sketched below:

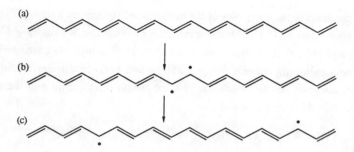

As can be seen, one of the solitons is centered on a carbon atom located on the upper side of the chain while the other, which is usually referred to as an antisoliton, is centered on a carbon located on the lower side of the chain. Topologically, two solitons need always to be separated by an antisoliton. We note that in the case of trans-polyacetylene, a soliton and an antisoliton possess the same characteristics since they are centered on carbon sites that are identical.

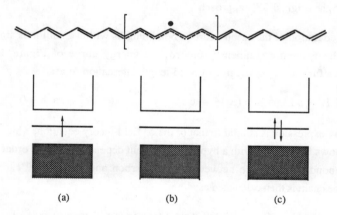

Figure 3.4   Top: Illustration of the geometric structure of a neutral soliton on a trans-polyacetylene chain.
Bottom: Schematic band structure for a trans-polyacetylene chain containing: (a) a neutral soliton; (b) a positively charged soliton; (c) a negatively charged soliton

In order to study the energetics of the formation of solitons on a trans-polyacetylene chain, we have followed [3.34, 3.35] the approaches of Pople and Walmsley [3.33] and Su, Schrieffer, and Heeger (hereafter denoted as SSH) [3.30] which are based on a simple Hückel model including σ-bond compressibility and bond-length dependent transfer (resonance) integrals. In the bond-order / bond-length relationship that we used, the transfer integral values are expressed as:

$$t(r) = A \times \exp(-r/B)$$

where r is the bond length; the energy of the σ framework is:

$$f(r) = C \times t(r) \times (r - r_0 + B)$$

where A, B, and C are parameters to be optimized and r_0 is usually taken as the length of a pure sp^2–sp^2 single bond (about 1.51 Å) but can also be optimized. The parameters have been adjusted in order to reproduce the trans-polyacetylene bandgap of 1.4 eV, a total width for the π and π^* bands of 10 eV, and a degree of bond-length alternation of 0.14 Å, in analogy to the original SSH approach.

The modulation of the carbon-carbon bond lengths within a soliton is simulated by an hyperbolic tangent dependence. Indeed, the energy curve of Figure 3.3 can be expressed as a function of the degree of bond-length alternation Δr as:

$$E = a (\Delta r)^4 - b (\Delta r)^2 + c \qquad\qquad a, b > 0$$

Such an expression is similar to that of the ϕ^4 field-theory equation, which among its solutions allows for solitons with a hyperbolic tangent dependence on the order parameter [3.30]. The bond-length value between the nth carbon and the $(n+1)^{th}$ carbon on the polyacetylene chain is then calculated as:

$$r \text{ (in Å)} = 1.40 + (-1)^n \, \Delta r/4 \, [\, \tanh(n/\ell) + \tanh(n+1/\ell) \,]$$

where Δr is the degree of bond-length alternation far away from the defect, n is the site location from the defect center located at site 0, and ℓ is a modulation factor to be optimized and corresponding to the soliton half-width.

In this framework, we calculate the energy of creation of a soliton by minimizing the total ($\pi + \sigma$) energy of a long chain (containing over 100 carbons) with respect to the ℓ value. The energy of creation is found to be lowest, 0.45 eV, when ℓ is equal to 7, implying a soliton full width of 15 carbons. These results are in excellent agreement with those of SSH. The presence of a soliton induces the appearance of a localized electronic state at the Fermi energy, i.e., 0.70 eV above (below) the valence (conduction) band edge. Localization of the electronic state associated to the soliton is confirmed by the fact that the total probability density of the radical is 0.855 over the 15 sites covered by the topological defect. It is most important to stress that *the energy to create a radical defect when it takes the form of a soliton is 0.25 (= 0.70 − 0.45) eV lower than the 0.70 eV that would be required on the basis of a rigid lattice model.* This feature, which intimately combines both the electronic structure and the geometric structure, constitutes the basis of the fascinating physics encountered in conjugated polymers, as will be fully illustrated in the next Sections.

Electron spin resonance experiments on undoped trans-polyacetylene indicate the presence in traditional Shirakawa samples of about one radical (neutral soliton) per 3000 carbon atoms [3.36]. This ratio is much larger than that to be expected on the basis of simple thermodynamics. This implies, as does the fact that this spin concentration shows little temperature dependence, that these neutral solitons are for the most part extrinsic defects which are created during the cis-trans isomerization process and trapped with respect to recombination between crosslinks.

Finally, we note that many authors have gone beyond the simple Hückel model namely by considering explicitly the repulsion between electrons. This is in order to account, at least partly, for electron-electron interaction effects and/or electron correlation effects (i.e., when going beyond the Hartree-Fock limit) and to investigate the influence of these effects on bond alternation and bandgap [3.37-3.46]. The conclusion which can be given is that, for parameters that are well adapted to polyacetylene, electron-electron interactions do enhance the bond alternation value whereas electron correlation effects slightly reduce it (a longer discussion of this topic will be given in Section 3.5). Generally speaking, it is striking that the SSH approach, which undoubtedly affords too naive a picture because it privileges the electron-lattice coupling at the expense of the electron-

electron interaction, remains in most instances qualitatively valid. It is therefore nowadays still widely used when some first-order theoretical insight is needed for a problem involving conjugated polymers.

3.3.2 Polyparaphenylene, polypyrrole, and polythiophene

The geometric structures of polyparaphenylene, polypyrrole, and polythiophene are sketched in Figure 3.1. In all cases, the chains appear to be mostly linear. In polyparaphenylene, the benzene rings are linked in the para positions whereas in polypyrrole and polythiophene, the connections between rings are mainly in $\alpha-\alpha'$ positions. Contrary to the situation in trans-polyacetylene, the ground state of these aromatic polymers is nondegenerate since the bond isomer form which can be envisioned, of quinoid type, corresponds to a higher total energy:

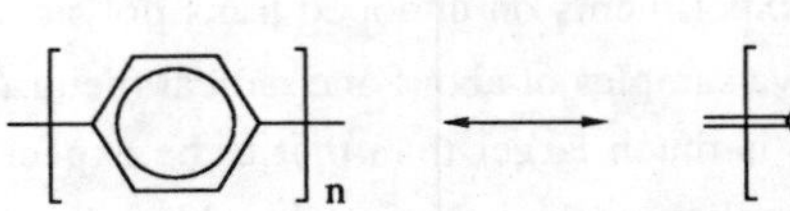

Crystallographic data on polyparaphenylene oligomers [3.47-3.49], from biphenyl to sexiphenyl, indicate that the carbon-carbon bond lengths within the rings are on the order of 1.406 Å while those connecting the rings are about 1.507 Å long. The same data also strongly suggest that adjacent rings are twisted with respect to one another by some 22.7°, every other ring being in the same plane. This torsion angle constitutes a compromise between the effect of conjugation and crystal-packing energy, which favor a planar structure, and the steric repulsion between ortho-hydrogen atoms, which favors a nonplanar structure. The translational unit cell therefore contains two benzene rings.

The VEH band structure and density of states of polyparaphenylene [3.50] are presented in Figure 3.5.

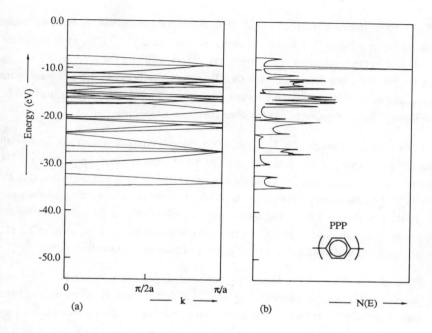

Figure 3.5 VEH band structure (a) and valence electronic density of states (b) of polyparaphenylene

Because of the glide plane symmetry running along the chain axis, all the bands are degenerate two by two at the end of the Brillouin zone. Taking that glide plane symmetry into account, we identify three π bands. The lowest occupied π band (band 11-11') is 1.6 eV wide; it corresponds, as usual, to the π wave function with no nodes and directly derives from the lowest π level of benzene. The other two occupied π bands derive from the two degenerate highest occupied electronic levels of benzene. One of those two π bands (band 14-14') is very flat because by symmetry it possesses almost no contribution from the π-atomic orbitals located on the carbons in para positions, i.e., those that connect the rings. Such a flat band is therefore obtained in all the polymers where aromatic rings are connected in the para or at the α–α' positions; it leads in the density of states to a very intense peak, see Figure 3.5, which usually dominates the low binding energy side of the

XPS or UPS photoemission spectra (and often obscures the contributions from the upper occupied, wider π band). The highest occupied π band (band 13-13'), the most important with respect to the conduction mechanism, has a large width on the order of 3.5 eV. Its wave function characteristics are such (bonding character between the ortho and para carbons and antibonding character between the ortho and meta carbons) that it would be destabilized if quinoid contributions to the geometry were to increase.

The calculated bandgap is 3.5 eV. Such a value is consistent with the very small conductivity of undoped polyparaphenylene ($< 10^{-15}$ S/cm at room temperature) [3.51]. The ionization potential is found to be significantly larger than in polyacetylene (5.6 eV vs. 4.7 eV); this allows one to rationalize that the p-type doping of polyparaphenylene can only proceed with strong oxidizing agents and that mild oxidants such as iodine do not lead to any appreciable conductivity increase. The electron affinity, on the order of 2.1 eV, is large enough to lead to effective n-doping with electron donors like alkali metals.

The shapes of the VEH band structures calculated for polypyrrole [3.52] and polythiophene [3.53] are qualitatively very similar to that of the polyparaphenylene band structure, especially as far as the π bands are concerned. In polypyrrole, we obtain a very low ionization potential on the order of 3.9 eV. This value is 0.8 eV lower than in polyacetylene and in excellent agreement with the 0.7-0.8 V oxidation potential difference measured between the two compounds [3.54]. This accounts for the fact that polypyrrole can be effectively doped with iodine but is also very unstable in air in the undoped state. The large bandgap (about 3.2 eV) combined with the low ionization potential results in an exceedingly small electron affinity, lower than 1 eV. This low value provides the reason why all the efforts to n-dope polypyrrole have failed.

In polythiophene, both the ionization potential and bandgap values are slightly larger than in polyacetylene, around 5.0 eV and 2.0 eV, respectively [3.53]. The ionization potential remains low enough to allow for p-type doping with iodine. The bandgap, in the middle of the visible, is responsible for the red color of polythiophene films. Recently, an enormous interest has been devoted to 3-alkyl substituted polythiophenes since, as mentioned in Section 3.2, such polymers are soluble in most common organic solvents. Provided the chain backbone remains coplanar (we will thoroughly discuss this aspect in the next Section), we calculate that the influence of the alkyl chains on the most important parameters of the electronic structure is not significant [3.55].

It is also interesting to investigate the influence of connecting the aromatic rings via vinylene linkages as in polyparaphenylene vinylene or polythiophene vinylene. These polymers are especially attractive because they can be synthesized via a soluble precursor route leading to highly oriented samples with excellent mechanical, electrical, and optical properties [3.56, 3.57].

The band structure of polyparaphenylene vinylene is displayed in Figure 3.6.

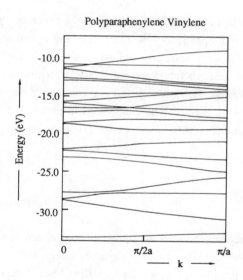

Polyparaphenylene Vinylene

Figure 3.6   VEH band structure of polyparaphenylene vinylene (only the valence occupied levels are shown)

In terms of the π bands, the electronic structure can be simply viewed as coming from the superposition of the polyacetylene and polyparaphenylene π bands. The VEH ionization potential and bandgap values of polyparaphenylene vinylene [3.58] are calculated to be intermediate between those of the parent polymers (5.1 eV for the ionization potential, to be compared to 4.7 eV in polyacetylene and 5.6 eV in polyparaphenylene; 2.5 eV for the bandgap, with respect to 1.4 and 3.5 eV). On the other hand, the width of the highest occupied valence band (2.8 eV) is significantly smaller than in both parent polymers (6.5 and 3.5 eV for polyacetylene and polyparaphenylene, respectively). However, it is worth noting that there occurs no localization of the upper occupied and lower unoccupied electronic levels on either the phenylene or vinylene moieties. The LCAO coefficients for those levels have indeed large values on all carbon atoms, implying delocalization of the corresponding electronic wave functions over the whole chain. This feature is important with regard to the transport or nonlinear optical properties of the material.

By and large, similar evolutions are found in polythiophene vinylene with respect to polyacetylene and polythiophene [3.53]. There also, the upper occupied and lower unoccupied levels are found to be delocalized over the whole chain. Consequently, the ionization potential and bandgap values are again calculated to be intermediate between those of the parent polymers. Generally speaking, it can be said that the presence of the vinylene linkages between the aromatic rings helps in preventing localization of the electronic effects within the aromatic moieties and favors delocalization over the whole chain.

3.3.3 Polyaniline

Polyaniline has recently become the subject of considerable scientific interest as a conducting polymer [3.10]. It is not a new material, however, and it has been known for over a hundred years namely as a precursor to aniline black. The redox chemistry of polyaniline is complex due to the various possible chemical structures for the backbone. Furthermore, the structures are not only affected by electrochemical oxidation and reduction but also by possible concomitant protonation or deprotonation of the nitrogens in the polymer backbone, especially in aqueous media. The various polyanilines can be expressed by using the following structure formula:

$$\{ \ (- C_6H_4 - NH - C_6H_4 - NH -)_y \ (- C_6H_4 - N = C_6H_4 = N -)_{1-y} \ \}_x$$

In the fully reduced leucoemeraldine form, all nitrogen atoms are amine nitrogens and y is equal to 1.0. In the semioxidized emeraldine form, half of the nitrogens are oxidized to the imine form and y is equal to 0.5; one phenyl ring out of every four adopts a quinoid geometric structure. In the fully oxidized pernigraniline form, y is equal to 0.0 since only imine nitrogens are present and every other ring is quinoid-like. In all cases, steric interactions between hydrogens on adjacent phenyl rings prevent the chains from adopting a coplanar conformation. Precise structural data are not available yet on polymer chains; in model molecules, the measured dihedral angles between adjacent rings lie in the range of 25°-65°.

At this stage, our interest is simply in discussing the evolution of the electronic structure in going from leucoemeraldine to emeraldine and then to pernigraniline. A thorough discussion on emeraldine will be presented in Section 3.7. The VEH band structures presented hereafter are based on chain geometries obtained from MNDO geometry optimizations on oligomers [3.60]. In the absence of structural data on polyaniline chains, a torsion angle of 30° has been chosen as a reasonable value.

The VEH band structures of leucoemeraldine, emeraldine, and pernigraniline in their base (unprotonated) form are shown in Figure 3.7. The unit cell contains two rings and two nitrogens in the leucoemeraldine and pernigraniline polymers and is twice as large for emeraldine (since one ring out of four adopts a quinoid geometry).

We first focus on leucoemeraldine. This polymer is isoelectronic to polyparaphenylene oxide, $(-C_6H_4-O-)$, and (if only valence electrons are considered) polyparaphenylene sulfide, $(-C_6H_4-S-)$. Because of the glide plane running along the chain axis, all the bands are degenerate two by two at the end of the Brillouin zone and can be considered as having two branches. Leucoemeraldine is calculated to have a wide bandgap, on the order of 4 eV, in excellent agreement with the experimental data indicating a gap of about 3.8 eV (hence, the prefix "leuco", i.e., white, in the name of the compound). The ionization potential is around 4.4 eV, slightly lower than in polyacetylene. The highest occupied band (with two branches denoted a and b in Figure 3.7) is very wide, about 3 eV for a 30° torsion angle value between adjacent rings. Even if the rings are considered to be perpendicular to one another (as it is the case experimentally

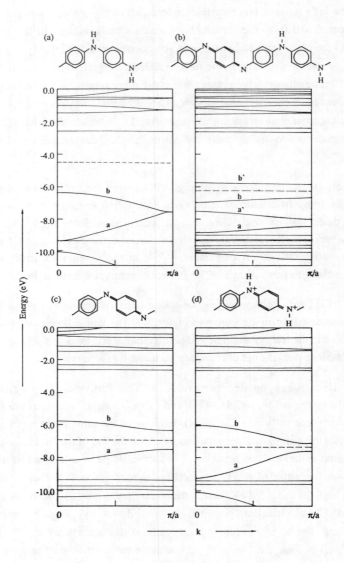

Figure 3.7 VEH band structures of (a) leucoemeraldine; (b) emeraldine; (c) pernigraniline; (d) protonated pernigraniline

in polyparaphenylene oxide and polyparaphenylene sulfide [3.61, 3.62]), the bandwidth of the HOMO is calculated to remain on the order of 2 eV while it is only 0.8 and 1.2 eV in polyparaphenylene oxide [3.63] and polyparaphenylene sulfide [3.50], respectively.

The most peculiar feature of the band structure of leucoemeraldine is, however, the fact that the shapes and wave function characteristics of the upper occupied bands are very different from those of the lower unoccupied bands, contrary to the situation found in polyacetylene and the polyaromatic systems (polyparaphenylene, polythiophene) and their vinylene derivatives. In condensed-matter physics terminology, such a situation corresponds to a breakdown of electron-hole symmetry. Indeed, while the upper occupied band is very wide and has contributions from all the backbone atoms, the lower unoccupied band, i.e., the conduction band, is totally flat as there is no contribution to that band from the para carbons or the nitrogens. The conduction band has actually a symmetry analogous to that of the flat occupied band located below branch a of the upper valence band.

When we switch to the emeraldine polymer, as mentioned above, the translational unit cell is twice as large as before and the glide plane symmetry is lost. As a result, the first Brillouin zone of the polymer is twice smaller as in leucoemeraldine, there appear twice as many bands, and the degeneracy at the end of the Brillouin zone is lost. For instance, if we focus on the evolution of the two branches a and b of the leucoemeraldine HOMO band, we observe that there are now four branches (denoted a, a', b, and b' on the Figure) that are all separated by gaps. Furthermore, because of the oxidation process that transforms half of the amine nitrogens into imine nitrogens, there are two electrons less per four-ring unit cell than in leucoemeraldine. The LUMO conduction band therefore becomes the upper branch b' of the former HOMO band. This is the reason why emeraldine is sometimes referred to a 3/4 filled band system from the electronic structure standpoint (a, a', and b are filled; b' is empty).

This evolution allows us to rationalize the much lower bandgap that exists in emeraldine, since the bandgap here corresponds to the gap opening between branches b and b' (in powder form, the compound, absorbing in the red, appears emerald-like, i.e., green in color, hence the origine of its name). The ionization potential is correspondingly larger since it is defined by the top of branch b. An analysis of the wave function characteristics of the a, a', b, and b' branches, indicates that branches a, a', and b are all constructed from a combination of atomic orbitals belonging to the three phenyl rings with an aromatic geometry. Branch b', i.e., the LUMO band of emeraldine, is very flat because its contributions come essentially from the quinoid rings that are widely separated from one another along the chain. The MNDO-optimized geometry together with torsion angles of 30° between adjacent rings, leads to a VEH calculated value for the bandgap which is on the order of 0.8 eV. This is too low with respect to the observed gap, peaking at about 2 eV [3.59-3.64]. We will come back to that point later in Section 3.7; the essential aspect here is, however, to show that the HOMO-LUMO transition in emeraldine results in localizing the excited electron on a quinoid ring while leaving a hole on the surrounding aromatic rings. This feature has been confirmed by INDO/CI semiempirical calculations on emeraldine oligomers [3.65] and has led condensed-matter physicists to describe this transition in terms of the formation of a molecular exciton [3.66].

Turning now to pernigraniline, the oxidation process has turned all the amine nitrogens into imine nitrogens. Every other phenyl ring here adopts a quinoid geometry and the unit cell contains two rings and two nitrogens as in the case of leucoemeraldine. With respect to leucoemeraldine, two electrons have been extracted from each unit cell and branch b is now fully empty. The bandgap is calculated to increase by about 0.4 eV relative to emeraldine, in good agreement with the evolution observed experimentally (pernigraniline absorbing in most of the visible region appears black). Branch b, i.e., the LUMO conduction band, is found to have contributions mostly from the N=quinoid=N fragments and branch a, the HOMO band, from the aromatic rings. In Figure 3.7, we also present the VEH band structure obtained when the pernigraniline chain is fully protonated in strongly acidic medium. The overall band structure is similar to that in unprotonated pernigraniline; however, the HOMO and LUMO bands are wider and the bandgap is accordingly found to be significantly smaller.

It is most interesting to realize that, like trans-polyacetylene, the pernigraniline chain has a degenerate ground state. This is easily observed when exchanging the aromatic and quinoid characters of the phenyl rings. However, there is an important difference with respect to polyacetylene. In the latter compound, all the atoms along the chains are identical and correspond to sp²-hybridized carbons linked to two other carbons and an hydrogen. In pernigraniline, there are nitrogen and carbon atoms which possess different electronegativities. Furthermore, all the carbons are not in a similar environment, the para carbons being connected to nitrogens and carbons, the ortho and meta carbons to carbons and hydrogens. We can therefore expect that nonlinear elementary excitations such as solitons possess peculiar characteristics.

This recently led dos Santos and Brédas to investigate the ground-state properties and the energetics of defect formation in pernigraniline on the basis of an SSH-type Hamiltonian [3.67]. A representative set of parameters for pernigraniline was optimized in order to reproduce the VEH band structure of Figure 3.7 and the ground-state geometry (N-C bonds are 1.30 and 1.41 Å long; aromatic rings have six equal bonds of 1.407 Å; quinoid rings have four single bonds of 1.46 Å and two double bonds of 1.36 Å). The site energies (i.e., the α Coulomb integrals) are fixed in relation to the carbons not connected to nitrogens and are optimized to be –5.5 eV for nitrogens and +1.0 eV for carbons in the para position.

The resulting π charge distribution leads to a net charge of –0.535 |e| on nitrogens, +0.948 |e| on quinoid rings, and +0.121 |e| on aromatic rings. These values are in agreement with the fact that phenyl rings tend to adopt a more quinoid structure when they are charged [3.68].

We searched for an undimerized structure and found it to be such that the N–C bonds are uniform and 1.355 Å long while the rings adopt a slightly quinoid geometry (four bonds of 1.432 Å and two of 1.390 Å). The corresponding band structure is metallic. The dimerized, Peierls distorted, ground-state geometry is stabilized by 0.12 eV per two N-ring units relative to the undimerized case.

An infinite pernigraniline chain is calculated to support two kinds of soliton distortions. In both cases, the center of the distortion corresponds to a nitrogen atom. The solitons have the following characteristics. The type-1 soliton has rings around the defect center which possess an aromatic geometry. It introduces a localized electronic state 0.11 eV below the Fermi level (of the chain carrying no defect). The corresponding wave function is symmetric, having a maximum at the central nitrogen. This electronic state is formed by 74% of a state coming from the top of the valence band and 26% of a state from the conduction band [3.67]. In the type-2 soliton, the rings around the defect center have a quinoid structure. The associated electronic localized level appears 0.06 eV above the original Fermi energy, with an antisymmetric wave function that has a node at the central nitrogen. It has contributions from the valence and conduction bands that exactly compensate those of the type-1 soliton. The optimized soliton geometries are shown in Figure 3.8. We note that both defects are rather localized as they are mostly confined within four rings.

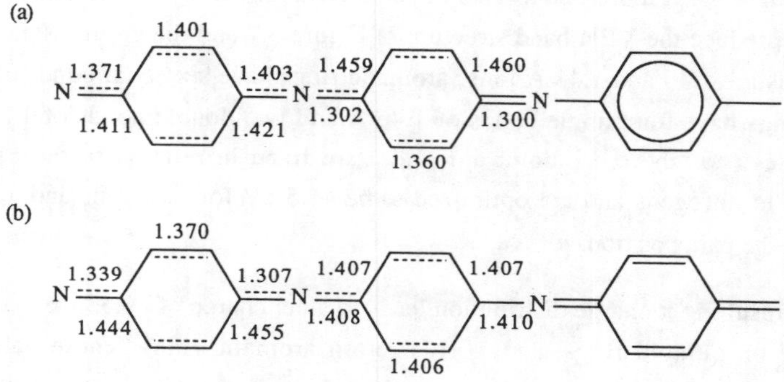

Figure 3.8 (a) Optimized type-1 and (b) type-2 soliton geometries in pernigraniline

The existence of two types of solitons has an important consequence. The most stable configuration for a soliton-antisoliton pair corresponds to the type-1 soliton level being doubly occupied and the type-2 soliton level left empty, with a total creation energy of 0.64 eV. This results in a net charge of $-0.52\,|e|$ around the type-1 soliton and an opposite charge around the type-2 defect. We thus observe *charge fractionalization of the solitons*, a feature which is simply driven by the difference in electronegativity between the carbon and nitrogen atoms. Furthermore, the most stable electronic configuration for a soliton-antisoliton pair is nonmagnetic, contrary to the situation in trans-polyacetylene. The excitation of a magnetic pair (corresponding to the case where two well-separated solitons are each occupied with a single electron) increases the above creation energy by 0.17 eV, i.e., the energy separation between the two localized levels. We note that the excess negative charge around the type-1 soliton mainly goes to the central nitrogen and to the second rings from the center, while the excess positive charge around the other soliton concentrates on the central rings; i.e., the rings having a quinoid geometry.

Recent experimental data on pernigraniline [3.69] tend to support the predictions of these calculations.

3.3.4 Chain-length dependence of the electronic properties of conjugated polymers

An interesting question in the area of conjugated systems is to understand the extent to which the properties of the polymer can be obtained on the basis of an extrapolation from oligomer data. Here, we decided to focus exclusively on oligomers and infinite chains of polyacetylene but the results that will be discussed do apply to the other conjugated compounds, such as polyparaphenylene, polypyrrole, or polythiophene [3.70]. We present VEH calculations of the electronic properties of polyacetylene oligomers containing from 1 up to 10 double bonds as well as the polymer, and diphenylpolyenes ranging in size from 1 to 8 double bonds in the polyenic segment. All VEH calculations are based on fully optimized geometries obtained with the MNDO technique.

In Figure 3.9, we have plotted the experimental and theoretical data for the bandgap of the polyenes vs. reciprocal chain length (n^{-1}), which is expected to give a linear relationship for large n values. Also included in Figure 3.9 are the data for the diphenylpolyenes (DPP), $C_6H_5-(HC=CH)_x-C_6H_5$. Since these molecules are more stable than H-terminated polyenes, $H-(HC=CH)_x-H$, there are more experimental studies of DPP molecules. When considering DPP as a model for the polyacetylene system, the effect of the phenyl end groups on the electronic properties must be considered. To do this, we define an effective conjugation length, $n_{eff} = x + A$, so that A describes the extension of the conjugation length by the phenyl rings beyond the x-unit polyene sequence. The value of A is established by adjustment until the experimental bandgap of a DPP with x double bonds is equal to that of a polyene with x + A double bonds. The results shown in Figure 3.9 use A = 2.7 and provide an excellent correspondence between the polyene and DPP experimental data.

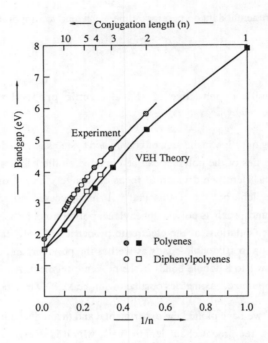

Figure 3.9   Bandgap for polyenes and diphenylpolyenes plotted against reciprocal conjugation length. For the diphenylpolyenes, n is the effective conjugation length defined in the text. All experimental data refer to absorption peaks in nonpolar organic solvents

We have carried out VEH calculations on two DPP molecules (x = 1,2). The computed bandgaps for these two molecules are plotted in Figure 3.9 with the same value of A used to define the effective conjugation lengths; the agreement with the calculations on polyenes is good. The VEH calculated ionization potentials for these DPP molecules agree nearly perfectly with the VEH calculations on polyenes if the effective conjugation lengths are again defined in the same way. In addition, the theoretical values of ionization potential (IP) for these DPP molecules (as well as those inferred from the VEH calculations on polyenes, again with A = 2.7) are in excellent agreement with the experimental gas-phase IP data of Hudson et al. [3.71].

Another reason for considering the DPP molecules in such detail is that a fairly complete set of experimental oxidation and reduction potentials is available. Our interest in these theoretical predictions and their implications, particularly regarding the electrochemical properties of polyacetylene, is fueled by the work on battery applications of polyacetylene and other conjugated polymers.

In Figure 3.10, we have plotted the theoretical values of the ionization potentials for the polyenes and DPP molecules vs. n^{-1}; on the same plot are experimental oxidation potentials, E_{ox}, for the DPP molecules. An approximately linear correlation between IP and E_{ox} is expected and found.

Furthermore, we have plotted the theoretical values of the electron affinity (where $EA = IP - E_g$) vs. n^{-1} for these compounds and compared these to their reduction potentials, E_{red}. E_{ox} and E_{red} values are given with respect to the saturated calomel electrode (SCE). In order to make quantitative comparisons to gas-phase IP and EA values, we have taken the zero of the SCE scale to be 6.3 V with respect to zero on the gas-phase scale. This same scale factor has been found previously and also gives the best agreement between theory and experiment in our case. For reference to the solid state, this 6.3 eV scale factor can be viewed as being composed of two components: a polarization energy correction to adjust gas-phase IP values to solid state values (typically 1.5-2.0 eV) and a scale factor relating SCE to vacuum (typically 4.7 eV).

The agreement between experiment and theory shown in Figure 3.10 is remarkably good, especially since our use of a constant 6.3 eV scale factor inherently involves an assumption of a chain-length-independent polarization energy correction to the theoretical values of IP. Other work showing a nearly linear relation between IP and E_{ox} also suggests a relatively constant polarization energy for a wide variety of organic molecules. The extrapolation in going from polyenes to polymer gives an oxidation potential of 0.4 V vs. SCE and a reduction potential of -1.1 V vs. SCE for polyacetylene. (The difference is precisely the theoretical E_g by construction). Note that these values for the polymer correspond to the first oxidation or reduction potential (removal or addition of one electron). Therefore, the E_{ox} and E_{red} values would correspond to onset values in terms of the observed oxidation or reduction of polyacetylene. Careful measurements by Shacklette in a battery cell configuration yield onset values of $E_{ox} = 0.2$ V and $E_{red} = -1.3$ V vs.SCE with uncertainties of 0.2 V. Rough estimates of the onset values from cyclic voltammetry are consistent with these values.

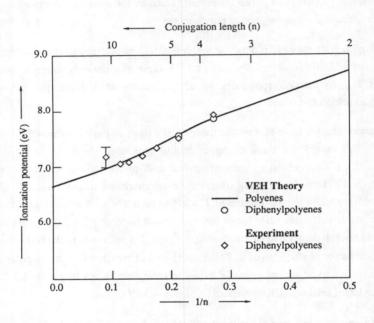

Figure 3.10 VEH computed ionization potentials for polyenes and diphenylpolyenes compared to oxidation potential experimental data

We believe the agreement between theory and experiment for the redox properties of polyacetylene is satisfactory, since we have not explicitly considered effects such as chemical and structural disorder, dopant (counterion) intercalation, and dopant diffusion, all of which can be of more importance in the energetics of polymer ionization than for the electrochemistry of oligomers in solution. It is also possible that a solvent shift would provide a difference between actual and oligomer-extrapolated redox energies, though none is indicated in the comparison to experiment.

In summary, we observe very generally that the electronic and electrochemical properties of conjugated systems such as polyacetylenes, *have a chain-length dependence that goes as 1/n*. In other words, 90% of the total evolution observed in going from the monomer to the infinite polymer chain, is already obtained for a chain which is ten monomer-units long. This will be further confirmed in the case of polythiophenes in the following Section. Note, however, that even if the presence of relatively short chains does not significantly modify the electronic or electrochemical properties, it can severely alter the transport properties.

3.4 INFLUENCE OF THE FLEXIBILITY OF CONJUGATED POLYMER CHAINS ON THE ELECTRONIC STRUCTURE

One of the major advances in the field of conducting polymers has been the recent advent of conducting polymer solutions. Solutions of poly(p-phenylene sulfide) were first obtained by Frommer and co-workers [3.72]; however, the solvents were rather exotic, such as an arsenic trifluoride-arsenic pentafluoride mixture. More recently, Elsenbaumer et al. [3.11], Sato et al. [3.73], and Hotta et al. [3.74] turned their interest to polythiophene and found that the addition of long flexible alkyl chains (containing four or more carbons) on position 3 of the thiophene rings allows the polymer chains to become soluble in common organic solvents such as chloroform, dichloromethane, or tetrahydrofuran. The addition of an ethane or butane sulfonate group, as considered by Wudl et al. or Meijer et al. [3.12], renders the polymer chains soluble in water and leads to the concept of self-doped polymers when the chains are oxidized. Intrinsic solubility (i.e., without side-chain addition) has been discovered for polyaniline [3.75]. This polymer is soluble in concentrated sulfuric acid and partially (or in some cases completely) soluble in organic solvents. These discoveries open the way to a more complete characterization of conducting polymers as macromolecules (e.g., determination of the molecular weight).

Furthermore, the dramatic improvements in the processibility of conducting polymers pave the way towards the commercialization of stable, intrinsically conducting polymers.

The availability of conducting polymer solutions has not only boosted the application potentialities of this class of compounds but has also opened up new areas of fundamental research in the field, such as for instance: (i) the investigation of solvatochromic and (solution and solid state) thermochromic effects and their dependence upon alkyl chain length [3.76]; or (ii) spectroscopic studies aimed at understanding the possible differences between charge storage in a conducting polymer in solution as compared to the solid state [3.77]. In this framework, the studies performed on alkyl-substituted polythiophenes expand on the works devoted since the end of the 70's to polydiacetylene, another conjugated polymer that is soluble in organic solvents. Polydiacetylenes are not dopable to any significant level of electrical conductivity but present interesting thermochromic effects exploited in time-temperature indicators [3.78].

In the solid state, it is well established that in the absence of strong steric interactions, crystal packing and the electronic energy associated with π-electron conjugation favor the appearance of coplanar conformations. In solution, the conformation is determined by the competition between the increased entropy of a disordered chain and the lower electronic energy of a fully conjugated coplanar chain. Thus, the very flexibility added to ensure solubilization can lead to significant deviations from planarity. Such deviations result in a lesser degree of conjugation and therefore directly influence those properties that are dominated by the π-electron framework. In early studies, we have examined the influence of the presence of a torsion angle on the electronic properties (such as bandgap, bandwidth, or ionization potential) of aromatic polymers; these properties have been found to follow a simple cosine law as a function of the values of the torsion angle between rings [3.79]. In this Section, we discuss recent calculations that have allowed us: (i) to rationalize the origin of the thermochromic effects that appear in the solid state in polyalkylthiophenes and (ii) to gain some information about the torsion potential curves associated with conjugated macromolecules; the knowledge of the energies involved in the torsion process is essential if a meaningful discussion of persistence and conjugation lengths in solution is to be addressed [3.80].

3.4.1 Influence of thermochromic effects on the electronic structure of poly(alkylthiophenes) in the solid state

Recently, several research groups have observed thermochromism in alkyl-substituted polythiophenes in the solid state or in solution [3.76, 3.77, 3.81]. In the solid state, the phenomenon has been monitored via optical absorption and XPS/UPS spectroscopies [3.76]. The results show reversible modifications with the variation of temperature. Heating films of poly(3-hexylthiophene), P3HT, from 21°C to 190°C leads to a shift in the optical absorption maximum by over 0.5 eV (from 2.5 eV up to 3.1 eV) and a decrease in the intensity of the peaks at 2.1 and 4.7 eV (that do not shift in energy). The temperature dependence of the UPS spectra indicates a narrowing of the highest occupied π band and an increase of the ionization potential values. To these experimental modifications correspond changes in the electronic properties, which we now address.

As we know that the electronic and geometric properties are closely connected in conjugated systems, we investigate theoretically the electronic structure evolution of poly(3-hexylthiophene), on the one hand as a function of chain length and, on the other hand as a function of changes in the conformation of the backbone or of the alkyl tail. The theoretical studies discussed here have been carried out with the VEH and MNDO methods.

3.4.1.1 Geometric and electronic properties of thiophene oligomers

The electronic properties of a conjugated polymer chain broken into mutually perpendicular segments (which confines π-electron delocalization within the segments) can be compared to those of oligomers of the same size as the segments themselves. Therefore, the evolution of the geometric and electronic properties as a function of oligomer (segment) length is first investigated. The thiophene rings are considered to be coplanar; adjacent rings are connected at the α positions and alternate in such a way that the sulfur atoms point in opposite directions. This conformation is called anti, while the situation where all heteroatoms point to the same side corresponds to a syn conformation.

Geometry optimizations on the oligomers indicate that while weak differences still exist between the central part of bithienyl and the middle ring of terthienyl, the inner rings of terthienyl and quaterthienyl are identical. Therefore, the geometries for the longer oligomers and the polymer chain of thiophene can be deduced from that of the central part of the trimer.

In Figure 3.11, the VEH electronic properties are presented for the oligomers up to sexithienyl and the polymer chain.

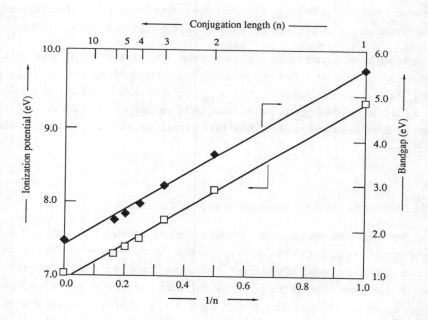

Figure 3.11 VEH ionization potential (IP) and bandgap (Eg) values vs. 1/n for the thiophene oligomers and polymer

The IP value decreases with chain length so that the value for sexithienyl is only 0.1 eV higher than that for the polymer. A plot of Eg as a function of 1/n (with n is equal to the number of thiophene rings) results in a *straight line* between all calculated points. It can be observed that a shift of 0.5 eV in the π–π* transition can originate in a total break of conjugation every 7 to 8 thiophene units along the chain (i.e., segments containing 7 to 8 thiophene rings and mutually perpendicular).

The comparison of the anti and syn conformations is indicative of very weak differences. Bond lengths differ by less than 0.01 Å and bond angles, by 0.1° or less, except for the angles between the C-S and interring bonds that vary by about 1°. The distance between two hydrogens on adjacent rings (2.6 Å) does not prevent planarity of the chain. The similarity in the geometries lead to equivalent electronic properties for the two coplanar conformations of quaterthienyl. Therefore, the thermochromic effects cannot be explained by a reversible transformation from one conformation to the other. At the MNDO level, we obtain an estimate of the energy difference between the two forms of the order of 1 kcal/mol, the anti conformation being more stable. Such a difference is hardly significant.

3.4.1.2 Poly(3-hexylthiophene)

Now we analyze the effects of hexyl substituents on the geometric structure and therefore on the electronic structure of polythiophene. When the aliphatic groups adopt an all-trans conformation and remain in the plane of the backbone, VEH calculations indicate that alkyl-substituted polythiophenes possess electronic properties very similar to those of polythiophene. The geometry of 3-hexylterthienyl shows the same features as those of the 3-alkylterthienyls with smaller alkyl substitutents. With respect to terthienyl, the inner ring displays a very weak dissymmetry in bond lengths and angles due to the hexyl tail, whereas the outer rings remain unaffected by the alkyl group. In this context, the size of the alkyl substituents has no influence on geometry.

Comparison of the VEH band structure of polythiophene and P3HT indicates that the HOMO and LUMO π bands are almost identical. The bands due to the presence of the hexyl group are located lower than the π-HOMO bands since they correspond to σ bonds between carbon atoms or hydrogen and carbon atoms. In P3HT, the IP (5.0 eV) and HOMO bandwidth (2.2 eV) values are slightly lower than in polythiophene, while the E_g values (1.7 eV) are identical for the two polymers.

We have also studied the effects on the thiophene ring geometry of different conformations where the hexyl substituent is bent relative to the rings (i.e., out of the plane of the backbone). MNDO optimizations carried out on 3-hexylthiophene indicate a maximum change of 0.01 Å in the bond lengths inside the ring; the steric repulsions between the ring and the alkyl tail are in fact minimized by modifications of bond angles within the tail.

Our results thus show that the influence of the alkyl tails on the geometry of the thiophene *rings* is very small and cannot account for the observed thermochromic effects. Furthermore, syn and anti *coplanar* polythiophene chains possess the same electronic structure. Therefore, we have to consider that, upon increasing the temperature, disordering effects appear in the alkyl chains which result in forcing a departure from planarity of the *polythiophene backbone*.

3.4.1.3 Evolution of electronic properties of polythiophene as a function of torsions between rings

We investigate three model poly(hexylthiophene) chains where departure from planarity occurs after every ring, every other ring, and every three rings, respectively (segments with full conjugation length containing one ring, two rings, and three rings, respectively). The torsion angle between two consecutive segments is varied from 0° (anti conformation) to 90° (perpendicular conformation). We consider two situations: (i) all the torsions correspond to a + angle; (ii) the torsions are alternatively + and - so that every other segment lies in the same plane. Both situations lead to identical results. In Figure 3.12, the IP and Eg evolutions as a function of torsion angle are presented for the three model chains. The evolution of these electronic properties can be well fitted by a cosine function. As the length of the conjugated segments increases, the curves in Figure 3.12

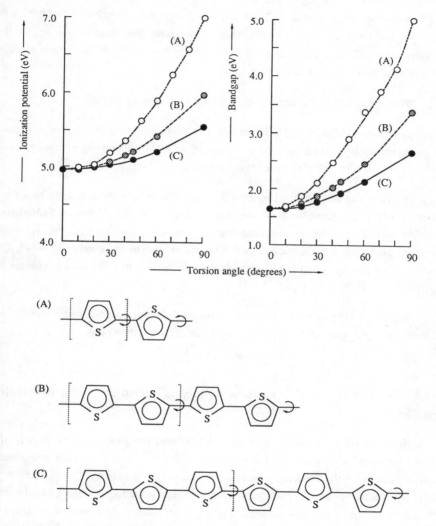

Figure 3.12 Evolution, as a function of the torsion angle between the rings (in deg.), of VEH ionization potential (IP) and bandgap (Eg) (in eV) for the monomer, dimer, and trimer models of polythiophene. (IP values are given after substracting 1.9 eV to correct approximately for the polarization energy of the lattice)

become flatter. From the Eg curves, a shift of about 0.5 eV in the first optical transition can be explained by a twist of 34°, 45°, and 57° for the model chains corresponding to one, two, and three thiophene ring segments, respectively. For these torsion angles, the IP values are 0.32, 0.28, and 0.27 eV larger for the monomer, dimer, and trimer models, respectively, than in the coplanar chain.

We can summarize these results by stressing the following points:

(i) The thermochromic effects occurring in polyalkylthiophenes in the solid state originate neither in alkyl tail induced modifications of the thiophene *ring* geometry, nor in an anti-to-syn conformation change of the backbone.

(ii) They can be explained by disordering in the tails leading to *departure from planarity* of the *polythiophene backbone*; model calculations indicate that a 0.5 eV shift of the first optical transition is consistent with a break up of the backbone into segments containing e.g., one, two, or three rings and rotated relative to one another by $\approx 34°$, $45°$, or 57°, respectively. A statistical model based on these results has been presented by Salaneck et al. [3.76].

(iii) Alkyl tails of different lengths present different order-disorder transitions, thereby modifying the backbone planarity in different temperature ranges.

3.4.2 Quantum-chemical evaluation of torsion potentials along conjugated backbones

Here, we are interested in gaining some information about the torsion potential curves associated with conjugated macromolecules. We have chosen to examine oligomers representative of polyacetylene, polydiacetylene, and polythiophene [3.82]. Till recently, polyacetylene had not been obtained in solution except in the form of block copolymers, despite being the most studied conducting polymer. Nevertheless, recent advances in synthetic pathways, e.g., the ring opening metathesis polymerization [3.20, 3.21], have provided polymer chemists with the tools to achieve solubilization of side-group substituted polyacetylenes. Our approach is to perform *ab initio* calculations either at the Hartree-Fock level or beyond Hartree-Fock by including correlation effects. Oligomers containing between 6 and 12 carbon atoms are considered and geometries as well as total energies are optimized as functions of torsion angle along the chains.

Two features of these calculations need to be carefully borne in mind. First, the calculations are carried out on discrete molecules in the gas phase and not on polymers in solution. Therefore, for instance, all solvent effects are neglected. This can lead to some discrepancies with experimental data, especially when the torsion process generates a dipole moment, as in the case of bithiophene. Second, our calculations deal with torsions occurring around only one bond along the molecule. It is clear that actual polymer conformations involve multiple torsions all along the chains. The results of our calculations do, however, provide a reliable basis from which to address the problems of persistence/conjugation lengths in solution with statistical mechanics tools [3.80].

3.4.2.1 Polyacetylene oligomers

We first investigate the polyacetylene oligomers since precise estimates of the barrier for s-trans to s-cis isomerization have been reported by Ackerman and Kohler on octatetraene [3.83]. We are thus in a position to evaluate the reliability of our theoretical approach on such systems.

From the optimized geometries of the coplanar conformations of octatetraene and dodecahexaene, we observe as expected that the degree of bond length alternation Δr (i.e., the difference between the length of a single bond and that of an adjacent double bond) slightly decreases as one proceeds towards the center of the molecule: $\Delta r = 0.139$ Å at the ends, $\Delta r = 0.129$ Å in the center of octatetraene, and $\Delta r = 0.125$ Å in the center of dodecahexaene. The very small evolution in bond lengths and bond angles in going from octatetraene to dodecahexaene indicates that only very small geometry variations are to be expected on further increase of the chain length. We note that the degree of bond alternation furnished by the 3-21G basis set is slightly larger than the X-ray scattering and nutation NMR experimental estimates [3.26, 3.27], which are on the order of 0.08-0.10 Å.

In octatetraene, we have also optimized the geometries for the case where there is a 90° or a 0° torsion around the middle single bond. In the 90° situation, the torsion causes marked changes in the bond length values, especially that of the twisted single bond which elongates by about 0.03 Å to reach 1.486 Å. The double bonds adjacent to it shorten by 0.006 Å to reach the same length as that of terminal double bonds in the 180° coplanar conformation. In the 0° case, the bond length modifications are much smaller, the central bond elongating by some 0.01 Å with respect to the all-trans conformation. The major evolution is that of the C=C-C angles around the central bond which increase to 127.0° from 124.1°. This opening allows the molecule to avoid any significant steric interactions between those hydrogen atoms that face one another.

At the Hartree-Fock 3-21G level, for a rotation around the central single bond and keeping the geometry frozen at its all-trans conformation value, we calculate the total energy to increase by 2.3 kcal/mol for a 150° angle and by 6.3 kcal/mol for a 120° angle, to reach a barrier of 6.9 kcal/mol for the 90° torsion. When relaxation of the geometry is allowed, the barrier is lowered, but merely by 0.4 kcal/mol to reach 6.5 kcal/mol. Using frozen geometries, we have also calculated the barrier that appears when the rotation takes place around an external single bond. The reason for this calculation is that there is experimental evidence that such a rotation is the lowest in energy [3.83]. We indeed find a barrier which is 0.6 kcal/mol smaller, i.e., 6.3 kcal/mol. Assuming that allowing for geometry relaxation would further lower the barrier by some 0.4 kcal/mol (as in the previous case), we obtain a theoretical estimate of about 5.9 kcal/mol. This result is in excellent agreement with the experimental measurements which locate the barrier at 6.04 kcal/mol [3.83].

The influence of including correlation effects and using polarization functions is, as expected, to decrease the barrier height. The reduction we calculate is slightly less than 1 kcal/mol, thus not affecting the barrier qualitatively and keeping it at the same order of magnitude.

At the Hartree-Fock 3-21G level, the cis conformation in its optimal geometry is calculated to be 3.4 kcal/mol less stable than the all-trans conformation for a rotation around the central bond. If we suppose that rotating around an external single bond would cost, as for the 90° case, around 0.6 kcal/mol less, we obtain a stability difference of the order of 2.8 kcal/mol., to be compared with an experimental value of 2.9 kcal/mol [3.83]. We note that in the case of the cis conformation, relaxation of the geometry exerts a larger influence on the total energy of the compound (lowering by about 0.9 kcal/mol, from 4.3 down to 3.4 kcal/mol) in order to avoid steric interactions as mentioned previously.

The total energy results obtained on octatetraene confirm that the Hartree-Fock 3-21G methodology is capable of reliably predicting the trends in relative stabilities for different conformations of conjugated oligomers, which is our main goal here.

It is interesting to pay some attention to the one-electron energy levels. Twisting the octatetraene molecule around the central bond (which, from the π-electron standpoint, results in the formation of two butadiene subunits) leads to a near degeneracy of the upper two occupied π levels, at -8.74 and -8.76 eV. These values are exactly midway between those for the upper two occupied π levels in all-trans octatetraene, -9.94 and -7.60 eV. The near degenerate levels in 90° twisted octatetraene correspond to having the π systems of each of the separate butadienic subunits weakly interacting with the σ levels of the other subunit. When the torsion is performed around an outer single bond, the upper π level remains closer to the value it has in the coplanar conformation (-7.95 eV with respect to -7.60 eV) since it now originates from an hexatrienic subunit.

The π levels come out to be almost identical in the coplanar (0° and 180°) conformations. This confirms that the π levels are almost unaffected by the conformation as long as the chain remains coplanar. Such a result was also observed on polyacetylene itself for which the π-band characteristics are very similar when considering the all-trans, cis-transoid, and trans-cisoid conformations.

In dodecahexaene, we have optimized the geometry for the all-trans conformation and kept that geometry frozen to calculate an "unrelaxed" barrier height. When twisting by 90° around the central single bond (which cuts the π network into two hexatriene units), the barrier is calculated to be 7.3 kcal/mol, that is on the same order but slightly larger by 0.5 kcal/mol than that in octatetraene. As a result, we expect that as the chain grows longer, it becomes more and more favorable to perform rotations around outer than around inner single bonds. This behaviour is consistent with the fact that, when a conjugated chain is interrupted by an sp^3 defect, it is energetically more favorable to locate the defect toward a chain end [3.84].

In 90° twisted dodecahexaene, the highest two occupied levels are almost degenerate (-7.96 and -7.97 eV). These values are similar to that of the HOMO level in end-twisted octatetraene since in both cases they originate in π hexatrienic-like subunits. The next four levels are also degenerate two by two and appear at energies similar to those of two levels of end-twisted octatetraene.

We note that our estimate for the torsional barrier in polyacetylene oligomers is about one order of magnitude lower than that obtained on the basis of Hückel-type calculations where a 90° rotation around a single bond was considered to fully decouple the two chain subunits [3.85]. Actually, a Mulliken population analysis of the *ab initio* wave functions indicates that the total (π plus σ) bond orders are essentially unaffected by the torsion, even that for the single bond around which the rotation is carried out. At the *ab initio* level, even though π conjugation is broken, there remains a strong coupling between the two chain subunits (the π framework of one subunit interacting with the σ framework of the other subunit and vice versa), which stabilizes the total energy of the chain. Such a coupling is of course not taken into account at the simple Hückel level.

3.4.2.2 Polydiacetylene oligomers

In the coplanar conformation of hexadienyne, the 3-21G optimized bond lengths (1.193 Å, 1.322 Å, and 1.429 Å for the triple, double, and single bonds, respectively) compare very well with those optimized for the polymer by Karpfen, using a similar basis set (1.194 Å, 1.321 Å, and 1.425 Å) [3.86] as well as those experimentally determined, e.g., for the paratoluenesulfonate (PTS) polymeric derivative (1.191 Å, 1.356 Å, and 1.428 Å) [3.87]. The double bond is experimentally observed to be about 0.03 Å longer than the calculated value.

For hexadienyne, no significant difference in the geometry is found when a torsion angle of 90° or 0° is imposed around one of the single bonds. With respect to the all-trans conformation, the bond lengths are modified by a most 0.002 Å in the perpendicular conformation and remain identical in the cis conformation. We note, however, that in the latter case, steric interactions would be present if bulky side groups, of the kind of paratoluenesulfonate, were explicit taken into account. In the 90° conformation, the bond which evolves most is as expected that around which the torsion occurs. This lack of significant geometry evolution as a function of torsion results in very small total energy differences between relaxed and unrelaxed geometries at 90°. Increasing the size of the oligomer up to decatrienediyne also leads to negligible geometry evolutions. The inner double (single) bonds slightly elongate (shorten) by 0.007 Å (0.005 Å), the C=C-C angles remaining almost identical.

In terms of the total energy evolution as a function of torsion angle, the most striking result is the very small height of the barrier calculated to perform a 90° rotation around a single bond. For hexadienyne, this barrier is obtained to be around 0.5 kcal/mol, with slight deviations depending on the level of calculation. At the Hartree-Fock 3-21G level, the barrier is 0.55 kcal/mol if the geometry is kept identical to that optimized for the 180° situation. The barrier is modified by an insignificant amount (lower by 0.01 kcal/mol) when the geometry is allowed to relax, consistent with the very small geometry evolution mentioned above. Increase of the basis set size to include polarization functions (6-31G*) and inclusion of correlation effects through Møller-Plesset second order perturbation theory sets the barrier value at 0.51 kcal/mol. In decatrienediyne, the unrelaxed barrier rises only slightly to 0.71 kcal/mol at the Hartree-Fock 3-21 G level.

The fact that the barrier in polydiacetylene oligomers is one order of magnitude lower than in corresponding polyacetylene oligomers can be rationalized in the following way. As rotation around a single bond occurs, the double bond on one side of that single bond remains in conjugation with the triple bond on the other side. The feature originates from the two perpendicular π systems carried by the triple bond. In that way, conjugation persists through the twisted single bond. The evolution of the one-electron energy levels is consistent with this picture. In the case of total rupture of conjugation, the HOMO level for 90° twisted decatrienediyne should appear at the same energy as that of the HOMO level for coplanar hexadienyne, i.e., at -8.42 eV. In fact, it is slightly less stable and is located at -8.24 eV, due to conjugation with the triple bond. As mentioned before, such an evolution is not present in the polyacetylene oligomers where we have documented a complete rupture of conjugation upon 90° rotation.

Such a small value of the barrier implies that a polydiacetylene chain can easily rotate around single bonds in good solvents, in particular in the absence of bulky side groups. Such groups, however, are almost always present and can hinder the rotation not only because of strong steric interactions but also due to the presence of hydrogen-bonding through the lateral groups as in the 3-BCMU or 4-BCMU urethane-like side groups.

As pointed out by Rossi et al. [3.80], it should be stressed that whenever the rotation around the single bonds is easy, this strongly effects the conjugation length. However, the persistence length is much less influenced; indeed, any rotation occurring around the [C–C≡C] segment has only a modest effect on the end-to-end distance, since the trans conformation of the double bonds imposes parallelism among all of these segments.

3.4.3 Polythiophene oligomers

Due to the size of the monomer units, optimization of the geometry has been restricted to the case of bithiophene. Some experimental data on the conformational preference of bithiophene have been reported by Bucci et al. [3.88]. These authors have investigated bithiophene as partially oriented in the nematic phase of a liquid crystalline solvent. NMR measurements indicate the anti conformation to be the most stable with a barrier for the conversion to the syn form of about 5±2 kcal/mol. A qualitative analysis of the data suggests that the stability difference between the anti and syn forms is very small, on the order of 0.2 kcal/mol.

The Hartree-Fock 3-21G relaxed geometries for the anti, perpendicular, and syn conformations indicate that rotations around the central bond lead to small evolutions of the geometry. In the perpendicular conformation, the interring bond length elongates from 1.441 Å up to 1.447 Å. In the syn conformation, the interring bond remains intermediate between the previous two values (1.445 Å) and the angle between this bond and the C-S bond increases to 121.5°, a value 1° larger than in the anti conformation.

The barrier we calculate at the perpendicular conformation is 2.6 kcal/mol (2.7 kcal/mol, unrelaxed value), i.e., at the low side of the experimental data of Bucci et al. [3.88]. On the other hand, the anti conformation is more stable than the syn conformation by 2 kcal/mol, i.e., a difference one order of magnitude larger than the 0.2 kcal/mol proposed experimentally. At this stage, it is difficult to discriminate between the theoretical and experimental data. From the theoretical standpoint, the full geometry optimizations that we have performed for all three conformations should provide a reliable framework. However, as is the case when using a 3-21G basis set, we calculate the C-S bonds to be almost 0.1 Å too long. This could lead to an enhancement of steric interactions in the syn conformation. Furthermore, as mentioned above, the solvent effects are neglected in our approach. Polar solvents could preferentially stabilize the bithiophene syn conformation (for which we calculate a dipole moment of 2.4 Debye) with respect to the anti conformation (which has no permanent dipole moment). From the experimental standpoint, on the other hand, it is not an easy task to evaluate how the nematic liquid crystal can influence the measured conformation of the bithiophene molecule.

Rossi et al. have theoretically investigated the influence of the shape of the torsion potential on the persistence and conjugation lengths of polythiophene chains in solution [3.80]. The differences between the potentials derived on the one hand from the NMR data of Bucci et al. [3.88] and, on the other hand, from our calculations [3.82] are large enough that precise experimental measurements of these lengths should enable one to ascertain the relative quality of the two approaches. For instance, the persistence length at room temperature is calculated to be of the order of 11 polymer repeat units in the former case and about 60 in the latter case. To the best of our knowledge, no relevant experimental data are available on unsubstituted polythiophene yet.

Finally, we note that the rigidity of the polymer backbone and therefore the persistence length should dramatically increase when the polymer is doped. Indeed, the doping process results in strong geometry modifications inducing quinoidic structures and thus double-bond character of the interring bonds. At the Hartree-Fock (HF) 3-21G and semiempirical AM1 levels, we have recently calculated that the barrier to rotation of adjacent rings gets to be about 25-30 times larger when a quinoid-like charged conformation is taken into account [3.89]. Such a dramatic evolution of the persistence length has been recently observed by Aimé et al. through small-angle neutron scattering experiments on poly(3-butylthiophene) [3.90]. These authors have found an increase in statistical length from about 55 Å up to > 850 Å upon doping. Doped poly(3-butylthiophene) solutions display properties which are most interesting because they are markedly different from those of conventional polyelectrolytes. In that context, the applicability of models, such as the conformon model proposed by Pincus et al. [3.91], should be an active field of investigation.

The major message of this Section is thus that, contrary to the conventional belief, conjugated molecules actually constitute flexible systems, at least in their undoped (uncharged) state. The coupling of the ring motion with the electronic structure leads to fascinating effects, such as the solid state thermochromism of polyalkylthiophenes.

3.5 MOLECULAR ENGINEERING OF ORGANIC POLYMERS WITH A VERY SMALL INTRINSIC BANDGAP

Major efforts are currently devoted to the search for organic polymers that would have very small or even vanishing bandgaps. This means looking for systems that would be *intrinsically* good electrical conductors or semiconductors without the need of doping. So far, two routes have mainly been explored. The first consists in trying to construct fully fused-ring hydrocarbon structures that would more or less correspond to a one-dimensional-like graphite. Examples are furnished by polyperinaphthalene or polyacene-like systems. The fused-ring nature of those systems all along the chain, however, hinders the flexibility of the macromolecules and hence could negatively affect the mechanical properties of the resulting polymeric material unless some derivatization of the chains is achieved.

Another way to accomplish the same goal is to consider some of the polymers that are already well known in the field of synthetic metals and to try and modify their electronic properties by the action of substituents. In that framework, a reasonable approach is to consider as parent polymers those which already have a rather small bandgap and whose chemical nature leads to facile substitution reactions. Polythiophene, which has an intrinsic bandgap of about 2 eV was an early target in this game because of the ease of substitution on the carbons in the 3 positions (without necessarily the appearance of strong steric interactions). These characteristics have led Wudl et al. [3.92] to prepare polyisothianaphthene, a polymer of a "nonclassical" thiophene. The monomer molecule, isothianaphthene, can be thought of as a thiophene unit with a benzene ring fused along the β-β' bond. Wudl et al. have found that polyisothianapthene has a 1.0 eV energy gap, which is one full eV lower than that of polythiophene itself.

In the first part of this Section, we set the theoretical framework in which it is possible to rationalize a bandgap lowering by establishing the relationship that exists between the bandgap value and the degree of bond length alternation (Δr) along the chain. In the case of polyacetylene, we describe the effects of electron-electron interactions and electron correlation on Δr. We then investigate, in the case of aromatic polymers such as polythiophene, polyparaphenylene, polypyrrole, or their derivatives, the influence of switching from an "aromatic" to a "quinoid-like" geometry. The concepts that will be illustrated in the course of these studies, are applied in the following parts of the Section, first to some polypyrrole derivatives and then to polypyrenylene vinylenes.

3.5.1 Relationship between bandgap and bond length alternation in conjugated polymers

Since the early days of quantum chemistry, the question of bandgap vs. bond length alternation has been a matter of controversy, especially in the case of polyene (polyacetylene) chains. On the basis of one-electron models, bond length alternation (i.e., charge-density-wave of bond-order type) is believed to be the primary cause for the existence of the 1.5-2.0 eV bandgap in the ground state of trans-polyacetylene [3.25]. However, inclusion of correlation effects via Unrestricted Hartree-Fock (UHF) methods leads to a spin-density-wave ground state where all carbon-carbon bonds have equal length; in this framework, the gap is accounted for exclusively by correlation effects [3.93, 3.94].

Interest in this problem was renewed by the discovery of high conductivity in doped polyacetylene. As discussed in Section 3.3, many of the peculiar properties of trans-polyacetylene have been explained on the basis of the presence of topological solitons on the polyacetylene chains. The existence of solitons implies the existence of a degenerate ground state and therefore bond length alternation along the chains.

There has been considerable discussion (and even confusion) in the physics community about the effect of electron-electron interactions and (when going beyond the Hartree-Fock limit) the effect of electron correlation on the magnitude of bond alternation in polymers such as trans-polyacetylene. For instance, the results of extended Hubbard calculations imply that weak electron-electron interactions enhance bond dimerization [3.37, 3.38] in contrast to results inferred from the addition of bond charge repulsion in the Hubbard model [3.40]. Recently, Wu, Sun, and Nasu have reported calculations combining the effects of both electron-electron interaction and electron correlation in trans-polyacetylene [3.41]. Their results indicate that for short-range (screened) electron-electron repulsion, bond alternation actually decreases with increasing electron-electron interaction strength (even in the weak interaction limit). On the other hand, they find that bond alternation increases when long-range e-e repulsion is considered. It is therefore informative to distinguish between the effects of electron-electron interaction and electron correlation and focus on the influence of the latter on bond alternation.

Paldus and co-workers have treated the case of large polyene rings (mimicking polyacetylene) using a Pariser-Parr-Pople (PPP) Hamiltonian with unscreened e-e repulsion and exchange [3.42]. For values of the parameters suitable to polyacetylene, they obtain at the Hartree-Fock PPP level a bond alternation on the order of 0.10 Å. From the Hamiltonian, it can be estimated that the contribution of the electron-lattice term to the bond alternation is much smaller, on the order of 0.01 Å. Therefore, the work of Paldus demonstrates that long-range electron-electron interaction increases bond alternation, in agreement with the results of Wu et al. [3.41]. We note that the PPP bond alternation value is on the same order as that derived at the Hückel Su-Schrieffer-Heeger Hamiltonian level [3.30]. This points to the fact that the origin of bond alternation is very different at the Hückel and PPP levels.

Using either delocalized Bloch functions or localized Wannier functions, Paldus and co-workers have then introduced electron correlation effects by following different approaches: (i) the unrestricted Hartree-Fock (UHF) method where they relax the constraint of doubly occupying the molecular orbitals; (ii) the alternant molecular orbital (AMO) technique where they also use different orbitals for different spins; (iii) many-body perturbation theory (MBPT) up to fourth order; and (iv) the coupled-cluster approach (CCA) [3.42]. In all cases, they find that the effect of electron correlation is to depress bond dimerization; the stronger the electron-electron interaction (at the Hartree-Fock level), the larger the reduction. For physical values of the parameters, the effect of electron correlation is to decrease the bond dimerization only slightly to 0.09 Å (down from the 0.10 Å RHF value). The reason is that for weak to intermediate electron-electron interaction strength, the correlation energy is calculated to be almost constant for bond dimerizations ranging between 0.05 and 0.15 Å. Note that the absence of alternation obtained when using UHF methods as mentioned above, is actually due to an unbalanced treatment of correlation in that type of techniques. Indeed, UHF overestimates correlation effects for small Δr values and underestimates them for large Δr values, resulting in an artificially favored equal bond length geometry.

Similar results have been obtained with the PPP Hamiltonian on long polyene chains. Inclusion of electron correlation via configuration interaction results in a slight lowering of bond alternation and, for fixed magnitude of dimerization, in a reduction of the single-particle gap. Suhai has performed *ab initio* quantum-chemical calculations with an extended basis set on an infinite chain of trans-polyacetylene [3.44]. Geometry optimizations lead to a 0.103 Å bond dimerization at the *ab initio* RHF limit. Again, this value is lowered to 0.084 Å by electron correlation introduced via a perturbation approach estimated to recover at least 75% of the total correlation energy.

All these results indicate that *electron correlation always tends to depress bond dimerization*. For values of the parameters appropriate to polyacetylene, the effect of electron correlation is to reduce the absolute value of bond dimerization by an amount on the order of 10-20% relative to the value obtained including all electron-electron interactions at the RHF level limit [3.45].

The existence of bond length alternation in polyacetylene is thus now firmly assessed both on experimental and theoretical grounds, its origin coming from both e-lattice coupling and electron-electron interaction effects. A number of calculations indicates that in systems which, like trans-polyacetylene, possess a degenerate ground state, the bandgap value increases as a function of the degree of bond length alternation. The question we address now is to examine which kind of relationship might exist in systems with a nondegenerate ground state, in particular aromatic polymers such as polythiophene, polyparaphenylene, or polypyrrole.

We have performed VEH calculations on those three compounds [3.95], first considering the neutral polymers and then varying their geometries from aromatic (which corresponds to the ground-state optimal structure) to quinoid-like. The geometries between which variations are made are taken from *ab initio* Hartree-Fock optimizations on undoped (aromatic) and 50% n-doped (quinoid) structures for quaterthienyl, quaterphenyl, and quaterpyrryl [3.68]. For each geometry, the degree of bond length alternation, Δr, is calculated as the largest difference between the carbon-carbon bond lengths. By definition, we set the Δr values to be negative (positive) for a strongly aromatic (quinoid) structures.

The evolution of the bandgap as a function of increasing quinoid contributions to the geometry is depicted in Figure 3.13 for polythiophene.

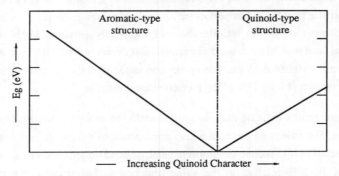

Figure 3.13 Sketch of the general evolution of the bandgap in aromatic polymers as a function of increasing quinoid character of the backbone geometry.

The bandgap is a direct bandgap at the Brillouin zone center (k=0). As can be observed from Figure 3.13, a most important result is that the bandgap evolves linearly as a function of Δr. This result is actually in agreement with the solid state physics concepts developed by Brazovskii and Kirova [3.96]. These authors define the bandgap (E_g) in compounds with a nondegenerate ground state in the following way:

$$E_g = 2 \Delta_o = 2 (\Delta_e + \Delta_i)$$

where Δ_e is a fixed "external" contribution to the gap due to the σ skeleton (which vanishes in the case of a compound with a degenerate ground state). Δ_i refers to the "internal" contribution to the gap from the π-electron framework; it corresponds to a Peierls-type contribution related to the degree of bond length alternation (or the degree of ring torsion alternation [3.97]).

In the case of polythiophene, the evolution is such that, at first, the bandgap *decreases linearly* when Δr increases. As the situation where Δr is equal to +0.06 Å is approached, the bandgap is calculated to become very small. A careful analysis of the symmetries of the HOMO and LUMO bands indicates that they belong to the same irreducible representation. Therefore, the bandgap is not calculated at the VEH level to completely vanish. However, an accidental degeneracy might occur at a point of high symmetry in the Brillouin zone, such as the zone center (k=0) or the zone edge (k=π/a). For Δr values larger than +0.06 Å, the bandgap then *increases linearly* with Δr.

An interesting aspect is that, with respect to the situation prevailing for Δr lower than +0.06 Å, the HOMO and LUMO characteristics have exchanged for Δr larger than +0.06 Å. Thus, the HOMO [LUMO] electronic pattern for what we can define as an aromatic-type geometry (Δr < +0.06 Å) is similar to the LUMO [HOMO] electronic pattern for a strongly quinoid-type geometry (Δr > +0.06 Å): there occurs an (avoided) crossing between electronic levels. In this framework, it becomes easy to understand the evolution of the bandgap when starting from an aromatic geometry. Indeed, as the quinoid contributions are increased, the HOMO becomes destabilized since it is equivalent to the LUMO of the quinoid structure whereas the LUMO becomes stabilized. This evolution results in a net decrease of the bandgap.

On Figure 3.13, we can thus divide the V-shaped curve into two parts: (i) the left part (for $\Delta r < +0.06$ Å) which corresponds to what we define as an aromatic-type electronic structure and (ii) the right part (for $\Delta r > +0.06$ Å) which corresponds to a quinoid-type electronic structure. The bandgap value for $\Delta r = 0$ allows us to estimate the "external" contribution Δ_e to the gap, since the Δ_i term then vanishes. In the case of polythiophene, we calculate Δ_e to be about 0.3 eV.

The evolution calculated for polyparaphenylene and polypyrrole is qualitatively similar to that of polythiophene [3.95]. There is, however, a major difference related to the fact that the intrinsic bandgap values of polyparaphenylene and polypyrrole in the ground state are 1.0-1.5 eV larger than that of polythiophene under the same conditions. As a result, the electronic structure of polyparaphenylene or polypyrrole always remains on the aromatic side (left side of the V-shaped curve), even for very strong quinoid geometries. We have calculated, at the Hückel level, that the level crossing in polyparaphenylene would occur for bond length values that are not reasonable: ≈ 1.30 Å double bonds and ≈ 1.50 Å single bonds (i.e., the bottom of the V curve would correspond to $\Delta r \approx +0.20$ Å).

The bandgap-bond length alternation relationship that we have derived can be used as a *qualitative guideline to design new organic polymers with a small intrinsic bandgap*. What is needed is a compound in which some quinoid contributions to the geometric structure or electronic structure are stabilized in the ground state. Note, however, that this relationship predicts that the corresponding polymeric chains will generally not have a metallic behaviour (except as a result of accidental degeneracy) but would constitute semiconductors with possibly (very) small bandgaps. Furthermore, it must be borne in mind that, for systems lying near the bottom of the V-shaped curve of Figure 3.13, correlation and nuclear relaxation phenomena may become important and lead to a larger bandgap than that predicted at the VEH level. Polysulfurnitride [3.6, 3.7], however, definitely demonstrates that a conjugated polymer can be intrinsically highly conducting namely as a result of three-dimensional interactions which suppress any tendency towards Peierls distortion.

The decrease in bandgap when going from polythiophene to polyisothianaphthene can therefore be rationalized by realizing that the fusion of a benzene ring onto the thiophene ring effectively increases the quinoid contributions to the electronic structure. As described above, this destabilizes the HOMO and stabilizes the LUMO. In polyisothianaphthene, the increased quinoid contributions to the electronic structure cut the bandgap in half with respect to that of the parent polymer, polythiophene.

We should stress that MNDO geometry optimizations carried out on the infinite polymer chain and Hückel calculations on long oligomers suggest that the geometric structure of a polyisothianaphthene chain tends to become quinoid-like, at least away from the chain ends [3.98]. Such a trend is, however, not found in Hartree-Fock *ab initio* calculations performed on oligomers up to the pentamer.

3.5.2 Polypyrrole derivatives

From the relationship that has been established between the energy gap and the bond length alternation in aromatic polymers, we know that promising compounds are those in which some quinoid contributions are stabilized in the ground state.

We have therefore investigated [3.99] the geometric and electronic structures of a polypyrrole derivative containing 50% of quinoid form in the ground state: poly(5,5'-bipyrrole methenylene), PBPM, Figure 3.14.

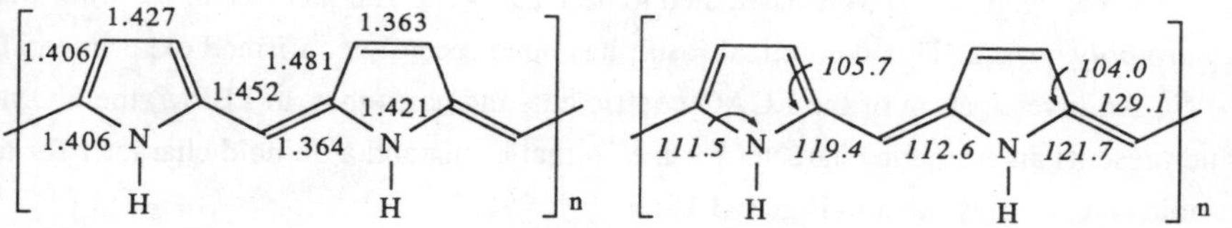

Figure 3.14 MNDO-optimized geometric structure of poly(5,5'-bipyrrole methenylene). Bond lengths are given in Å and angles in degrees. The system is fully coplanar

The presence of an odd number of conjugated carbons between the rings forces those to alternate between aromatic and quinoid geometries. This polymer is interesting in its own right since it has a degenerate ground state and thus can be expected to present topological excitations such as solitons; moreover, the recent report of the synthesis of PBPM [3.100] is a further motivation for studying theoretically this compound. The electronic structure of PBPM has also been reported by Lee and Kertesz at the simple Hückel level [3.101].

The geometry of the five-ring oligomer of PBPM is optimized with MNDO and then used as input for a VEH band structure calculation. The optimized geometry is depicted in Figure 3.14 (we note that the geometries taken separately for the aromatic and quinoid rings of PBPM are very similar to those optimized for polypyrrole with entirely aromatic or quinoid structures, respectively).

We have also calculated the VEH electronic band structures of pure polypyrrole for two different geometries: (i) where the pyrrole rings adopt the geometry of the PBPM aromatic rings and (ii) where they adopt the geometry of the PBPM quinoid rings. In the aromatic geometry, the bandgap of polypyrrole is 3.18 eV; in the quinoid geometry, it decreases to 0.86 eV. In agreement with what we mentioned above, it is important to recall that both geometric structures of polypyrrole result in an aromatic-like electronic structure: i.e., the bonding-antibonding patterns for the highest occupied valence band and the lowest unoccupied conduction band correspond to aromatic and quinoid characters, respectively.

The bandgap of PBPM is calculated to be 1.10 eV, i.e., about *three times lower than in pure polypyrrole*. This theoretical result has been recently confirmed experimentally [3.100]. An investigation of the LCAO coefficients indicates that, in PBPM, the HOMO band presents an aromatic character for the aromatic ring and a quinoid character for the quinoid ring, as illustrated in Figure 3.15.

The influence of the bridging carbon is thus such that quinoid electronic character is maintained for the quinoid ring. In the case of the LUMO band, there occurs an inversion of the electronic patterns relative to the HOMO band, so that here the aromatic ring takes on a quinoid electronic character and the quinoid ring, an aromatic character.

Examining the largest LCAO coefficients, we find that the HOMO band is dominated by the aromatic ring (with an aromatic electronic pattern) and the LUMO band, by the quinoid ring (also with an aromatic electronic pattern). The important consequence is that the top of the HOMO band of PBPM lies near (0.21 eV above) the top of the HOMO band of the aromatic polypyrrole chain and the bottom of the LUMO band of PBPM lies near (0.18 eV above) the top of the HOMO band of the quinoid polypyrrole chain. The ≈ 0.2 eV upward shifts are a consequence of the destabilization coming from the other ring present in the unit cell. Thus, the bandgap in PBPM, 1.10 eV, is comparable to the destabilization (1.13 eV) of the HOMO band in polypyrrole when it evolves from an aromatic geometry (corresponding to that of the PBPM aromatic rings) to a quinoid geometry (corresponding to that of the PBPM quinoid rings).

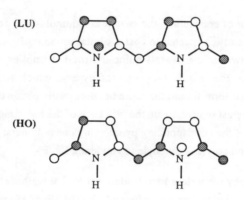

Figure 3.15 HOMO and LUMO band bonding-antibonding patterns for PBPM

It follows from that analysis that an even lower bandgap could be found if the two kinds of PBPM rings had more similar geometries. This could be achieved, for instance by introducing substituents located exclusively on one kind of rings [3.99].

3.5.3 Application to poly(pyrenylene vinylenes)

In this section, we examine the possibility of designing novel low bandgap polymeric materials by following a slightly different route. We investigate the electronic structure of polymers based on an aromatic skeleton where the switch to a quinoid-type geometry is expected to cost a minimal amount of total energy because aromaticity is preserved in some part of the system.

In this framework, pyrene appears to be an excellent candidate. The idea, based on simple chemical intuition, is as follows [3.102]. In the ground state, pyrene has an aromatic-like geometry and can be best described as a biphenyl molecule to which two ethylenic units are connected. However, when we consider vinylene derivatives of pyrene, two different situations can be envisioned depending on whether the vinylene substitution takes place on positions 1 and 6 or on positions 2 and 7, see Figure 3.16.

In the 1-6 mode of connection, the switch to a quinoid-type geometry is similar to that in pyrene itself and is expected to cost a similar amount of total energy. In the 2-7 connection case, however, an overall quinoid structure implies the formation of an aromatic structure in the middle two rings of pyrene, which acquire a naphthalene character. The quinoid form should therefore be more easily obtainable in the latter case. We thus found of interest to investigate the influence of the two connection modes on the electronic structure of the corresponding pyrene vinylene polymers: poly(1,6-pyrenylene vinylene), P16PV, and poly(2,7-pyrenylene vinylene), P27PV.

The methodology we follow here is identical to that we have used in the previous instances: geometry optimizations are performed with the MNDO semiempirical technique; the nonempirical VEH pseudopotential method is employed for the band structure calculations.

From the geometry optimizations on P16PV and P27PV, we obtain that the maximum degrees of bond alternation (i.e., the maximum differences between the lengths of carbon-carbon bonds) are almost identical for the P16PV and P27PV systems: 0.123 and 0.120 Å, respectively. These maxima are found within the vinylene links. Within the pyrene rings, the largest bond length differences are 0.082 and 0.095 Å for the P16PV and P27PV, respectively. They correspond in the latter case, to the difference between an "ethylenic" unit double bond and the single bond in the middle of the "biphenyl" unit.

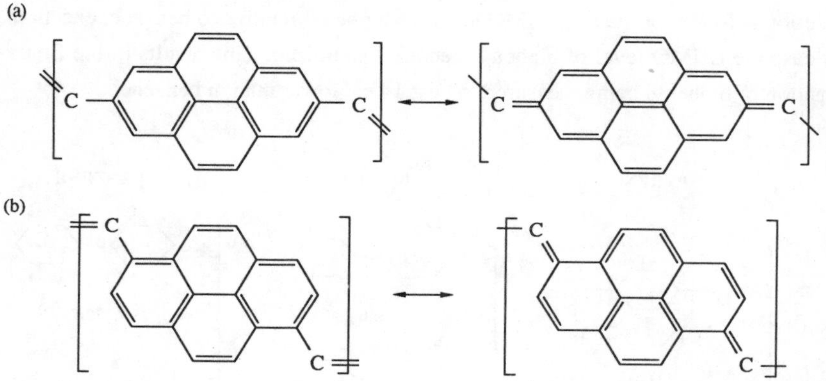

(a)

(b)

Figure 3.16 Resonance structures of (a) poly (1,6-pyrenylene vinylene) and (b) poly (2,7-pyrenylene vinylene)

3.5.3.1 The pyrene molecule

We first investigate the electronic structure of the pyrene molecule since it will serve as a useful reference to rationalize the polymeric results. The two highest occupied molecular orbitals and the two lowest unoccupied molecular orbitals of pyrene are depicted in Figure 3.17 on the basis of their bonding-antibonding electronic patterns. They are compared to those of benzene and biphenyl.

The four energy levels of the pyrene and biphenyl molecules illustrated in Figure 3.17 are reminiscent of the corresponding levels in benzene. In the latter compound, the HOMO and LUMO levels are both doubly degenerate. The lower symmetry present in biphenyl and pyrene leads to a removal of this degeneracy. In biphenyl, the equivalent to the benzene e_{1g} HOMO level is found at nearly the same energy, due to the lack of

222

contributions from the carbons located in para positions (i.e., those carbons located along the long axis of the molecule). The b_{2g} HOMO level of biphenyl is equivalent to the benzene e_{1g} HOMO level and is strongly destabilized (by 1.08 eV) with respect to it, due to antibonding interactions between the two phenyl rings, see Figure 3.17. A similar evolution is found for the b_{1u} LUMO levels of biphenyl relative to benzene, except that in this case the LUMO level of biphenyl becomes stabilized. This results in the first $\pi-\pi^*$ transition of biphenyl being measured to be 2.14 eV lower than in benzene.

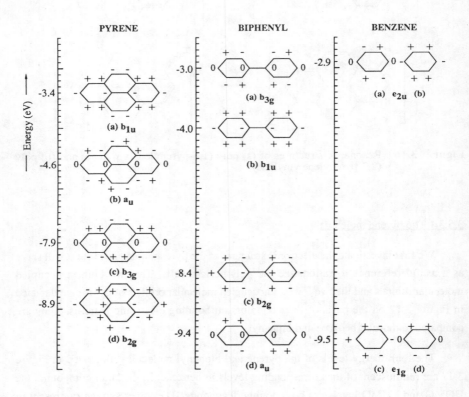

Figure 3.17 Bonding-antibonding electronic patterns of (a) LUMO + 1, (b) LUMO, (c) HOMO, and (d) HOMO - 1 orbitals (with their respective energies in eV) of pyrene, biphenyl, and benzene molecules

Very interestingly, the evolution is markedly different in pyrene due to the presence of the ethylenic units which bridge the biphenyl unit. The b_{3g} HOMO level is in pyrene equivalent to the benzene e_{1g} HOMO level (and therefore to the biphenyl HOMO–1 level) because of antibonding interactions with the ethylenic units. The pyrene HOMO level has thus no contributions coming from the carbons located along the long axis of the molecule. The pyrene b_{2g} HOMO–1 level corresponds to the biphenyl b_{2g} HOMO level due to a stabilization originating in bonding interactions with the ethylenic units, see Figure 3.17.

This difference in the locations of the pyrene b_{3g} and b_{2g} levels with respect to biphenyl is actually the clue to rationalizing the polymer results.

3.5.3.2 Poly(1,6-pyrenylene vinylene) and poly(2,7-pyrenylene vinylene)

The band structures of P16PV and P27PV are displayed in Figure 3.18. As done above, we focus on the highest two occupied bands and the lowest two unoccupied bands. The overall shapes of these bands are very similar in both polymers. We note that among the two highest occupied (lowest unoccupied) bands, there occurs a very flat band (width on the order of 0.1 eV) and a significantly more disperse band (width on the order of 1.1 eV). Going from P16PV to P27PV, the upper flat band becomes destabilized (by about 0.6 eV). Similar trends are observed for the LUMO levels. In the case of P27PV, this evolution leads to the absence of any crossing between the highest two valence bands (lowest two conduction bands). It also results in a very significant decrease in the bandgap value, Eg ≈ 1.7 eV, relative to P16PV, Eg ≈ 3.0 eV [3.102].

In order to rationalize the Eg differences between P16PV and P27PV, it is best to investigate first the situation in P16PV. Here, the vinylene linkages appear in the "para" positions of the biphenyl unit of pyrene. Since (as in the case of the pyrene HOMO level) there is no contribution to the HOMO from the carbons located on the long axis of the molecule, namely carbons 1 and 6, the interactions between the vinylene linkages and the pyrene unit are very weak and lead to a dispersionless band located at the same energy level (±7.92 eV) as in pyrene itself. On the other hand, strong interactions between the vinylene and pyrene moieties occur for the HOMO-1 band of pyrene: bonding interactions at $k=0$ (center of the Brillouin zone) and antibonding interactions at $k=\pi/a$ (end of the first Brillouin zone), producing a 1.05 eV wide band. Thus, the HOMO and LUMO bands of P16PV (at $k=\pi/a$) mostly derive from the HOMO–1 and LUMO+1 levels of pyrene interacting with vinylene linkages.

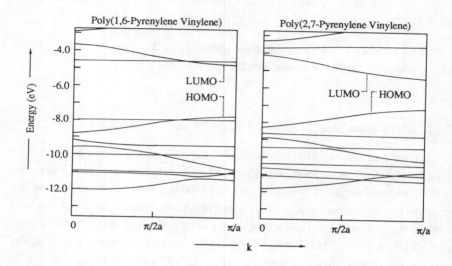

Figure 3.18 Valence effective Hamiltonian (highest occupied and lowest unoccupied bands) band structure of (a) P16PV and (b) P27PV

The relevance of this analysis is confirmed in the following way. In P16PV, we can consider that the bandgap (HOMO-LUMO separation) is determined by the electronic interplay between a vinylene moiety and a biphenyl-like moiety. This holds especially true as the LCAO coefficients on the ethylenic units of pyrene are extremely low for the HOMO and LUMO bands. In such a case, the polymer chain (as far as the HOMO and LUMO levels are concerned) can be modeled as being formed by a regular "copolymer" of poly(paraphenylene vinylene), PPV, and polyparaphenylene, PPP. It turns out that in such regular "copolymers", the bandgap value usually corresponds to the average of the bandgap values in the parent polymers. Here, we find that the bandgap of P16PV is 3.0 eV, intermediate between those of PPP (3.5 eV) and PPV (2.5 eV).

The 2–7 mode of connection leads to a very different picture, due to the strong modification of the symmetry of the interactions between the vinylene and pyrene moieties. The electronic overlap between the vinylene units and the pyrene HOMO is here effective and leads to the P27PV highest valence band with a width of 1.2 eV. For this band (as well as for the LUMO band), large value of LCAO coefficients are found all over the periphery of the pyrene unit, only the central two carbons having negligible contributions. On the other hand, the interaction with the pyrene HOMO–1 level is weak, resulting in the appearance of a flat band in the polymer, centered around the energy of the pyrene HOMO–1 level, at about - 8.8 eV.

This evolution allows us easily to understand the strong reduction in bandgap value when going from P16PV to P27PV. In the pyrene molecule, the HOMO–1 (LUMO+1) level is located about 1.0 eV below (above) the HOMO (LUMO) level. For the 1-6 polymer mode of connection, through the interaction with the vinylene links, it is the HOMO–1 (LUMO+1) level of pyrene which gains some width to produce the polymer HOMO (LUMO) band. As a result, the highest and lowest electronic levels in the polymer are located at energies near those in the pyrene molecule; the first $\pi-\pi^*$ transition calculated to be 3.35 eV in the molecule is reduced by only 0.38 eV to reach 2.97 eV in the polymer.

For the 2-7 mode of connection, the most effective interaction of the vinylene linkages is with the pyrene HOMO (LUMO) band itself, producing a strong destabilization of the highest occupied electronic level and a strong stabilization of the quinoid-like lowest unoccupied level relative to pyrene, in contrast to the 1-6 connection. Quinoid contributions to the electronic structure are thus favored in the 2-7 compound.

It has been claimed that only the degree of bond length alternation along the polymer chain is important to determine the bandgap value of a conjugated polymer. However, our results demonstrate that, in these pyrene polymeric derivatives which possess an almost identical alternation degree, important differences can actually be rationalized only in the framework of the model we have developed in the first part of this Section and which is based on the importance of the stabilization of quinoid contributions to the electronic ground state.

We find it most interesting that the chemist's intuitive idea which forms the basis of this work – i.e., that strong quinoid contributions to the electronic structure are expected in the 2,7 compound but not in the 1,6 compound – is in agreement with a detailed analysis of the electronic characteristics of the highest occupied and lowest unoccupied bands in these polymers. On that basis, an active search can be made for novel polymer structures with possibly even lower bandgaps.

3.6 NATURE OF THE CHARGES APPEARING ON THE POLYMER CHAINS UPON DOPING

One of the most challenging tasks in the field of conducting polymers has been to rationalize the evolution of the electronic, optical, and transport properties which take place upon "doping". Quantum chemical calculations have greatly helped in providing a detailed microscopic understanding of the modifications of the polymer characteristics when excess charges are added to the chains. In this Section, we will try to provide a description of the fascinating phenomena that occur and which have led to the formulation of novel concepts in the chemistry and physics of the organic solid state. First, we present the general features which make the doped organic polymers so different from the more traditional doped inorganic semiconductors such as silicon.

In a polymer, the interaction of a polymer unit cell with all its neighbours leads to the formation of electronic bands, as described in detail in the first part of the book. The highest occupied electronic levels constitute the valence band (VB) and the lowest unoccupied levels, the conduction band (CB). The width of the forbidden band, or bandgap (E_g), between the VB and CB determines the intrinsic electrical properties of the material. If E_g is larger than ~2eV, the material is an insulator because, at room temperature, very few electrons have the required energy to jump from the VB to unoccupied levels in the CB and carry a current upon application of an electric field. If E_g is lower than 2 eV, the material is usually referred to as a semiconductor, and as a metal or a semimetal if the Eg value is zero. For all the organic conjugated polymers we have discussed so far, the bandgap is larger than 1.5 eV so that these materials are intrinsically insulating.

Initially, the high conductivity increase observed upon doping organic polymers was thought to result from the formation of unfilled electronic bands. It was simply assumed than upon p-type or n-type doping, electrons were, respectively, removed from the top of the VB or added to the bottom of the CB, in analogy to the mechanism of generation of charge carriers in doped inorganic semiconductors. This assumption was however quickly challenged by the discovery (first reported in the case of polyacetylene and later of polyparaphenylene and polypyrrole) that conducting polymers can display conductivity which does not seem to be associated with unpaired electrons (spin 1/2) but rather with spinless charge carriers. The first question we address here is the following: What happens on the polymer chain when it is oxidized or reduced, or in other words, when the chain becomes ionized by removing or adding an electron?

In *organic molecules*, it is usually the case that the equilibrium geometry in the ionized state is different from that in the ground state, e.g., the geometry of neutral biphenyl is benzenoid-like but becomes quinoid-like in the ionized state, for example in the Rb^+-biphenyl$^-$ complex. The energies involved in the ionization process of a molecule are schematically depicted in Figure 3.19. (Note that in the rest of this Section, we will mainly consider the process of removing electrons from the system, i.e., oxidation or p-type doping, but the concepts that will be developed apply equally well to reduction or n-type doping). A vertical, Franck-Condon-like ionization process costs an energy E_{IP-v}. If a geometry relaxation then takes place in the ionized state, we gain back a relaxation energy E_{rel}. Conceptually, going from the ground state to the relaxed ionized state can also be thought of in the following way. The geometry of the molecule is first distorted in the ground state in such a way that the molecule adopts the equilibrium geometry of the ionized state. This costs a distortion (elastic) energy E_{dis} (Figure 3.19). If we consider the one-electron energy levels of the molecule, this distortion leads to an upward shift $\Delta\varepsilon$ of the HOMO and a downward shift of the LUMO, as illustrated in Figure 3.20. If we then proceed to the ionization of the distorted molecule, it requires an energy E_{IP-d}.

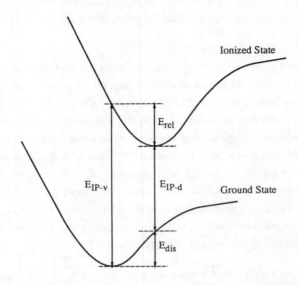

Figure 3.19 Illustration of the energies involved in a molecular ionization process. $E_{IP\text{-}v}$ is the vertical ionization energy; E_{rel}, the relaxation energy gained in the ionized state; E_{dis}, the distortion energy to be paid in the ground state in order that the molecule adopts the equilibrium geometry of the ionized state; $E_{IP\text{-}d}$, the ionization energy of the distorted molecule

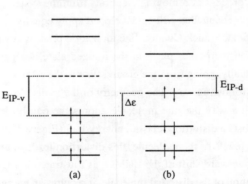

Figure 3.20 Schematic illustration of the one-electron energy levels for an organic molecule (e.g., biphenyl) in its ground-state electronic configuration adopting: (a) the equilibrium geometry of the ground state; (b) the equilibrium geometry of the first ionized state

From Figure 3.19, it is clear that it is energetically favorable to have a geometry relaxation in the ionized state when the quantity [E_{IP-v} minus E_{IP-d}] (which actually corresponds to ΔE as can be inferred from Figure 3.20) is larger than the distortion energy E_{dis}; or, in other words, when the reduction ΔE in ionization energy upon distortion is larger than the energy E_{dis} required to make that distortion.

In a polymer, or any solid, a vertical ionization process E_{IP-v} results in creating a hole on top of the valence band, see Figure 3.21 (a). In this case, three remarks can be made. First, by the very definition of the process, no geometry relaxation (lattice distortion) takes place on the chain. Second, the positive charge on the chain is *delocalized* over the whole polymer chain. Third, the presence of a hole (unfilled level) on top of the VB leads, as we discussed above, to the appearance of a metallic character. This situation corresponds to the initial assumption made about the conduction mechanism in doped organic polymers. However, in an *organic polymer chain*, it can be energetically favorable to *localize* the charge that appears on the chain and to have, around the charge, a local distortion (relaxation) of the lattice. This process causes the presence of localized electronic states in the gap due to a local upward shift Δ_e of the HOMO and downward shift of the LUMO (Figure 3.21 (b)). Considering the case of oxidation, i.e., the removal of an electron from the chain, we lower the ionization energy by an amount $\Delta\epsilon$. If $\Delta\epsilon$ is larger than the energy E_{dis} necessary to distort the lattice locally around the charge, this charge localization process is favorable. We then obtain the formation of what condensed-matter physicists call a polaron. In chemical terminology, the polaron is just a radical-ion (spin 1/2) associated with a lattice distortion and the presence of localized electronic states in the gap referred to as polaron states. The ability of a charge to deform significantly the lattice around it is the manifestation of a strong electron-phonon coupling.

The quantity $\Delta\epsilon$-E_{dis} (= E_{rel}) corresponds to the polaron binding energy. Our calculations (based on Hückel theory with σ-bond compressibility) have indicated that polaron formation is energetically favorable in all the organic conjugated polymers we have studied. The polaron binding energy is of the order of 0.05 eV in polyacetylene, 0.03 eV in polyparaphenylene, and 0.12 eV in polypyrrole. It must be stressed that in the case of polaron formation, the VB remains full and the CB empty. There is no appearance of metallic character since the half-occupied level is localized in the gap, Figure 3.21 (b).

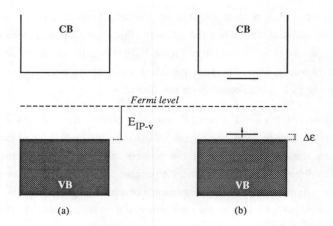

Figure 3.21 Illustration of the band structure of a polymeric chain in the case of: (a) a vertical ionization process; (b) the formation of a polaron. The chemical potential, or Fermi level, is taken as reference level

Next, we consider what happens when a second electron is removed from the polymer chain: is it more favorable to take the second electron from the polaron or from anywhere else on the chain (in which case we have two polarons). In the former case, we have bipolaron formation. A bipolaron is defined as a pair of like charges (here a dication) associated with a strong local lattice distortion. The bipolaron can be thought of as analogous to the Cooper pair in the BCS theory of superconductivity, which consists of two electrons coupled through a lattice vibration, i.e., a phonon. The formation of a bipolaron implies that the energy gained by the interaction with the lattice is larger than the Coulomb repulsion between the two charges of same sign confined in the same location.

The electronic band structure corresponding to the presence of two polarons and that of one bipolaron is depicted in Figure 3.22. Since the lattice relaxation around two charges is stronger than around only one charge, E_{dis} for the bipolaron is larger than E_{dis} for the polaron and the electronic states appearing in the gap for a bipolaron are further away from the band edges than for a polaron.

If we compare the creation of a bipolaron relative to that of two polarons, our calculations for polyacetylene, polyparaphenylene, and polypyrrole indicate that the distortion energy E_{dis} to form one bipolaron is roughly equal to that to form two polarons. On the other hand, the decrease in ionization energy is much more important in the bipolaron case (2 x $\Delta\varepsilon^{bip}$) than for two polarons (2 x $\Delta\varepsilon^{pol}$), see Figure 3.22. This is the reason why usually one bipolaron is thermodynamically more stable than two polarons in these systems despite the Coulomb repulsion between two similar charges. Furthermore, the latter is also largely screened by the presence of dopants (counterions) with opposite charges.

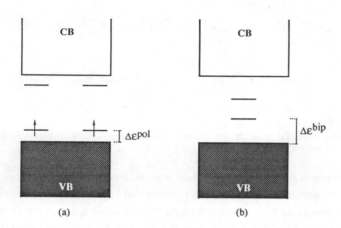

Figure 3.22 Band structure of a polymer chain containing: (a) two polarons; (b) one bipolaron

In case of p-(n-)type doping, the bipolaron levels in the gap are empty (fully occupied), see Figure 3.22 (b). The bipolarons are thus *spinless*. The presence of bipolarons on polymer chains results in the possibility of two optical transitions below the bandgap transition: for p-type doping, from the VB to the lower bipolaron level and from the VB to the upper bipolaron level. The oscillator strength for the lowest energy optical bipolaron transition has been calculated to be larger than for the highest energy one. In the case of polarons, a third absorption is possible below the gap, corresponding to an optical transition between the two polaron levels (Figure 3.22 (a)).

In the case of trans-polyacetylene, we have shown previously that there exists a degenerate ground state, i.e., two geometric structures corresponding exactly to the same total energy. As a result, the two charges forming a would-be bipolaron in trans-polyacetylene can readily separate, Figure 3.23.

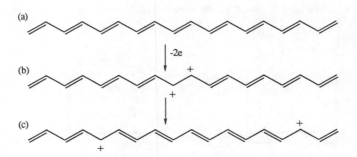

Figure 3.23   Illustration of the formation of two charged solitons on a chain of trans-polyacetylene

This process is favorable because there is no increase in distortion energy when the two charges separate, since the geometric structure that appears between the two charges has the same energy as the geometric structure on the other sides of the charges. This leads to the appearance of two isolated charged solitons on the trans-polyacetylene chain.

Unlike trans-polyacetylene, systems such as polyparaphenylene, polypyrrole, or polythiophene possess a nondegenerate ground state since their ground state corresponds to a single geometric structure which, in their case, is aromatic-like. A quinoid-like bond isomer ("resonance") structure can be envisioned but has a higher total energy. We have performed *ab initio* calculations which show that the quinoid structure has a lower ionization potential and larger electron affinity than the aromatic structure. This explains why, on doping, the chain geometry in these compounds relaxes locally around the charges towards the quinoid structure which has a larger affinity for charges.

The charge generation process outlined for doped organic polymers can be contrasted to that in inorganic doped semiconductors. In the latter compounds, the atoms are covalently bonded in all three dimensions. The lattice is therefore much less deformable than along a polymer chain and rigid-band models usually provide a good description of their solid state properties. The process of doping is achieved by introducing ppm quantities of atoms, such as boron or phosphorus atoms, which fit into the lattice either substitutionally or interstitially. In the example of boron-doped silicon, boron has one less valence electron (three) than silicon (four). This yields empty electronic levels associated with the boron slightly above the top of the silicon VB. These empty dopant levels can be thermally populated with electrons from the Si VB, which results in the creation of holes on top of the VB and the possibility of electronic conduction. By comparison, in polymer chains, the electronic levels coming into the gap are intrinsic levels of the polymer chains, the dopant levels lying usually well outside the polymer forbidden band.

3.7 EVOLUTION OF THE ELECTRONIC STRUCTURE AT HIGH DOPING LEVEL

As mentioned earlier, high electrical conductivity was first observed in AsF_5- doped polyacetylene [3.9]. After this breakthrough, a number of doped conjugated organic polymers have been shown to exhibit high conductivity, e.g., polypyrrole, polythiophene, polyparaphenylene, polyparaphenylene vinylene, and polyaniline. Polyacetylene is still one of the most studied conducting polymers and has during the past few years received additional attention because of the extremely high electrical conductivity ($\sigma > 1.5 \times 10^5$ S/cm), comparable to that of copper, obtained for iodine doped samples prepared by the method of Naarmann [3.19].

A large number of physical and chemical methods applied to the studies of conducting polymers have naturally been concentrated on the doped, conducting state. The undoped state of the polymers is also interesting, namely as a reference state. Quantities such as optical bandgap, first ionization potential, electron affinity, and density of valence states have been obtained from optical absorption and photoemission studies on the undoped polymers. For the doped polymers, a variety of physical properties have been measured: optical absorption, photoinduced optical absorption, and Electron Energy Loss

Spectroscopy (EELS) are used to study the energy levels associated with the defect states. Vibrational studies focus on the effects of the defects upon the phonon spectra. Various types of magnetic studies are applied to investigate the spin properties of the materials. Conductivity is measured as a function of temperature, electric field strength, or frequency in order to obtain information on the transport properties. From electron and X-ray diffraction studies, information on the structure of the polymers is obtained. The combined knowledge collected from all these studies have produced a rather clear picture of the physical properties of the conducting polymers.

Due to the differences in chemical composition and morphology, a totally unified picture of the effects of doping for all conducting polymers is difficult to reach. However, despite dissimilarities between conducting polymers, there are classes of materials for which the overall features are the same. To a first class of polymers, we can assign those that present structural degeneracy in the ground state, such as trans-polyacetylene or pernigraniline. In this case, nonlinear excitations of soliton-type can be observed. Another class consists of most conjugated systems that do not exhibit a structural degeneracy. Here, we find e.g., polythiophene, polypyrrole, polyparaphenylene, polyparaphenylene vinylene. We have chosen to present polythiophene (PT) as a representative for this kind of conducting polymer, but we should bear in mind that the properties for this material are qualitatively reproduced also for the other polymers listed above. Since the structural degeneracy is removed, the energetically unfavorable phase is suppressed, leading to the creation of confined soliton-antisoliton pairs, i.e., bipolarons.

At high doping levels, many of these conducting polymers exhibit a transition into a metallic state. The metallic properties are verified by the finding of high Pauli susceptibility, a linear dependence of the thermopower with temperature, far-IR absorption, i.e., properties which all originate in the appearance of a finite density of states at the Fermi level. Polyacetylene was the first polymer for which a high Pauli susceptibility was found in the highly doped state [3.102]. Several attempts to explain this phenomenon have been presented. Mele and Rice [3.103] and Epstein and co-workers [3.104] showed that the band tailing due to disorder and three-dimensional interchain couplings can lead to gap closure (due to the formation of an incommensurate gapless Peierls system). Kivelson and Heeger [3.105] have pointed out that the sharp increase in

Pauli susceptibility at about 5% doping could be understood in terms of a first order phase transition from a soliton lattice to a polaron lattice. More recently, Jeyadev and Conwell have proposed a scenario in which dopants progressively provoke the introduction of electronic states in the gap up to a point where the gap closes at about 5% doping level [3.106]. Following polyacetylene, metallic-like properties were found in polyparaphenylene [3.107], polythiophene [3.108, 3.109], and polypyrrole [3.108] but only for certain dopant substances; e.g., polypyrrole exhibits a high Pauli susceptibility for doping with BF_4^- [3.108] whereas there is no sign of Pauli susceptibility for doping with ClO_4^- [3.110].

A third class of conducting polymers is exemplified by the emeraldine form of polyaniline. Due to its high stability, polyaniline has received considerable attention during the last few years. It is, up to now, one of the few examples of conducting polymers used in industrial applications, in this case as an electrode in a lithium-polyaniline battery [3.111]. The emeraldine base form of polyaniline corresponds to the oxidation state where half of the nitrogens are of imine type and the other half of amine type [3.10]. Emeraldine base (EB) is doped by means of proton transfer to the imine nitrogens of the polymer. In contrast to the commonly used electron transfer doping, EB is thus chemically modified upon proton doping and undergoes an insulator-to-conductor transition without any change in its number of electrons. The magnetic properties of protonated EB base, termed emeraldine salt (ES), show a linear increase in Pauli susceptibility starting from low or intermediate doping levels [3.112]. This feature has been explained in terms of a phase segregation into undoped and fully doped regions [3.112, 3.113]. In the fully doped state where all imine nitrogens are protonated, the polymer exhibits a very high Pauli susceptibility indicating that a metallic state has been reached.

In this Section, we survey some of the band structure calculations we have performed on highly conducting polymers. Results are presented for systems at low, intermediate, and high doping levels for the three polymers introduced above, namely, trans-$(CH)_x$, PT, and EB. To be able to study the evolution of the effects of doping on the electronic structure, we first discuss the VEH band structures of the undoped systems.

3.7.1 Undoped Polymers

<u>3.7.1.1 Trans-polyacetylene</u>

The Valence Effective Hamiltonian (VEH) band structure for trans-polyacetylene, trans-$(CH)_x$, was already discussed previously and illustrated in Figure 3.2. Due to the Peierls distortion, the π-band splits into two bands, termed π (lower) and π^* (upper) bands. Considering an MNDO optimized geometry, the energy gap between these two bands is calculated to be 1.43 eV and the total π-π^* bandwidth, 13.6 eV [3.114]. The single particle bandgap agrees very well with experimentally observed frequency dependent conductivity data for trans-$(CH)_x$, which exhibit a maximum at 1.4 eV interpreted as due to the π-π^* transition [3.115]. Slightly different results are found from optical absorption data, which as mentioned previously onset at ~1.4 eV and peak at ~1.8-1.9 eV [3.116]. As clearly seen in Figure 3.2, the π and π^* bands appear very symmetric about the energy gap. This shows that the electron-hole symmetry, which is complete in the widely used Su-Schrieffer-Heeger (SSH model [3.30], is well preserved even at the VEH level. The π^* band is slightly wider than the π band. We calculate bandwidths of 6.49 eV and 5.71 eV, for the π^* and π bands, respectively.

The 1-dimensional band structure for trans-$(CH)_x$, as calculated by the VEH method, shows large similarities with the band structure calculated by the Extended Tight-Binding (ETB) method [3.117] and by the local density approach [3.118, 3.119]. All three methods produce a very similar structure for the π-band. Some differences are observed in the σ part of the band structure but the overall agreement is rather good. In Figure 3.24, we have compared the VEH theoretical and experimental XPS results. Note, as a result of the atomic XPS cross sections, that the XPS intensity of the π-band is strongly suppressed as compared with the bands having large carbon 2s contributions. In Figure 3.24, we have also included the result of a band structure calculation using the density functional method [3.120]. Both the VEH and density functional methods provide results in close agreement with experimental data.

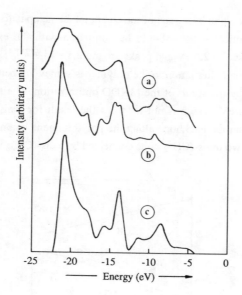

Figure 3.24 The lower curve shows the XPS density of states for trans-$(CH)_x$ as calculated by the VEH method. It is compared with the experimental XPS spectrum (due to C.R. Brundle as reported in Ref. 3.49) shown in the uppermost curve and the result obtained by Mintmire et al. [3.120] from a local density type of calculation (middle curve)

3.7.1.2 Polythiophene

A part of the VEH band structure for PT is presented in Figure 3.25. The unit cell in this calculation is one thiophene unit which is half the true translationally invariant unit cell. The band structure for the dithiophene unit cell is obtained in Figure 3.25, simply by folding the bands about $\pi/2a$, into the k-space region between k=0 and k=$\pi/2a$. At the VEH level, undoped polythiophene presents a gap between the highest occupied and lowest unoccupied bands of 1.72 eV on the basis of an MNDO-optimized geometry. We can compare this value with that obtained from a VEH calculation using an Hartree-Fock STO-3G geometry. Since the STO-3G-optimized geometry shows a stronger bond length alternation along the carbon backbone of the chain, the bandgap is found to be larger, 2.2 eV [3.68]. The theoretical values should be compared with the experimental optical absorption, which onsets at 2.0 eV and peaks around 2.5 eV [3.121]. This indicates that the effective degree of bond alternation in the polymer is somewhat better described by the STO-3G result than by the result of the MNDO optimization. The total π bandwidth is calculated to be 14.3 eV, which is not far from the result for trans-$(CH)_x$. This is not surprising since the carbon backbone along the polythiophene chain can be viewed as a pseudopolyene with two trans-$(CH)_2$ units connected by a cis-$(CH)_2$ unit.

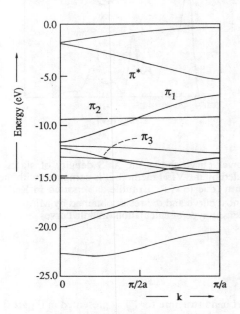

Figure 3.25 VEH band structure for undoped polythiophene

Out of the five π-bands, built up from the five p_z atomic orbitals of the thiophene unit, three are occupied. The π_2 band (see Figure 3.25) contains most of the sulfur $2p_\pi$ electron density. Due to a very small contribution from the $C\alpha$ $2p_\pi$ atomic orbitals, this band appears almost flat. A considerable contribution from the sulfur $2p_\pi$ is also found in the lowest unoccupied band (π_1*), e.g., at k=π/a the ratios $S(2p_\pi)/C\alpha(2p_\pi)$ and

$S(2p_\pi)/C\beta(2p_\pi)$ are 1.35 and 1.03, respectively. In comparison, we obtain zero values for the same ratios of the wave function coefficients in the highest occupied band (band π_1) at $k=\pi/a$. Thus, as far as the wave functions are concerned, the electron-hole symmetry about the Fermi level is far from being obeyed. However, despite the dissimilarities in the wave functions, the bands appear rather symmetric in the plot of the band structure.

The VEH result for the Density of Valence States (DOVS) spectra including XPS cross sections is shown in Figure 3.26 together with the experimental result for undoped polythiophene [3.122].

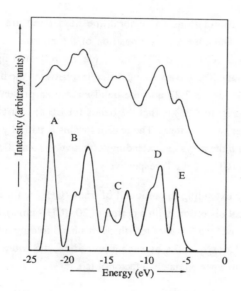

Figure 3.26 The lower curve shows the XPS density of states for polythiophene as calculated by the VEH method. It is compared with the experimental XPS spectrum [3.122] (upper curve)

Again, the comparison between the theoretical and experimental results is very good. The main peaks in the spectra, denoted by A, B, C, D, and E (see Figure 3.26), can be assigned to the respective atomic orbitals of the unit cell as follows: A) $S(3s)-C(2s)$ σ

states; B) $S(3s)$-$C(2s)$ σ states with some contributions from $C(2p_\sigma)$ and $H(1s)$; C) mixed σ states dominated by the flat part of the σ-band structure just above -20 eV (see Figure 3.25); D) a mixture of σ bands, including $S(3p_\sigma)$ and $C(2p_\sigma)$ orbitals and the π_3 band; E) the sharp peak at -9.4 eV is a result of the nondispersive character of the π_2 band with zero contributions from the $C\alpha(2p_\pi)$ atomic orbitals (see above). The tail at the low binding energy side is dominated by the flat part of the π_1 band closest to the energy gap.

3.7.1.3 Emeraldine base

Band structure calculations for emeraldine base (EB) are performed using a tetraaniline unit cell. The chemical composition of the polymer is considerably more complex than for trans-$(CH)_x$ and PT. This is clearly reflected in the band structure obtained for this polymer. The most striking difference between the band structure for EB which is presented in Figure 3.27 and the band structures presented for trans-$(CH)_x$ (Figure 3.2) and PT (Figure 3.25) is that the bands for EB appear to be much more flat than those for the other two polymers. The reason for this is partially due to the large size of the EB unit cell, with the π-system extending (effectively) over 20 sites, as compared to 2 and 4 sites for trans-$(CH)_x$ and PT, respectively.

For instance, by extending the unit cell of trans-$(CH)_x$ ten times (by allowing for some small geometrical distortion, with a period of 20 -(CH)- units), the width of the first Brillouin zone is reduced by a factor of ten (through a band folding into the k-space region between 0 to 1/10a). Consequently, by such a procedure, the width of the individual π-bands is also reduced by approximately a factor of ten.

The quinoid ring in the EB unit cell is connected by double bonds to imine nitrogen atoms. Since twists around a double bond are energetically very costly (and thus very unlikely to appear), the quinoid ring and the two imine nitrogens are expected to be almost coplanar (using the AM1 method, the quinoid ring is found to be twisted by at most 5-15° out of the plane of the C-N=C group). In order to decrease the steric interactions between hydrogens on neighbouring rings, a large twist occurs around the C-N bond linking the imine nitrogen to the benzenoid ring. This twist angle is, using the AM1 method, optimized to be around 40°-45°, a value for which the p_π orbitals of the carbon and nitrogen atoms forming this bond overlap very weakly. As a result, some π-states are strongly localized within the diimine-quinoid unit. In the polymer, such states on neighbouring unit cells overlap very weakly and form an essentially nondispersive band. Such a band, denoted $\underline{c}'$ in Figure 3.27, is the lowest unoccupied band for EB base. As depicted before, the three bands lying just below this band (bands $\underline{c}$, $\underline{b}'$ and $\underline{b}$ in Figure 3.27) are formed from the π-systems of the three benzenoid rings of the EB unit cell. The bandwidths are quite small, around 0.3 eV, even for bands $\underline{c}$, $\underline{b}'$ and $\underline{b}$. Below these bands, we observe four very flat bands (the bandwidth is calculated to be ~0.03 eV), denoted by $\underline{a}$ in Figure 3.27. The same group of four nondispersive bands are found in the unoccupied part of the band structure (band $\underline{d}$ in Figure 3.27). The states in these bands all originate from the benzene molecular orbital with nodes in the para positions of the aromatic ring. Since no contribution to these states is found in the C-N-C interring link, all these states are localized to individual aromatic rings, which explains the nondispersive character of the bands.

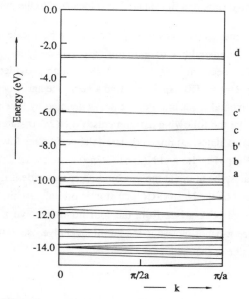

Figure 3.27 VEH band structure for emeraldine base

Aside from the flatness of the electronic bands of EB base, we also observe another important difference as compared to the band structures of trans-(CH)x and PT, namely, the *complete lack of electron-hole symmetry in the electronic structure for EB base*, which has already been stressed in Section 3.3.3. The fact that the highest occupied band (band c in Figure 3.27) and the lowest unoccupied band (band c' in Figure 3.27) are localized to different regions in space is the clearest example of this completely nonsymmetric arrangement of the states below and above the Fermi level. This has important consequences as concerns the formation of defect states.

The bandgap between the occupied and unoccupied bands is calculated to be 1.0 eV, which is considerably smaller than the lowest peak in the optical absorption spectrum for PEB, which is observed at 2.1 eV. This indicates that the electronic transitions involved in this absorption are not a simple, single particle, band-to-band transition. Calculations allowing for configuration interaction (CI) have shown that the low energy absorption has an excitonic character and involves an excitation from a state localized to rings surrounding the quinoid ring, to a final state that is localized to the quinoid ring [3.66, 123].

In the same manner as for trans-(CH)$_x$ and PT, we have calculated the theoretical DOVS spectrum for the emeraldine base. In this case, however, comparison is made with an experimental UPS spectrum [3.24] (Figure 3.28), which provides a better resolution than existing XPS spectra for EB. Again, we find a very nice agreement between the VEH result and the experimental spectrum. For the peaks denoted by letters A, B, C, D, E, and F, we are able to make the following assignments: A) N(2s) σ states; B) N(2s)-C(2s) σ states; C) C(2s)-C(2s) σ states; D) mixed σ states, the peak at the low binding energy side is due to localized N(2pσ)-H(1s) and N(2pσ)-C(2pσ) σ states; E) C(2pσ)-H(1s) and ring C(2pσ)-C(2pσ) σ states, the bump at the low binding energy side is due to mixed σ and π states mostly localized on the N(2pπ)-C(2pσ) atomic orbitals; F) nitrogen lone pair states and π states. The large peak arises from the four nondispersive π bands displayed in Figure 3.27. The three π-bands closest to the Fermi level give rise to the small bump at the low binding energy side of the main peak.

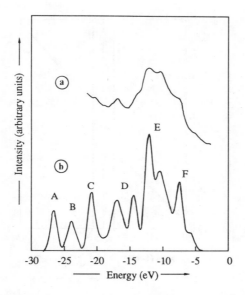

Figure 3.28 The UPS density of state for emeraldine base, as calculated by the VEH method (lower curve), is compared with the experimental UPS spectrum of Nilsson and Salaneck [3.124]

3.7.2 Low and Intermediate Doping Level

3.7.2.1 Trans-polyacetylene

To determine the electronic properties at low and intermediate doping levels, we have performed a VEH band structure calculation for trans-$(CH)_x$ in the soliton lattice conformation. The distance between two solitons is in this case taken to be 29 -(CH)- units and defines the unit cell of the polymer. This choice of unit cell corresponds to a doping level y=1/29=0.034. The band structure in the extended zone scheme presentation is displayed in Figure 3.29 together with the band structure of the undoped polymer (dashed line). The midgap soliton band is 0.16 eV wide; the subgaps appearing on both sides of the soliton band are equivalent, which means that the electron-hole symmetry is well preserved at the VEH level of calculation.

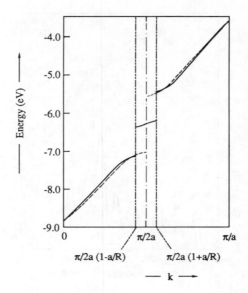

Figure 3.29 VEH band structure for the soliton lattice conformation of trans- $(CH)_x$ represented in the extended zone scheme. The distance between neighbouring solitons, R, is 29 -(CH)- units, corresponding to a doping level of y=0.034. (The dashed lines refer to the band structure of the undoped system)

Except for the region near the gap, the band structure for the undoped and doped trans-$(CH)_x$ chains are practically indistinguishable. No additional gaps can be detected at multiples of π/R (where R corresponds to the lattice constant), which is due to the fact that opening of gaps inside the occupied band does not lead to any lowering of the electronic energy. The energy gap between the valence band and the unoccupied soliton band is found to be 0.75 eV. This value compares favourably with the experimentally observed optical absorption [3.116] and frequency dependent conductivity results [3.115] for doped trans-$(CH)_x$, which both peak at 0.7 eV. Furthermore, we calculate the π-π^* transition to onset at 1.68 eV. Again, we find very good agreement with the experimentally observed value of 1.6 eV obtained from the optical conductivity studies.

3.7.2.2 Polythiophene

Since the bipolaron is doubly charged, the unit cell is twice as large as in the polaron case for a given doping level, e.g., to represent the bipolaron lattice at y=0.05 we need a unit cell of 40 thiophene rings containing 960 electrons. This is clearly a difficult problem to solve. However, from geometry optimization calculations, we have observed that the structural width of the bipolaron is roughly six monomers. Assuming that the tails of the electronic state associated with the bipolaron extend somewhat outside this region, we expect the onset of bipolaron-bipolaron overlap to occur at roughly y=0.20, corresponding to a bipolaron spacing of 10 thiophene units. The band structure for this doping level is shown in Figure 3.30.

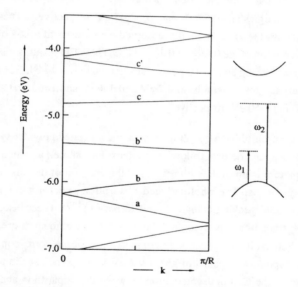

Figure 3.30 VEH band structure for the bipolaron lattice conformation of polythiophene. The distance between neighbouring bipolarons, R, is 10 thiophene units, corresponding to a doping level of y=0.20. To the right is shown a schematic picture of the electronic transitions involving the bipolaron levels

The bipolaron bandwidths are 0.11 eV and 0.08 eV for the lower and upper bipolaron bands, respectively. These results show that y=0.20 is indeed almost within the noninteracting regime and significant comparisons can be made with experimental data reported for much lower doping levels. In this context, it is worthwhile to point out again that our theoretical studies are performed on perfectly regular lattices whereas, experimentally, there can exist a more or less random structural distribution of the doping-induced defects. The optical absorption spectra naturally show some signature of this randomness as well as other perturbing effects not present in our calculations. This makes a direct translation of our ideal y value of the onedimensional lattice into the experimental doping level difficult and only qualitative agreement is expected.

We find transitions from the valence band (band $\underline{b}$) to the bipolaron bands (bands $\underline{b}'$ and $\underline{c}$) to occur at energies ω_1=0.57 eV and ω_2=1.47 eV, respectively. These values are in good agreement with the experimentally observed absorptions at ~0.65 eV and ~1.5 eV for PT doped within the range 0.05<y< 0.10 [3.116]. The VEH result is also in agreement with results obtained using an SSH-type of Hamiltonian [3.125, 3.126]. Bertho and Jouanin [3.125] find the ω_1 and ω_2 transitions at 0.7 eV and 1.4 eV and Brédas et al. [3.126] locate them at 0.50 eV and 1.60 eV, respectively.

The large electron-phonon coupling present in conjugated polymer systems also lead to formation of localized defects following photoexcitation, i.e., an electron-hole pair in the rigid band picture is unstable with respect to the creation of defects such as solitons, polarons, or bipolarons. The photoinduced absorption spectrum for PT shows two symmetric absorptions peaking at 0.45 eV and 1.25 eV [3.127]. These peaks are attributed to the ω_1 and ω_2 transitions involving bipolaron states, since no spins are observed to be created in the photoexcitation process. In comparison with the corresponding values for doping-induced bipolarons, a shift of about 0.2 eV is observed for the two peaks. This shift is an effect of the Coulomb interaction between the dopant ion and the bipolaron, which is present only in the case of doping induced defects. As noted above, it is more appropriate to study the difference between the two absorption energies due to bipolarons. It is an important feature of our results that this difference is observed to be almost identical in the three cases, namely ω_1-ω_2=0.90 eV in our theoretical spectra and 0.85 eV for both photo- and doping-induced bipolarons. The SSH-type calculations of Brédas et al. [3.126] lead to slightly larger separation of the bipolaron states, $\omega_1 - \omega_2 = 1.1$ eV. This is expected since, as noted above, the geometrical distortion associated with the bipolaron is more shallow at the Hückel level, which produces bipolaron states closer to the band edges.

We can point out that the VEH band structures for moderate soliton (in trans-$(CH)_x$) or bipolaron (in PT) concentrations lead to very good agreement between experimental and theoretical optical transitions. Thus, our results support the interpretation of the absorption data at low and intermediate doping levels in terms of a soliton model in trans-$(CH)_x$ and of bipolarons in PT.

3.7.3 High Doping Levels

The discussion now focusses on the band structures obtained in the highly doped regime, i.e., when the defects interact strongly. This regime is of special interest because it exhibits metallic properties.

3.7.3.1 Trans-polyacetylene

The VEH band structure of highly doped trans-$(CH)_x$ is presented in Figure 3.31(a) for the soliton lattice conformation (y=0.058) and in Figure 3.31(b) for the polaron lattice conformation (y=0.0625). Note that the reciprocal unit cell of the polaron lattice is about 6% larger than the one of the soliton lattice. The soliton band (band $\underline{c}$ in Figure 3.31a) is 0.74 eV wide and the subgaps appearing on both sides of this band are 0.62 eV and 0.60 eV for the lower and upper gaps, respectively. In the polaron band structure, there appear two bands within the gap, bands $\underline{b}$ and $\underline{c}$ in Figure 3.31(b). The widths of the lower and upper polaron bands are 0.95 eV and 0.94 eV, respectively. The two polaron bands are separated (at k=0) by 0.85 eV and are 0.09 eV away (at k=π/a) from the conduction or valence band edges. It is notable that the VEH method gives polaron bands that are *wider* than the soliton band. The appearance of highly dispersive bands in the metallic regime is also experimentally verified from EELS studies on doped highly oriented trans-$(CH)_x$

248

[3.128]. In earlier, more qualitative theoretical studies, the polaron bands have been presented as very narrow, essentially dispersionless, bands [3.105]. The disagreement between this result and the experimental findings has been used as an argument against the polaron lattice model. It is therefore important to point out that the VEH results for the polaron lattice are in qualitative agreement with the EELS data.

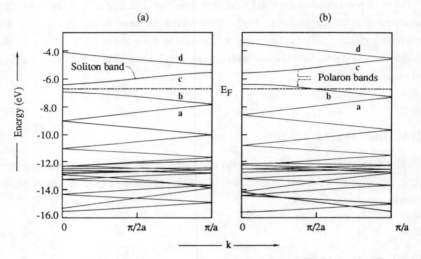

Figure 3.31 VEH band structure for trans-(CH)$_x$: (a) the soliton lattice at doping level y=0.058 and (b) the polaron lattice at y=0.0625

When comparing Figure 3.31(a) and 3.31(b), an interesting feature is that the half-filled polaron band neatly fits into the gap (between the valence band and the soliton band) at the Fermi level of the soliton lattice. Therefore, excitations across this gap in the soliton lattice become comparable in energy to *intraband* excitations in the half-filled polaron band. This feature has important consequences for the comparison with experimental optical absorption data, as is further discussed below.

The first interband transition in the soliton lattice (between bands b and c of Figure 3.31(a)) is calculated to onset at 0.62 eV. In the polaron lattice, we must take into account the intraband transition within the half-filled polaron band (band b in Figure 3.31(b)). Optical absorption within this band is expected in the far-IR region up to an energy roughly corresponding to the width of the polaron band, i.e., ~1.0 eV. We may recall that optical transitions are strictly allowed only if k vector (momentum) can be conserved, i.e., intraband transitions within the half-filled band are forbidden. However, we can expect some disorder effects to be present, which will destroy the symmetry of the lattice and thus the importance of the k-conservation rule.

Another important aspect of these results is the large difference in the π-π^* energy gap. Taking into account the proper unit cell for the soliton lattice, we obtain a direct bandgap of 2.0 eV, compared to 2.9 eV for the polaron lattice. This difference is most easily understood by the fact that *there are twice as many bands (states) in the bandgap for the polaron lattice since each polaron brings two states into the gap while each soliton only brings one state into the gap.*

Optical conductivity data on doped trans-$(CH)_x$ (including the low frequency region) [3.115] show that above the critical doping concentration at which the polymer enters a metallic-like regime, there occurs a rounding of all optical transitions in the 0.5-2.5 eV region and a marked increase of absorption into the IR. The first absorption peak shifts slightly down to 0.6 eV and the absorption previously related to the π-π^* transition is no longer observed. Within our model, the IR absorption can be well understood as originating from intraband absorption within the half-filled polaron band in the polaron lattice, as described above. The disappearance of the π-π^* transition is in agreement with our calculations showing that in the polaron lattice the π-π^* transition shifts to ~2.9 eV. Furthermore, for the polaron lattice, besides intraband excitations, optical transitions at 1.3 eV (from band a to band b), 1.6 eV (from band b to band c), and 2.8 eV (from band b to band d) are also predicted. This is in qualitative agreement with the rounding of the absorption data between 0.5 eV and 2.5 eV. Disorder effects can also lead to further rounding of these absorptions since around the doping levels where the metallic transition takes place, the widths of the polaron (or soliton) bands are very dependent on the actual doping levels. Indeed, the widths of the defect wave functions being of the order of 15-20 sites, slight local modifications in the actual doping level can significantly change the overlap between defect wave functions.

3.7.3.2 Polythiophene

The band structure of polythiophene in the bipolaron and polaron lattice conformations for y=0.25 are shown in Figure 3.32(a) and 3.32(b), respectively. Note that the length of the reciprocal unit cell in the bipolaron lattice is about half the one of the polaron lattice and contains twice as many bands.

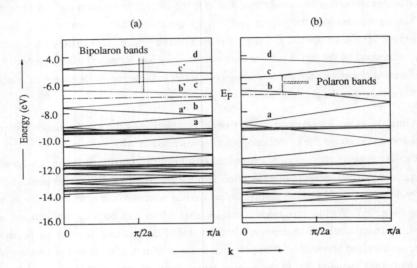

Figure 3.32 VEH band structures for polythiophene at doping level y=0.25: (a) the bipolaron lattice and (b) the polaron lattice

A comparison of the main features of the band structures for the two lattice conformations calls for the following remarks:

(i) Conceptually, if we go from the polaron lattice (Figure 3.32(b)) to the bipolaron lattice (Figure 3.32(a)), the two polaron bands (b and c in Figure 3.32(b)) are split into four bands (b, b', c, and c' in Figure 3.32(a)), the middle two (bands b' and c) residing in the gap as bipolaron bands; it is therefore apparent that the π-π* gap (between bands b and c') in the bipolaron lattice is smaller than that (between bands a and d) in the polaron lattice.

(ii) The Fermi level (or chemical potential) appears at about the same energy in both cases.

(iii) The bipolaron bands lie deeper into the intrinsic bandgap.

The differences in the band structure between the two lattices suggest a description in which *the bipolaron lattice is the result of a dimerization of the polaron lattice*.

The widths of the lower and upper bipolaron bands are 0.28 eV (band b' in Figure 3.32(a)) and 0.26 eV (band c), respectively, i.e., the electronic structure is rather symmetric around midgap. This feature also holds for the polaron bands, which are calculated to be 0.98 eV (band b) and 0.85 eV (band c) wide. For both lattice conformations, however, there is a quite large discrepancy as concerns the wave functions associated with the defect states. Whereas, in the upper defect band, the wave functions have a considerable contribution from the sulfur 2p atomic orbitals, there are only carbon contributions to the wave functions of the lower defect band.

The first interband transition in the bipolaron lattice, between bands b and b' (Figure 3.32(a)), appears at 0.5 eV. It is evident from the comparison between the polaron and bipolaron lattice band structures presented in Figure 3.32 that this transition has its correspondence in an intraband absorption of the half-filled polaron band. Such an absorption onsets at zero energy and can extend up to about 1.0 eV, i.e., the polaron bandwidth. These predictions for the polaron lattice are in agreement with the experimental results, which for y~0.25 indicate an optical absorption peaking at 0.6 eV, with a long tail into the IR regime [3.129].

For higher photon energies, we note that the bipolaron lattice would present a transition between bands b and c with an onset at 1.50 eV. It is interesting to study the evolution of this transition; for y=0.20, it is found at 1.47 eV, whereas for y=0.33, it onsets at 1.53 eV. Thus, this transition is almost independent of the doping level. A very similar behaviour is observed in optical absorption spectra on dissolved poly(3-hexylthiophene) doped up to y=0.36 [3.77]. Very interestingly, no metallic properties are found in this system suggesting that it remains in the bipolaron conformation even at very high doping levels. An absorption peaking at ~1.5 eV is observed for doping levels between 10 and 36%, in excellent agreement with the VEH value for the transition between the valence band and the upper bipolaron band. In contrast, for polythiophene in the solid state, which exhibits metallic properties in the highly doped state, a shift from 1.6 eV to 1.9 eV is observed in the doping range y=0.14 to y=0.20 [3.121]. This shift compares much more favourably with the evolution of the transition between the polaron bands (bands $\underline{b}$ and $\underline{c}$ in Figure 3.32(b)), which is located at 1.73 eV for y=0.10, 1.93 eV for y=0.25, and 2.18 eV for y=0.33. Furthermore, in the polaron lattice conformation there are also transitions, first between bands $\underline{a}$ and $\underline{b}$ (onset at 1.4 eV) and also between bands $\underline{a}$ and $\underline{c}$ (onset at 2.8 eV). These transitions explain qualitatively the finite absorption observed in the whole energy region up to ~2.5 eV. Finally, we observe that the π-π* transition in the polaron lattice has its correspondence in transitions between states deep into the valence and conduction bands of the bipolaron lattice, and consequently, appears at a much higher energy than in the bipolaron lattice. The transition is calculated to onset at 2.9 eV. No sharp transition is experimentally found in the 2-3 eV region [3.115, 3.129], in better agreement with the polaron lattice model.

A schematic representation of the soliton, polaron and bipolaron band structures in the limit of wide defect bands is given in Figure 3.33. In this Figure, we have qualitatively taken into account the properties of the detailed band structure description presented above. The closure of the single particle gap as the system is driven from a "dimerized" bipolaron lattice conformation towards a polaron lattice conformation is clearly illustrated, as well as the relation between the bandwidths and the bandgaps for the respective configurations.

We stress that the polaron bands in trans-$(CH)_x$ and PT are both calculated to be 1.0 eV wide. Actually, by comparing the shape of the polaron bands in these two polymers (Figure 3.31(b) for trans-$(CH)_x$ and Figure 3.32(b) for PT), it is hard to notice any significant differences, the narrow subgaps between bands $\underline{a}$ and $\underline{b}$, and bands $\underline{c}$ and $\underline{d}$ being also very similar. This similarity originates from the facts that: (i) there is almost no sulfur contribution to the lower polaron band in polythiophene; (ii) the carbon backbone in polythiophene resembles that of cis-polyacetylene; and (iii) the charge per carbon introduced by the doping is formally the same (0.0625) in both cases.

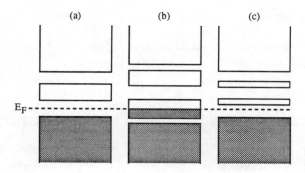

Figure 3.33 Schematic representation of the band structure at high doping level for (a) the soliton lattice, (b) the polaron lattice, and (c) the bipolaron lattice

As a result of the similarities pointed out above between the polaron lattice band structures in trans-$(CH)_x$ and PT, it is expected that, if the polaron lattice model is correct, the experimental absorption spectra of trans-$(CH)_x$ and PT should be very similar. This is actually the case if we compare, for example, the data given for trans-$(CH)_x$ [3.115] and for PT [3.121, 3.129]. At high doping, both polymers exhibit an absorption tail extending into the far-IR region and a rather symmetric peak at 0.6 eV (which can be understood in terms of an intraband transition within the half-filled polaron band) as well as very broad absorptions in the 1-3 eV region.

Finally, we note that UPS has recently been used to observe changes in the π-band structure of a PT derivative, poly(3-hexylthiophene), induced by doping using NOPF$_6$. A charge-induced movement of the Fermi energy and a finite density of states at the Fermi energy are seen unambiguously in the most highly doped material [3.130]. The results are consistent with the polaron lattice calculations and suggest the appearance of a polaron-lattice state at saturation-doping level [3.130].

254

3.7.3.3 Emeraldine salt

The band structures for the bipolaron lattice and the polaron lattice in emeraldine salt (ES) are presented in Figure 3.34(a) and 3.34(b), respectively.

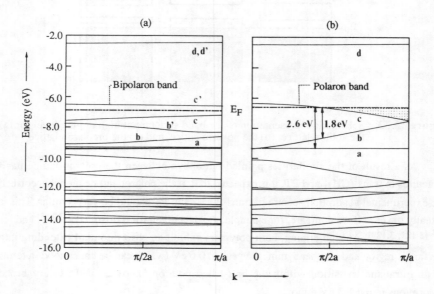

Figure 3.34 VEH band structure of the emeraldine salt for (a) the bipolaron lattice and (b) the polaron lattice. The unit cell for the polaron lattice, *a*, is half the one for the bipolaron lattice

The unit cell for the bipolaron lattice is twice that of the polaron lattice, as in the case of PT. Again, it is observed that the almost perfect electron-hole symmetry present in trans-$(CH)_x$ and PT is not present in ES. *Only one bipolaron and one polaron band* (bands c' and c in Figure 3.34(a) and 3.34(b), respectively) *appear deep in the gap, instead of the two deep defect bands found in the other conducting polymers studied so far.* Therefore, the schematic representation of the bipolaron and polaron band structures given in Figure 3.33 does not apply to emeraldine. Actually, the second conduction band in ES is totally flat and as a result, the defect (polaron or bipolaron) band originating from it remains very shallow and is hardly distinguishable from it. This has interesting consequences since the first optical transitions, which usually involve both defect bands, are expected here to involve only the lower defect band.

Even though the doping level is high, we observe that the lower bipolaron band is narrow, its width being merely 0.1 eV. This is due to the fact that the bipolaron defect is mostly localized on a single quinoid-like ring. For the polaron conformation, the polaron bandwidth is larger, in our calculation 1.1 eV. The first interband transition in the bipolaron lattice occurs between bands $\underline{c}$ and $\underline{c}'$ and onsets at 0.6 eV. Transitions including higher unoccupied bands do not appear for energies below 3 eV. Instead, there would be low lying absorptions due to the promotion of electrons from deeper valence bands into the lower bipolaron band. In particular, a strong absorption to be due to the transition between the very flat band $\underline{a}$ and the likewise flat bipolaron band is expected at 2.7 eV. As was discussed for trans-$(CH)_x$ and PT, we expect the half-filled polaron band to allow for intraband transitions up to roughly 1.1 eV, the polaron bandwidth. This transition should be followed by interband transitions between bands $\underline{b}$ and $\underline{c}$ (1.8 eV), and bands $\underline{a}$ and $\underline{c}$ (2.6 eV).

The experimental absorption data for ES show a broad asymmetric absorption peaked at 1.5 eV containing a long absorption tail extending into the IR region [3.131]. The shape of this absorption is different from the low energy peaks in trans-$(CH)_x$ and PT. Our interpretation is that the long tail at the low energy side of the peak originates from the intraband absorption in the half-filled polaron band. This absorption is smoothly connected with the $\underline{b}$ to $\underline{c}$ interband transition, which we calculate to be at 1.8 eV. The width of the IR absorption is experimentally determined to be ~1.0 eV, in very good agreement with our calculated width of 1.1 eV for the polaron band [3.131]. Contrary to what is observed in trans-$(CH)_x$ and PT, a second clear absorption is seen in emeraldine salt. This absorption peaks at 2.8 eV, i.e., in agreement with the $\underline{a}$ to $\underline{c}$ transition of the polaron lattice.

Up to now, there are no direct structural optimizations of protonated emeraldine which have been able to include the formation of a polaron lattice. It is questionable, if methods based on the Roothaan-Hartree-Fock equations (like the MNDO and AM1 methods we have used to optimize the geometry of oligomeric equivalents to the polymers) can be used in this context, since it is well known that the Hartree-Fock theory gives a nonphysical vanishing density of states at the Fermi level [3.132], i.e., in the case of a polaron lattice, at the center of the polaron band. Instead, the local density approach, which has been very successful in calculating electronic properties of ordinary metals, might be more promising for the calculation of structural properties of a true polaronic metal. However, despite the fact that no direct proof of a structural relaxation into a polaron lattice upon protonation of EB is available, the results above show that the electronic properties for a polaron lattice conformation of ES, based on the geometry of an oligomeric system, is in excellent agreement with both optical and magnetic data.

It should be stressed again that doping by means of protonation, like the conversion of EB into ES, does not alter the number of electrons in the polymeric system. Therefore, it is the response of the lattice to protonation which is the only cause of the distortion. If the polaron lattice model is correct, *this distortion is commensurate with respect to the nitrogen to proton (or counterion) ratio, i.e., 2 for protonated emeraldine, instead of commensurate to the bandfilling, i.e., 4 (for both EB ansd ES since the number of electrons is unchanged), as for ordinary Peierls systems.* Very interestingly, the same type of argument can also be applied to the formation of polaron lattices in polyacetylene and polythiophene. For these polymers the soliton and bipolaron lattices, respectively, constitute the (Peierls) conformations commensurate to the bandfilling. There is no chemical modification of the polymer which can set up a potential commensurate with the polaron lattice; however, several types of interactions which might cause an electronic transition into a polaron lattice conformation have already been discussed previously. In addition to these, we note that a periodic arrangement of the dopant ions can set up a potential commensurate to the polaron lattice. Therefore, in future theoretical studies of the polaron lattice, it is essential to investigate the effects of dopant ions on both the structural and electronic properties of highly doped conducting polymers, as done, for example, by Conwell et al. [3.106].

In this Section, we have discussed the electronic and structural properties of three categories of highly conducting polymers: trans-polyacetylene, which is the prototype of a system with a degenerate ground state; polythiophene, as one example of a system with a nondegenerate ground state; and emeraldine which has the special property that it can be doped by means of protonation. The electron transfer doping of trans-$(CH)_x$ results in soliton formation and a corresponding soliton band is formed in the electronic structure. For low doping levels, electronic excitations involving states in this band appear at roughly half the energy of the intrinsic bandgap of undoped trans-$(CH)_x$. The evolution calculated for this absorption, as a function of doping concentration, is found to be in very good agreement with experimental observations. The experimental findings in terms of optical absorption and magnetic properties for lightly doped PT are consistent with the electronic structure for an array of essentially noninteracting polarons. For higher doping levels, the optical spectrum for PT first changes in a way consistent with the recombination of pairs of polarons into bipolarons whereas at saturation doping, a polaron lattice seems to be present. Protonated emeraldine as well as highly doped trans-$(CH)_x$ and PT exhibit metallic properties. Our results indicate that the electronic structure of a polaron lattice is consistent with the electronic properties observed in the metallic state for all three polymers.

It must be stressed that, at the present stage, no definite consensus has been reached about the detailed nature of the metallic-like regime at very high doping levels. The polaron lattice model is one of the candidates but is far from being fully agreed upon for all polymers. In our opinion, elucidation of the transport phenomena at high doping is one of the most challenging tasks to be taken up in the 90's. One of the main problems facing the application of quantum chemical Hartree-Fock approaches in this respect is the difficulty of providing a correct description of metallic situations such as those corresponding to the formation of a polaron lattice. In this context, the future might be bright for the application of treating electron correlation effects explicitly.

3.8 EVOLUTION OF THE GEOMETRIC AND ELECTRONIC STRUCTURES IN THE FIRST B_u EXCITED STATE OF POLYACETYLENE CHAIN

Following the pioneering work of Ducuing and co-workers [3.133], it has been realized that conjugated polymers inherently possess very high nonlinear optical responses

[3.134, 3.135]. It has recently been pointed out that the influence of electron-lattice coupling and e.g., of the presence of nonlinear elementary excitations such as solitons could also be significant for the nonlinear optical properties [3.136, 3.139]. In all-trans-polyacetylene, photoexcitation across the gap into the $1B_u$ state produces electron-hole pairs that are found to decay very rapidly (in 10^{-13} sec.) into pairs of separated, positively and negatively charged, solitons [3.140-3.142]. Such an evolution is in agreement with the early theoretical predictions of Su and Schrieffer [3.143]. It is important to note that this process results in an efficient charge separation mechanism, a feature essential to provide large optical nonlinearities [3.144]. Furthermore, in the off-resonance regime, instantons (i.e., virtual soliton-antisoliton pairs) [3.145] have been suggested to yield enhanced optical nonlinearities with respect to a purely rigid lattice situation [3.136, 3.138].

In short polyenes, spectroscopic studies indicate that strong geometry relaxations also take place in the first B_u one-photon optically-allowed excited state [3.146, 3.147]. These geometry relaxations are traditionally modeled theoretically using Bond Order/Bond Length (BOBL) relationships [3.25]. In this framework, the geometry relaxations follow exactly the wave function characteristics in the excited state. They are calculated to become smaller as the chain length increases and tend to be neglibible for chains containing over 15-20 carbon atoms [3.148, 3.149]. Such an evolution is, however, inconsistent with the situation appearing in very long chains, i.e., polyacetylene, where the first B_u state strongly relaxes to produce a pair of charged solitons, as mentioned above [3.140, 3.141].

In this Section, we review the calculations we have performed in order to remove the inconsistency between the traditional understanding of the $1B_u$ relaxation process in short polyenes and the perception of the situation prevailing in all-trans-polyacetylene and provide a coherent picture of the evolution between short and long polyenes. First, we investigate the relaxation process in the $1B_u$ excited state of polyene chains by means of the methodology of Su, Schrieffer, and Heeger [3.30] , as adapted by Brédas et al. [3.34, 3.35]. We make use of a Hückel Hamiltonian with bond length dependent transfer integrals and σ-bond compressibility, as described in Section 3.3. We study chains ranging from 10 to 58 carbon atoms in order to understand the evolution of the relaxed geometry as a function of chain length. Second, calculations are carried out at a much higher level of sophistication (combined Restricted Hartree-Fock *ab initio*/Pariser-Parr-Pople

Configuration Interaction level) for three polyene molecules: hexatriene (C_6H_8), decapentaene ($C_{10}H_{12}$), and tetradecaheptaene ($C_{14}H_{16}$). We are in this way able to assess our results on a firmer ground.

3.8.1 Su-Schrieffer-Heeger Hamiltonian Approach

In order to be in a position, on the one hand, to examine the evolution in going from short to long polyenes and, on the other hand, to make a significant comparison to polyacetylene, we have chosen to work first at the Su-Schrieffer-Heeger Hamiltonian level (see Section 3.3.1). The parameters originally optimized for polyacetylene are applied to polyene chains ranging in size from 10 to 58 carbon atoms [3.139]. We recall that these parameters lead to: (i) a degree of bond length alternation of 0.14 Å in the ground state (single bond equal to 1.47 Å and double bond equal to 1.33 Å), (ii) a (somewhat underestimated) bandgap of 1.4 eV in trans-polyacetylene, and (iii) a creation energy of 0.92 eV for a pair of totally separated solitons [3.139].

We search the configuration space to find the geometry providing the lowest $1B_u$ excited-state total energy. Two types of situation are mostly investigated:

(i) We allow for geometries similar to those obtained with traditional BOBL relationships in the framework of Pariser-Parr-Pople (PPP) calculations including configuration interaction. The amplitude and the extent of the relaxation are both optimized.

(ii) We look for geometries resulting in the formation of a soliton pair (two-soliton geometries). Both the distance between the solitons and the soliton widths are optimized.

The results corresponding to the two-soliton geometries are given in Table 3.1 for chains containing from 10 to 58 carbon atoms. In the longest chain investigated here (58 carbon atoms), results are identical to those obtained for the infinite polyacetylene chain. The $1B_u$ excited-state geometry relaxes to form a pair of solitons (with a creation energy of 0.922 eV), which lowers the $1B_u$ excited-state total energy by 38% relative to the vertical excitation energy. The solitons are calculated to be optimally located on sites 19 and 40, i.e., they are as distant from one another as from the chain ends. Each soliton is found to extend over about 15 carbons, as in trans-polyacetylene.

As the chain length decreases, the total energy lowering due to the relaxation of the $1B_u$ excited state becomes smaller. However, even in the case of decapentaene, this energy lowering due to a soliton pair formation still amounts to 23% of the vertical transition energy.

Very importantly, the relaxed excitation energies for excited-state geometries calculated on the basis of BOBL relationships are 0.2-0.3 eV larger than in the two-soliton formation situation. The major result of these Hückel-like calculations is thus to find that the relaxation effects in the $1B_u$ excited state of polyene molecules are qualitatively similar to those in polyacetylene. *Even in the shortest chains considered here, the $1B_u$ excited-state is found to relax optimally in such a way as to produce a pair of solitons.* In order to be accommodated in shorter chains, the solitons shrink in size: their width decreases from $2\ell=15$ for trans-polyacetylene, to $2\ell=11$ for chains containing between 30 and 50 carbon atoms, $2\ell=7$ for chains between 20 and 30 carbon atoms, and $2\ell=3$ in chains containing less than 20 carbons. In all cases, the solitons tend to be centered on locations separating them equally from one another and from the chain ends. The evolution between short and long polyene chains is thus found to be fully coherent.

It is important to stress that the two-soliton formation in the $1B_u$ excited state of decapentaene appears to agree better with experimental data than does the situation where the $1B_u$ excited-state geometry is described in terms of BOBL relationships. In the former case, the average bond-length variations between the $1B_u$ excited state and the ground state are calculated to be 0.070 and 0.087 Å for the double and single bonds, respectively. In the latter case, the corresponding values are 0.043 and 0.047 Å [3.148]. These results are to be compared with the experimental estimates of Granville et al. [3.147] which provide average variations of 0.085 and 0.081 Å for double and single bonds, respectively. Furthermore, the relaxation energy experimentally measured in the $1B_u$ state of decapentaene (at 77 K) is on the order of 0.8 eV [3.147]. Such a relaxation represents 20% of the vertical excitation energy [3.147]. Within the two-soliton configuration, the relaxation is calculated to be about 23% of the vertical transition energy, while it is only 10-12% in the BOBL-derived $1B_u$ excited state [3.148].

Table 3.1 **Evolution as a function of chain length n of:**
(i) the vertical excitation energy (in eV) to the first B_u excited state, E_{vert};
(ii) the corresponding relaxed excitation energy (in eV), E_{rel};
(iii) the relative total energy lowering (in %) due to the relaxation of the $1B_u$ excited state;
(iv) the optimal .i.soliton width;, 2ℓ (in number of sites);
(v) the optimized site locations of the soliton defects, n_1 and n_2; and
(vi) the HOMO-LUMO separation (in eV) in the relaxed geometry.
The calculations are performed in the framework of the Su-Schrieffer-Heeger Hamiltonian

n	E_{vert}	E_{rel}	%	2ℓ	n_1,n_2	HOMO-LUMO
58	1.477	0.922	37.6	15	19,40	0.14
46	1.515	0.939	38.0	11	15,32	0.20
38	1.557	0.983	36.9	11	11,28	0.20
26	1.682	1.105	34.3	7	9,18	0.30
22	1.760	1.178	33.1	7	7,16	0.55
18	1.876	1.340	28.6	3	5,14	0.70
14	2.064	1.541	25.3	3	5,10	0.98
10	2.400	1.853	22.8	3	3,8	1.16

The overall picture obtained within the simple Su-Schrieffer-Heeger Hamiltonian is that there exists a smooth evolution of the relaxation of the $1B_u$ excited-state geometry in going from short polyenes to polyacetylene. This evolution is consistent with the strong geometry relaxations experimentally observed in short polyenes (e.g., decapentaene) as well as polyacetylene. In all cases, the relaxation is such as to produce the formation of a soliton-antisoliton pair on the chain.

It is however important to address the question of the $1B_u$ excited-state relaxation using a more sophisticated theoretical approach, in particular taking explicitly into account the effects of electron-electron interactions and electron correlation. The calculations have therefore been extended on short polyenes, using a combined PPP Configuration Interaction and Hartree-Fock *ab initio* approach.

3.8.2 PPP-CI/RHF *ab initio* Approach

PPP Hamiltonians are usually suitably parameterized to reproduce excited state transition energies. However, they do not provide reliable geometry optimizations in terms of total energy differences. On the other hand, if *ab initio* techniques (which can afford good total energy differences) can be used quite easily to optimize ground-state geometries, they prove to be difficult to apply for the optimization of excited-state geometries in large systems.

Therefore, we have seeked to combine these two techniques in order to be able to perform calculations on relatively large molecules and to assess more reliably the trends in the first B_u excited state geometry [3.150]. We have thus considered: (i) geometry optimizations at the Hartree-Fock 3-21G split valence basis set level, which allow for a correct sketch of the *ground-state potential energy curve;* and (ii) PPP Configuration Interaction (CI) calculations including single excitations from the ground-state reference framework and based on the various 3-21G optimized geometries. The PPP parameters we have used are chosen in such a way as to provide a degree of bond-length alternation in the ground state almost identical to that calculated at the RHF 3-21G level, see below. Although it does not predict a correct ordering of the excited states of polyene molecules, a single CI approach is known to describe adequately the first B_u excited state [3.148]. Combining the ground state total energies with the PPP vertical transition energies, leads to the *relative excited-state total energies* corresponding to various geometrical situations.

We have considered three linear polyene molecules: hexa-1,3,5-triene (C_6H_8), deca-1,3,5,7,9-pentaene ($C_{10}H_{12}$) and tetradeca-1,3,5,7,9,11,13-heptaene ($C_{14}H_{16}$). The RHF *ab initio* calculations are performed *in the ground state* for three different geometry situations:

(i) In the first case, we carry out a full (i.e., all bond lengths and valence angles) 3-21G optimization of the ground-state geometry assuming coplanar conformations.

(ii) In the second case, we consider the geometry corresponding to the relaxed $1B_u$ excited-state, obtained using BOBL relationships with the PPP Hamiltonian [3.24]. All bond angles are optimized at the *ab initio* 3-21G level, while the carbon bond lengths are kept at their PPP B_u-state values and the carbon-hydrogen bond lengths are set at their optimal *ab initio* ground-state values.

(iii) In the third case, we partly optimize the geometry corresponding to the formation of a pair of solitons, as obtained at the Hückel/SSH Hamiltonian level described previously. Here, we force the two carbon-carbon bonds surrounding the soliton centers to be of equal lengths. The values of these bond lengths and those of the bonds towards the ends of the molecules are optimized. The carbon-carbon bonds located between the solitons are chosen to be identical to the single and double bonds appearing in the 3-21G optimized ground-state geometry. All the bond angles are fully optimized, while the carbon-hydrogen bonds are fixed at the *ab initio* ground-state optimal values (Table 3.2).

Table 3.2 **Optimized Hartree-Fock 3-21G carbon-carbon bond lengths for the ground state of hexatriene, decapentaene, and tetradecaheptaene.(in Å)**
The numbers between parentheses for decapentaene refer to the PPP Hamiltonian-optimized values. The atoms are labeled starting from one end of the molecule

	C_6H_8	$C_{10}H_{12}$	$C_{14}H_{16}$
$R(C_1-C_2)$	1.322	1.322 (1.325)	1.322
$R(C_2-C_3)$	1.462	1.461 (1.465)	1.461
$R(C_3-C_4)$	1.327	1.329 (1.330)	1.329
$R(C_4-C_5)$		1.456 (1.463)	1.455
$R(C_5-C_6)$		1.330 (1.332)	1.331
$R(C_6-C_7)$			1.454
$R(C_7-C_8)$			1.331

In Table 3.2, we present the ground-state geometries of hexatriene, decapentaene, and tetradecaheptaene, as optimized at the Hartree-Fock 3-21G level. The decapentaene PPP values are also indicated. The PPP values are obtained by using the following bond order (l_{pq})/ bond length (R_{pq}, in Å) relationship: $R_{pq} = 1.51 - 0.19\ l_{pq}$.

The evolution of the degree of bond-length alternation along the polyene chains obtained on the basis of BOBL relationships are depicted in Figure 3.35.

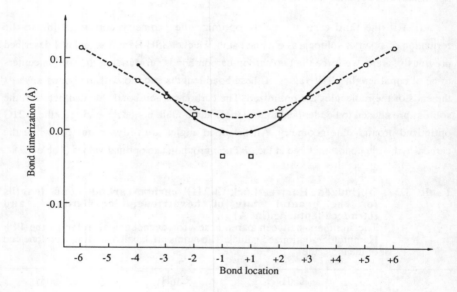

Figure 3.35   Ilustration of the evolution of the bond dimerization value (in Å) along the molecules of hexatriene (squares), decapentaene (closed circles, solid line), and tetradecaheptaene (open circles, dashed line), in the BOBL-derived relaxed geometry for the first B_u excited state. The sign of the dimerization is by convention taken to be positive if it is the same as in the ground state. The bonds are labeled starting from the central bond of the molecule

It is clearly observed that in this framework the geometry modifications in the first singlet B_u excited state strongly decreases with increasing chain length. In hexatriene, all bonds are significantly affected by the excitation; in the middle of the molecule, the sign of bond dimerization reverses. This is only slightly the case in decapentaene, where the two central bonds are almost equal and the bond alternation for the outer bonds reaches +0.009 Å. In tetradecaheptaene, however, the sign of the bond alternation remains positive all along the molecule, indicating no single bond-double bond character reversal. (Note that by convention, we consider the sign of the bond alternation to be positive when it is the same as in the ground state).

In Figure 3.36, we present the results obtained when considering the possibility of formation of a soliton-antisoliton pair as the $1B_u$ state relaxes.

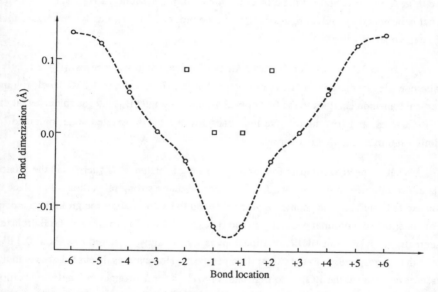

Figure 3.36   Ilustration of the evolution of the bond dimerization value (in Å) along the molecule of hexatriene (squares), decapentaene (closed circles and dashed line), and tetradecaheptaene (open circles, dashed line) in the two-soliton relaxed geometry for the first B_u excited state. The sign of the dimerization is by convention taken to be positive if it is the same as in the ground state. The bonds are labeled starting from the central bond of the molecule

A comparison of Figure 3.35 and 3.36 illustrates that the two-soliton geometry leads to bond length modifications relative to the BOBL-derived geometries, which are smaller in the hexatriene case but much larger in the longer chains. We note that for hexatriene, it is actually rather misleading to speak in terms of the formation of a two-soliton configuration since the soliton and the antisoliton are localized on adjacent sites. Hexatriene is therefore too short a chain to provide for a complete bond alternation reversal in the middle of the molecule when considering a two-soliton configuration. Such a reversal, however, clearly occurs in decapentaene and tetradecaheptaene. In decapentaene, the average bond-length modification with respect to the optimal ground-state geometry is 0.073 Å. We stress again that this value is in much better agreement with the 0.08 Å experimental estimate of Granville et al. for the relaxed B_u state [3.147] than what is provided by BOBL relationships (on the order of 0.04 Å, i.e., twice smaller than the experimental value).

In Table 3.3, we list for the three kinds of investigated geometries: (i) the relative total energies in the ground state, as obtained at the Hartree-Fock 3-21G level; (ii) the vertical transition energies to the first optically-allowed singlet B_u state, as calculated at the PPP-CI level, and (iii) the relative total energies in the $1B_u$ excited state, obtained by simply summing the first two terms.

What we need to compare for our purpose are the relative total energies in the excited state. For all three molecules, we find as expected that a strong relaxation takes place in the excited state, i.e., the geometry which is found to be optimal for the ground state does not constitute the optimal geometry in the B_u excited state. In hexatriene, the B_u relaxed geometry derived from BOBL relationships is significantly favoured (by about 0.3 eV) over that corresponding to the two-soliton geometry. However, as discussed above, in the case of hexatriene, the BOBL configuration provides for a stronger geometry relaxation and a two-soliton configuration is not truly achieved. In decapentaene, the two-soliton geometry becomes very close in energy (within 0.05 eV) with respect to the BOBL geometry for the excited state. In both configurations, the relaxation relative to a vertical process is found to be on the order of 0.9 eV, a value in excellent agreement with the 0.84 eV value which is experimentally measured [3.147]. Importantly, in the case of tetradecaheptaene, the two-soliton geometry corresponds to the most stable situation for the excited-state, being about 0.07 eV lower in energy than the BOBL geometry. The gain in stability of the two-soliton configuration with respect to the BOBL configuration as chain length increases is fully consistent with the experimental observation on the photogeneration of soliton pairs in the $1B_u$ excited state of polyacetylene. In long chains, a BOBL-derived configuration would in contrast lead to negligible relaxation.

Table 3.3   Comparison of the energies involved in hexatriene, decapentaene, and tetradecaheptaene for the three geometry situations investigated in this work:
(A) relative RHF *ab initio* total energies in the ground state (the fully optimized RHF *ab initio* 3-21G value being taken as reference);
(B) PPP-CI vertical transition energies to the first singlet B_u state;
(C) relative total energies in the $1B_u$ excited state (calculated by summing the first two terms). All energies are given in eV.

	A	B	C
hexatriene			
ab initio ground state	0.00	5.86	5.86
BOBL	0.63	4.14	4.77
two-soliton	0.27	4.83	5.10
decapentaene			
ab initio ground state	0.00	5.03	5.03
BOBL	0.51	3.56	4.07
two-soliton	1.19	2.93	4.12
tetradecaheptaene			
ab initio ground state	0.00	4.65	4.65
BOBL	0.43	3.37	3.80
two-soliton	1.08	2.65	3.73

An interesting feature is uncovered when (i) observing in Figures 3.35 and 3.36 the evolution in bond length alternation for decapentaene and tetradecaheptaene; and (ii) comparing it to the geometry relaxation process due to a photogenerated electron-hole pair in polyacetylene (see Figure 2a of Ref. 3.145). From this observation, we suggest that the configuration given by BOBL relationships corresponds to the *early stage* of the process of electron-hole separation into a pair of solitons whereas the two-soliton configuration relates to the *final stage* of the separation process. This separation is actually induced through the explicit coupling of the electronic structure to the lattice, an ingredient which is absent from the BOBL approach where the geometry has to follow the wave function characteristics. It is worth pointing out that in the two-soliton configuration, the phase of the bond alternation pattern between the locations of the soliton and the antisoliton is opposite to that of the wave function bonding-antibonding pattern.

These results obtained from the combined RHF *ab initio*/PPP-CI approach thus confirm that in linear polyenes containing at least ten carbon atoms, the electron-lattice coupling plays a significant role in the relaxation of the first B_u excited state. We stress that the fast relaxation of the $1B_u$ state related to the decay of a photoinduced electron-hole pair into a pair of charged solitons has been invoked to be an important factor in the large nonlinear optical response of polyacetylene. Such a decay indeed provides major shifts in oscillator strengths and a very effective charge separation mechanism, which are essential to large hyperpolarizabilities.

Our work thus suggests that going beyond frozen geometry models of the polarizabilities and hyperpolarizabilities by incorporating electron-lattice coupling could also prove to be essential to describe properly the optical nonlinearities in short and intermediate-sized polyenes [3.151, 3.152], as well as in other oligomers where electron-lattice coupling effects are known to be important (polypyrrole, polythiophene,...).

We note that the two-soliton geometry relaxation leads to the appearance of new electronic states in the gap. This feature leads to important shifts in oscillator strengths. It would be therefore most interesting to have photoinduced absorption experiments carried out in the subpicosecond regime. Such measurements indeed constitute an ideal means to probe the energies and fast time evolution of the new optical absorptions resulting from such a relaxation in linear polyenes.

As a general conclusion to this third part of the book, we would like to stress that quantum-chemical polymer methods have proven very useful in addressing the electronic properties of conjugated chains. In particular, they have allowed us to determine the very strong relationship existing between the geometric structure and the electronic structure in these systems, both in the ground state and in the excited states, to gain a deep insight into the mechanisms leading to the appearance of high electrical conductivity upon redox or protonic acid doping of conjugated polymers and to offer some predictions in terms of novel macromolecular architectures that could result in low bandgap characteristics.

270

References

[3.1] "Handbook of Conducting Polymers", (T.A. Skotheim, Ed.), Marcel Dekker, New York (1986).

[3.2] Proceedings of the International Conferences on Synthetic Metals (ICSM'86, ICSM'88, ICSM'90), published in Synthetic Metals.

[3.3] (a) "Conjugated Polymeric Materials: Opportunities in Electronics, Optoelectronics, and Molecular Electronics", (J.L. Brédas and R.R. Chance, Eds.), Kluwer, Dordrecht (1990).

(b) "Conjugated Polymers: The Novel Science and Technology of Highly Conducting and Nonlinear Optically Active Materials", (J.L. Brédas and R. Silbey, Eds.), Kluwer, Dordrecht (1991).

[3.4] A.J. Heeger, S. Kivelson, J.R. Schrieffer, and W.P. Su, Rev. Mod. Phys., **60**, 781 (1988); J.L. Brédas and G.B. Street, Acc. Chem. Res., **18**, 309 (1985).

[3.5] H. Naarmann, Naturwissenschaften, **56** Heft 6, 308 (1969).

[3.6] V.V. Walatka, M.M. Labes, and J.H. Perlstein, Phys. Rev. Lett., **31**, 1139 (1973).

[3.7] R.L. Greene, G.B. Street, and L.J. Suter, Phys. Rev. Lett., **34**, 577 (1975).

[3.8] W.D. Gill, W. Bludau, R.H. Geiss, P.M. Grant, R.L. Greene, J.J. Mayerle, and G.B. Street, Phys. Rev. Lett., **38**, 1305 (1977).

[3.9] H. Shirakawa, E.J. Louis, A.G. MacDiarmid, C.K. Chiang, and A.J. Heeger, J. Chem. Soc. Chem. Commun. 578 (1977); C.K. Chiang, C.R. Fincher, Y.W. Park, A.J. Heeger, H. Shirakawa, E.J. Louis, S.C. Gau, and A.G. MacDiarmid, Phys. Rev. Lett., **39**, 1098 (1977).

[3.10] J.C. Chiang and A.G. MacDiarmid, Synth. Met., **13**, 193 (1986).

[3.11]   R.L. Elsenbaumer, K.Y. Jen, and R. Oboodi, Synth. Met., **15**, 169 (1986).

[3.12]   E.E. Havinga, L.W. van Horssen, W. ten Hoeve, H. Wynberg, and E.W. Meijer, Polymer Bull., **18**, 277 (1987); A.O. Patil, Y. Ikenoue, F. Wudl, and A.J. Heeger, J. Am. Chem. Soc., **109**, 1858 (1987).

[3.13]   J.H. Edwards and W.J. Feast, Polymer Commun., **21**, 595 (1980); J.H. Edwards, W.J. Feast, and D.C. Bott, Polymer, **25**, 395 (1984).

[3.14]   J.H. Burroughes, C.A. Jones, and R.H. Friend, Nature, **335**, 137 (1988).

[3.15]   J.H. Burroughes, C.A. Jones, R.A. Lawrence, and R.H. Friend, in "Conjugated Polymeric Materials: Opportunities in Electronics, Optoelectronics and Molecular Electronics", (J.L. Brédas and R.R. Chance, Eds.), p.221, Kluwer Academic Publishers, Dordrecht (1990).

[3.16]   D. Ofer and M.S.Wrighton, J. Am. Chem. Soc., **110**, 4466 (1988).

[3.17]   T. Ito, H. Shirakawa, and S. Ikeda, J. Polym. Sci., Polym. Chem. Ed., **12**, 11 (1974).

[3.18]   K. Akagi, S. Katayama, M. Ito, H. Shirakawa, and K. Araya, Synth. Met., **28**, D51 (1989).

[3.19]   H. Naarmann and N. Theophilou, Synth. Met., **22**, 1 (1987); J. Tsukamoto, A. Takahashi, and K. Kawasaki, Jap. J. Appl. Phys., **29**, 125 (1990).

[3.20]   F.L. Klavetter and R.H. Grubbs, J. Am. Chem. Soc., **110**, 7807 (1988); E.J. Ginsburg, C.B. Gorman, S.R. Marder, and R.H. Grubbs, J. Am. Chem. Soc., **111**, 7621 (1989); E.J. Ginsburg, C.B. Gorman, R.H. Grubbs, F.L. Klavetter, N.S. Lewis, S.R. Marder, J.W. Perry, and M.J. Sailor, in "Conjugated Polymeric Materials: Opportunities in Electronics, Optoelectronics and Molecular Electronics", (J.L. Brédas and R.R. Chance, Eds.), p. 65, Kluwer Academic Publishers, Dordrecht (1990).

272

[3.21] C. Schaverien, J. Dewan, and R.R. Schrock, J. Am. Chem. Soc., **108**, 2771 (1986).

[3.22] J. Cheung, R.B. Rosner, M.F. Rubner, X.Q. Yang, J. Chen, and T.A. Skotheim, in "Conjugated Polymeric Materials: Opportunities in Electronics, Optoelectronics and Molecular Electronics", (J.L. Brédas and R.R. Chance, Eds.), p. 91, Kluwer Academic Publishers, Dordrecht (1990).

[3.23] T. Ito, H. Shirakawa, and S. Ikeda, J. Polym. Sci., Polym. Chem.Ed., **13**, 1943 (1975).

[3.24] J.L. Brédas, R.R. Chance, R. Silbey, G. Nicolas, and Ph. Durand, J. Chem. Phys., **75**, 255 (1981).

[3.25] L. Salem, "The Molecular Orbital Theory of Conjugated Systems", Benjamin, New York (1966).

[3.26] C.R. Fincher, C.E. Chen, A.J. Heeger, A.G. McDiarmid, and J.B. Hastings, Phys. Rev. Lett., **48**, 100 (1982).

[3.27] C.S. Yannoni and T.C. Clarke, Phys. Rev. Lett., **51**, 1191 (1983).

[3.28] R.E. Peierls, "Quantum Theory of Solids", Clarendon Press, Oxford (1955).

[3.29] M.J. Rice, Phys. Lett. A, **71**, 152 (1979).

[3.30] W.P. Su, J.R. Schrieffer, and A.J. Heeger, Phys. Rev. Lett., **42**, 1698 (1979); Phys. Rev. B, **22**, 2209 (1980).

[3.31] S. Kuroda and H. Shirakawa, Phys. Rev. B, **35**, 9380 (1987).

[3.32] M. Mehring, A. Grupp, P. Höfer, and H. Käss, Synth. Met., **28**, D399 (1989).

[3.33] J.A. Pople and S.H. Walmsley, Mol. Phys., **5**, 15 (1962).

[3.34] J.L. Brédas, R.R. Chance, and R. Silbey, Mol. Cryst. Liq. Cryst., **77**, 319 (1981).

[3.35] J.L. Brédas, R.R. Chance, and R. Silbey, Phys. Rev. B, **26**, 5843 (1982).

[3.36] A.G. MacDiarmid and A.J. Heeger, Synth. Met., **1**, 101 (1979).

[3.37] See the review by D. Baeriswyl, D.K. Campbell, and S. Mazumdar, in "Conducting Polymers", (H. Kiess, Ed.), Springer, New York (1990) and references therein.

[3.38] J.E. Hirsch, Phys. Rev. Lett., **51**, 296 (1983).

[3.39] Z.G. Soos and S. Ramasesha, Phys. Rev. Lett., **51**, 2374 (1983); Phys. Rev. B, **29**, 5410 (1984).

[3.40] S. Kivelson, W.P. Su, J.R. Schrieffer, and A.J. Heeger, Phys. Rev. Lett., **58**, 1899 (1987).

[3.41] C.Q. Wu, X. Sun, and K. Nasu, Phys. Rev. Lett., **59**, 831 (1987).

[3.42] J. Paldus and E. Chin, Int. J. Quantum Chem., **24**, 373 (1983); J. Paldus, E. Chin, and M.G. Grey, Int. J. Quantum Chem., **24**, 395 (1983); R. Pauncz and J. Paldus, Int. J. Quantum Chem., **24**, 411 (1983); J. Paldus and M. Takahashi, Int. J. Quantum Chem., **25**, 423 (1984); M. Takahashi and J. Paldus, Phys. Rev. B, **31**, 5121 (1985).

[3.43] H. Fukutome and M. Sasai, Progr. Theor. Phys., **67**, 41 (1982); **69**, 373 (1983).

[3.44] S. Suhai, Phys. Rev. B, **27**, 3506 (1983).

[3.45] J.L. Brédas and A.J. Heeger, Phys. Rev. Lett., **63**, 2534 (1989).

[3.46] G. König and G. Stollhoff, Phys. Rev. Lett., **65**, 1239 (1990).

274

[3.47] Y. Delugeard, J. Desuche, and J.L. Baudour, Acta Crystallogr., B**32**, 702 (1976).

[3.48] J.L. Baudour, H. Cailleau, and W.B. Yelon, Acta Crystallogr., B**33**, 1773 (1977).

[3.49] J.L. Baudour, Y. Delugeard, and P. Rivet, Acta Crystallogr., B**34**, 625 (1978).

[3.50] J.L. Brédas, R.R. Chance, R. Silbey, G. Nicolas, and Ph. Durand, J. Chem. Phys., **77**, 371 (1982).

[3.51] L.W. Shacklette, R.R. Chance, D.M. Ivory, G.G. Miller, and R.H. Baughman, Synth. Met., **1**, 307 (1979).

[3.52] J.L. Brédas, B. Thémans, and J.M. André, J. Chem. Phys., **78**, 6137 (1983).

[3.53] J.L. Brédas, R.L. Elsenbaumer, R.R. Chance, and R. Silbey, J. Chem. Phys., **78**, 5566 (1983).

[3.54] A.F. Diaz, J. Crowley, J. Bargon, G.P. Gardini, and J.B. Torrance, J. Electroanal. Chem., **121**, 355 (1981).

[3.55] B. Thémans, J.M. André, and J.L. Brédas, Synth. Met., **21**, 149 (1987).

[3.56] S. Shi and F. Wudl, in "Conjugated Polymeric Materials: Opportunities in Electronics, Optoelectronics and Molecular Electronics", (J.L. Brédas and R.R. Chance, Eds.), p.83, Kluwer Academic Publishers, Dordrecht (1990).

[3.57] H. Eckhardt, L.W. Shacklette, K.Y.J en, and R.L. Elsenbaumer, J. Chem. Phys., **91**, 1303 (1989); H. Eckhardt, K.Y.J en, L.W. Shacklette, and S. Lefrant, in "Conjugated Polymeric Materials: Opportunities in Electronics, Optoelectronics and Molecular Electronics", (J.L. Brédas and R.R. Chance, Eds.), p. 305, Kluwer Academic Publishers, Dordrecht (1990).

275

[3.58] J.L. Brédas, R.R. Chance, R.H. Baughman, and R. Silbey, J. Chem. Phys., **76**, 3673 (1982); R. Lazzaroni, M. Lögdlund, S. Stafström, W.R. Salaneck, D.D.C. Bradley, R.H. Friend, N. Sato, E. Orti, and J.L. Brédas, in "Conjugated Polymeric Materials: Opportunities in Electronics, Optoelectronics and Molecular Electronics", (J.L. Brédas and R.R. Chance, Eds.), p. 149, Kluwer Academic Publishers, Dordrecht (1990).

[3.59] A.G. MacDiarmid and A.J. Epstein, in "Conjugated Polymeric Materials: Opportunities in Electronics, Optoelectronics and Molecular Electronics", (J.L. Brédas and R.R. Chance, Eds.), p. 53, Kluwer Academic Publishers, Dordrecht (1990).

[3.60] D.S. Boudreaux, R.R. Chance, J.F. Wolf, L.W. Shacklette, J.L. Brédas, B. Thémans, J.M. André, and R. Silbey, J. Chem. Phys., **85**, 4584 (1986); R.R. Chance, D.S. Boudreaux, J.F. Wolf, L.W. Shacklette, R. Silbey, B. Thémans, J.M. André, and J.L. Brédas, Synth. Met., **15**, 105 (1986).

[3.61] J. Boon and E.P. Magre, Makromol. Chem., **126**, 130 (1969).

[3.62] B.J. Tabor, E.P. Magre, and J. Boon, Eur. Polym. J., **7**, 1127 (1971).

[3.63] B. Thémans, J.M. André, and J.L. Brédas, Mol. Cryst. Liq. Cryst., **118**, 121 (1985).

[3.64] S.D. Phillips, G. Yu, Y. Cao, and A.J. Heeger, Phys. Rev., **B39**, 10702 (1989).

[3.65] B. Sjögren and S. Stafström, J. Chem. Phys., **88**, 3840 (1988).

[3.66] C.B. Duke, E.M. Conwell, and A. Paton, Chem. Phys. Lett., **131**, 82 (1986).

[3.67] M.C. dos Santos and J.L. Brédas, Phys. Rev. Lett., **62**, 2499 (1989); M.C. dos Santos and J.L. Brédas, Synth. Met., **29**, E321 (1989).

[3.68] J.L. Brédas, B. Thémans, J.G. Fripiat, J.M. André, and R.R. Chance, Phys. Rev B **29**, 6761 (1984).

[3.69] Y. Cao, Synth. Met., **35**, 319 (1990).

[3.70] J.L. Brédas, R. Silbey, D.S. Boudreaux, and R.R. Chance, J. Am. Chem. Soc., **105**, 6555 (1983).

[3.71] B.S. Hudson, J. Ridyard, and J. Diamond, J. Am. Chem. Soc., **98**, 1126 (1976).

[3.72] J.E. Frommer, Acc. Chem. Res., **19**, 192 (1986).

[3.73] M. Sato, S. Tanaka, and K. Kaeriyama, J. Chem. Soc. Chem. Commun., 873 (1986).

[3.74] S. Hotta, S. Rughooputh, A.J. Heeger, and F. Wudl, Macromolecules, **20**, 212 (1987).

[3.75] A. Andreatta, Y. Cao, J.C. Chiang, P. Smith, and A.J. Heeger, Synth. Met., **26**, 383 (1988); Y. Cao, P. Smith, and A.J. Heeger, in "Conjugated Polymeric Materials: Opportunities in Electronics, Optoelectronics and Molecular Electronics", (J.L. Brédas and R.R. Chance, Eds.), p. 171, Kluwer Academic Publishers, Dordrecht (1990).

[3.76] W.R. Salaneck, O. Inganäs, B. Thémans, J.O. Nilsson, B. Sjögren, J.E. Osterholm, J.L. Brédas, and S. Svensson, J. Chem. Phys., **89**, 4613 (1988).

[3.77] M. Nowak, S.D.D.V. Roghooputh, S. Hotta, and A.J. Heeger, Macromolecules **20**, 965 (1987); M.J. Nowak, D. Spiegel, S. Hotta, A.J. Heeger, and P. Pincus, Macromolecules, **22**, 2917 (1989).

[3.78] R.H. Baughman, R.L. Elsenbaumer, Z. Iqbal, G.G. Miller, and H. Eckhardt, Springer Series in Solid State Sciences, **76**, 432 (1987).

[3.79] J.L. Brédas, G.B. Street, B. Thémans, and J.M. André, J. Chem. Phys., **83**, 1323 (1985).

[3.80] G. Rossi, R.R. Chance, and R. Silbey, J. Chem. Phys., **90**, 7594 (1989).

[3.81] S. Rughooputh, S. Hotta, A.J. Heeger, and F. Wudl, J. Polym. Sci., Polym. Phys. Ed., **25**, 1071 (1987).

[3.82] J.L. Brédas and A.J. Heeger, Macromolecules, **23**, 1150 (1990).

[3.83] J.R. Ackerman and B.E. Kohler, J. Chem. Phys., **80**, 45 (1984).

[3.84] J.L. Brédas, J.M. Toussaint, G. Hennico, J. Delhalle, J.M. André, A.J. Epstein, and A.G. McDiarmid, Springer Series in Solid State Sciences, **76**, 48 (1987).

[3.85] D. Spiegel, P. Pincus, and A.J. Heeger, Synth. Met., **28**, C385 (1989).

[3.86] A. Karpfen, J. Phys. C: Solid State Phys., **13**, 5673 (1980).

[3.87] D. Kobelt and E.F. Paulis, Acta Crystallogr. B, **30**, 232 (1973).

[3.88] P. Bucci, M. Longeri, C.A. Veracini, and L. Lunazzi, J. Am. Chem. Soc., **96**, 1305 (1974).

[3.89] R. Lazzaroni, S. Rachidi, and J.L. Brédas, Phys. Rev. B, submitted for publication.

[3.90] J.P. Aimé, F. Bargain, M. Schott, H. Eckhardt, G.G. Miller, and R.L. Elsenbaumer, Phys. Rev. Lett., **62**, 55 (1989).

[3.91] P.A. Pincus, G. Rossi, and M.E. Cates, Europhys. Lett., **4**, 41 (1987).

[3.92] F. Wudl, M. Kobayashi, and A.J. Heeger, J. Org. Chem., **49**, 3381 (1984); M. Kobayashi, N. Colaneri, M. Boysel, F. Wudl, and A.J. Heeger, J. Chem. Phys., **82**, 5717 (1985).

278

[3.93]   R.A. Harris and L.M. Falicov, J. Chem. Phys., **51**, 5034 (1969).

[3.94]   A.A. Ovchinnikov, I.I. Ukrainskii, and G.V. Kventsel, Usp. Fiz. Nauk., **108**, 81 (1972); Sov. Phys. Usp., **15**, 1575 (1973).

[3.95]   J.L. Brédas, J. Chem. Phys., **82**, 3808 (1985).

[3.96]   S.A. Brazovskii and N. Kirova, Zh. Eksp. Teor. Fis. Pis'ma, **33**, 6 (1983); JETP Lett., **33**, 4 (1981).

[3.97]   J.L. Brédas, J. Libert, C. Quatrocchi, A.G. MacDiarmid, J. Ginder, and A.J. Epstein, Phys. Rev. B, submitted for publication.

[3.98]   J. Kürti and P.R. Surjan, J. Chem. Phys., **92**, 3247 (1990).

[3.99]   J.M. Toussaint, B. Thémans, J.M. André, and J.L. Brédas, Synth. Met., **28**, C205 (1989).

[3.100] R. Becker, G. Blöchl, H. Bräunling, in "Conjugated Polymeric Materials: Opportunities in Electronics, Optoelectronics and Molecular Electronics", (J.L. Brédas and R.R. Chance, Eds.), p. 133, Kluwer Academic Publishers, Dordrecht (1990).

[3.101] Y.S. Lee and M. Kertesz, J. Chem. Phys., **88**, 2609 (1988).

[3.102] S. Ikehata, J. Kaufer, T. Woerner, A. Pron, M.A. Druy, A. Sivak, A.J. Heeger, and A.G. MacDiarmid, Phys. Rev. Lett., **45**, 1123 (1980).

[3.103] E.J. Mele and M.J. Rice, Phys. Rev. B, **23**, 5397 (1981)

[3.104] A.J. Epstein, in "Handbook of Conducting Polymers", (T.A. Skotheim, Ed.), p. 1041, Marcel Dekker, New York (1986).

[3.105] S. Kivelson and A.J. Heeger, Phys. Rev. Lett., **55**, 308 (1985).

[3.106] E.M. Conwell and S. Jeyadev, Phys. Rev. Lett., **61**, 361 (1988) .

[3.107] K. Kume, K. Mizuno, K. Mizoguchi, K. Nomura, Y. Naniwa, J. Tanaka, M. Tanaka, and H. Watanabe, Mol. Cryst. Liq. Cryst., **83**, 285 (1982).

[3.108] K. Mizoguchi, K. Misuo, K. Kume, K. Kaneto, T. Shiraishi, and K. Koshino, Synth. Met., **18**, 195 (1987).

[3.109] F. Moraes, D. Davidov, M. Kobayashi, T.C. Chung, J. Chen, A.J. Heeger, and F. Wudl, Synth. Met., **10**, 169 (1985).

[3.110] P. Pfluger, U.M. Gubler, and G.B. Street, Solid State Commun., **49**, 911 (1984).

[3.111] T. Nakajima, Bridgestone Corporation.

[3.112] A.J. Epstein, J.M. Ginder, F. Zuo, R.W. Bigelow, H.S. Woo, D.B. Tanner, A.F. Richter, W.S. Huang, and A.G. MacDiarmid, Synth. Met., **18**, 303 (1987).

[3.113] J.M. Ginder, A.F. Richter, A.G. MacDiarmid, and A.J. Epstein, Solid State Comm., **63**, 97 (1987); A.G. MacDiarmid, J.C. Chiang, A.F. Richter, and A.J. Epstein, Synth. Met., **17**, 285 (1987).

[3.114] S. Stafström and J.L. Brédas, Phys. Rev. B, **38**, 4180 (1988).

[3.115] X.Q. Yang, D.B. Tanner, M.J. Rice, H.W. Gibson, A. Feldblum, and A.J. Epstein, Solid State Commun., **61**, 335 (1987).

[3.116] T.C. Chung, F. Moraes, J.D. Flood, and A.J. Heeger, Phys. Rev. B, **29**, 2341 (1984).

[3.117] P.M. Grant and I.P. Batra, Solid State Commun., **29**, 225 (1979).

[3.118] J. von Boehm, P. Kuvalainen, and J.L. Calais, Solid State Commun., **48**, 1085 (1983); Phys. Rev. B, **35**, 8177 (1987).

[3.119] M. Springborg, Phys. Rev. B, **33**, 8475 (1986); **37**, 1218 (1988).

[3.120] J.W. Mintmire, F.W. Kutzler, and C.T. White, Phys. Rev., B **36**, 3312 (1987).

[3.121] T.C. Chung, J.H. Kaufman, A.J. Heeger, and F. Wudl, Phys. Rev., B **30**, 702 (1984).

[3.122] C.R. Wu, J.O. Nilsson, O. Inganäs, W.R. Salaneck, J.E. Österholm, and J.L. Brédas, Synth. Met., **21**, 197 (1987).

[3.123] S. Stafström, B. Sjögren, and J.L. Brédas, Synth. Met., **29**, E219 (1989).

[3.124] J.O. Nilsson and W.R. Salaneck, unpublished.

[3.125] D. Bertho and C. Jouanin, Phys. Rev., B **35**, 626 (1987).

[3.126] J.L. Brédas, F. Wudl, and A.J. Heeger, Solid State Commun., **63**, 577 (1987).

[3.127] Z. Vardeny, E. Ehrenfreund, O. Brafman, M. Nowak, H. Schaffer, A.J. Heeger, and F. Wudl, Phys. Rev. Lett., **56**, 671 (1986).

[3.128] J. Fink, N. Nücker, B. Scheerer, A. vom Felde, and G. Leising, Springer Series in Solid State Sciences, **76**, 84 (1987).

[3.129] S. Hasegawa, K. Kamiya, J. Tanaka, and M. Tanaka, Synth. Met., **18**, 225 (1987).

[3.130] M. Lögdlund, R. Lazzaroni, S. Stafström, W.R. Salaneck, and J.L. Brédas, Phys. Rev. Lett., **63**, 1840 (1989).

[3.131] S. Stafström, J.L. Brédas, A.J. Epstein, H.S. Woo, D.B. Tanner, W.S. Huang, and A.G. MacDiarmid, Phys. Rev. Lett., **59**, 1464 (1987).

[3.132] H. Monkhorst, Phys. Rev. B, **20**, 1504 (1979); J.M. André in "Large Finite Systems", (J. Jortner, A. Pullman, and B. Pullman, Eds.), p. 277, Reidel, Dordrecht (1987).

[3.133] J.P. Hermann and J. Ducuing, J. Appl. Phys., **45**, 5100 (1974); C. Sauteret, J.P. Hermann, R. Frey, F. Pradere, J. Ducuing, R.R. Chance, and R.H. Baughmann, Phys. Rev. Lett., **36**, 956 (1976).

[3.134] "Nonlinear Optical Properties of Polymers", (A.J. Heeger, J. Orenstein, and D.R. Ulrich, Eds.), Vol. 109, Materials Research Society, Symposium Proceedings, (1988).

[3.135] "Nonlinear Optical Properties of Organic Molecules and Crystals", (D.S. Chemla and J. Zyss, Eds.), Academic Press, New York (1987).

[3.136] M. Sinclair, D. Moses, K. Akagi, and A.J. Heeger, in "Nonlinear Optical Properties of Polymers", (A.J. Heeger, J. Orenstein, and D.R. Ulrich, Eds.), Vol. 109, p. 205, Materials Research Society, Symposium Proceedings, (1988).

[3.137] M. Sinclair, D. Moses, K. Akagi, and A.J. Heeger, Phys. Rev. B, **38**, 10724 (1988).

[3.138] M. Sinclair, D. Moses, D. McBranch, A.J. Heeger, J. Yu, and W.P. Su, Synth. Met., **28**, D655 (1989).

[3.139] J.L. Brédas and A.J. Heeger, Chem. Phys. Lett., **154**, 56 (1989).

[3.140] M. Sinclair, D. Moses, and A.J. Heeger, Solid State Commun., **57**, 343 (1986).

[3.141] L. Rothberg, T.M. Jedju, S. Etemad, and G.L. Baker, Phys. Rev. Lett., **57**, 3229 (1986); Phys.Rev. B, **36**, 7529 (1987).

[3.142] S. Roth and H. Bleier, Adv. Phys., **36**, 385 (1987).

[3.143] W.P. Su and J.R. Schrieffer, Proc. Natl. Acad. Sci., **77**, 5626 (1980).

[3.144] J.R. Heflin, K.Y. Wong, O. Zamani-Khamiri, and A.F. Garito, Phys. Rev., **B38**, 1573 (1988).

[3.145] M. Sinclair, D. Moses, D. McBranch, A.J. Heeger, J. Yu, and W.P. Su, Phys. Scr., **T27**, 144 (1989).

[3.146] B.S. Hudson, B.E. Kohler, and K. Schulten, in "Excited States", (K.C. Lim, Ed.), Vol. 6, p. 1, Academic Press, New York (1982).

[3.147] M.F. Granville, B.E. Kohler, and J. Bannon Snow, J. Chem. Phys., **75**, 3765 (1981).

[3.148] P. Tavan and K. Schulten, J. Chem. Phys., **85**, 6602 (1986).

[3.149] J.L. Brédas, M. Dory, and J.M. André, J. Chem. Phys., **83**, 5242 (1985).

[3.150] J.L. Brédas and J.M. Toussaint, J. Chem. Phys., **92**, 2624 (1990).

[3.151] C.P. DeMelo and R. Silbey, J. Chem. Phys., **88**, 2567 (1988).

[3.152] Z.G. Soos and S. Ramasesha, J. Chem. Phys., **90**, 1067 (1989).

4

QUANTUM CHEMISTRY AIDED DESIGN OF CHAINS FOR OPTOELECTRONICS

4.1 SUMMARY AND OBJECTIVES

The purpose of this chapter is

(1) to define the electrical polarizability and hyperpolarizabilities of molecules and polymers,

(2) to summarize the current methods of calculations of polarizability and hyperpolarizabilities of molecules, analyze their mutual advantages, and review their applications to oligomers,

(3) to state the difficulties met in the calculation of the polarizability of infinite chains and polymers, write down the perturbation expression in this case, and apply it to model polymeric chains in the simple Hückel approximation and within an *ab initio* approach,

(4) to exemplify by a set of selected examples the application of quantum chemistry to the molecular design of compounds having a high polarizability.

4.2 CALCULATIONS OF LINEAR POLARIZABILITY AND OF FIRST AND SECOND NONLINEAR HYPERPOLARIZABILITIES OF LONG OLIGOMERS

Materials which exhibit high linear and nonlinear responses are currently the subject of intense and worldwide research activities. The interest in organics not only lies in their enhanced electric responses over a wide frequency range and ultrafast response times, but also in the many ways of varying their molecular structures, the possibility of film forming and processing, and higher laser damage thresholds. High electric susceptibilities very much depend on the nature of the delocalized network of π-electrons and it is important to find molecular structures that yield the largest possible responses because, among other conditions and advantages, the larger the nonlinear response and the smaller the electric field required to achieve the desired nonlinear effect. The molecular structure of organic materials can easily be modified in order to maximize their electric responses, but this versatility has also its drawback because, due to cost and time, it is practically impossible to prepare and test all the interesting compounds. Moreover, since it is expected that the compounds used to form materials will depend on the particular application under consideration, there will be a continuous need for designing new molecules for optoelectronics.

The first and minimal condition for organic molecules to be of potential interest for optoelectronics is obviously to have high microscopic electric responses. At the microscopic level, polarization is due to a molecular dipole moment induced by the various electric fields. In the case of electric fields, the energy in presence of the external field, $E(F)$, and the dipole moment, $\mu(F)$, can be expanded as power series of the electric field strength F [4.1,4.2]:

$$E(F) = E_0 - \mu_0.F - \frac{1}{2!}\alpha{:}F^2 - \frac{1}{3!}\beta\vdots F^3 + \dots$$

$$\mu(F) = \mu_0 + \alpha.F + \frac{1}{2!}\beta{:}F^2 + \frac{1}{3!}\gamma\vdots F^3 + \dots$$

where μ_0 is the permanent dipole moment and α, β, and γ are known as the polarizability (rank 2 tensor) and the first and second order hyperpolarizabilities (rank 3 or rank 4 tensors). Note that the previous expansions of $E(F)$ and $\mu(F)$ are consistent with the definition of the dipole moment as the derivative of the energy with respect to the field components:

$$\mu(F) = - \frac{\delta E(F)}{\delta F}$$

If we suppose that the electric field is applied in the z-direction ($F = Fe_z$, where e_z is a unit vector in the z-direction) and we use the Taylor series:

$$E(F) = E_0 + (\frac{dE}{dF})_0 F + \frac{1}{2} (\frac{d^2E}{dF^2})_0 F^2 + \frac{1}{3!} (\frac{d^3E}{dF^3})_0 F^3 + \ldots$$

we can make the following identifications:

$$\mu_{0z} = - (\frac{dE}{dF})_0$$

$$\alpha_{zz} = - (\frac{d^2E}{dF^2})_0$$

$$\beta_{zzz} = - (\frac{d^3E}{dF^3})_0$$

but other conventions exist like $\mu(F) = \mu_0 + \alpha.F + \beta:F^2 + \gamma\vdots F^3 + \ldots$ In the latter case, correction factors $\frac{1}{2}$ and $\frac{1}{6}$ are introduced respectively into the numerical values of β_{ijk}, a component of the first hyperpolarizability and of γ_{ijkl}, a component of the second hyperpolarizability.

Large values of these coefficients are necessary conditions for high macroscopic electric responses. For organics, high electric responses have invariably been obtained in systems containing delocalized π-electrons. Out of possible conjugated structures, many can readily been discarded on the basis of simple considerations (number of π-electrons, chain length, etc.), but more quantitative assessments require direct quantum mechanical calculations of the coefficients α, β, and γ. For instance, factors such as the geometry and the resulting electron density distribution turn out to have a great influence on these properties and are hardly predicted from rules of thumb.

Furthermore, to be useful materials, these compounds must combine, in addition to high electric susceptibilities, many other properties such as a suitable organization at the molecular level with possible symmetry constraints, chemical stability, appropriate transparency regions, etc. Like drug design, the design of organic chains for optoelectronics is a complex task where many constraints on the prospective system must be met simultaneously before a good material can be claimed. Quantum chemistry can provide assistance in this endeavor by providing calculated molecular properties to discriminate among the interesting candidates. In Section 4.2.1, we shortly describe currently available methods of calculations of (hyper)polarizabilities of molecular systems. In Sections 4.2.2 and 4.2.3, a short and therefore nonexhaustive account of early and present contributions made in this domain on the basis of quantum mechanical calculations is given. They consist of a limited number of selected contributions to the quantum chemistry aided design for optoelectronics made by other groups; most of these contributions have been a source of inspiration for our own work described in Sections 4.2.4 and 4.2.5 and in a more systematic way in Part 4.4 of this Chapter. Part 4.3 states the difficulties met in calculating the polarizability of infinite polymeric systems and gives an introduction to the quantum chemical methods used in this field.

4.2.1 Currently Available Methods of Calculations

4.2.1.1 Finite Field-Self Consistent Field method

The Finite Field-Self Consistent Field (F.F.) method was originally proposed by Cohen and Roothaan [4.3] and is equivalent to an analytic Coupled Hartree-Fock scheme, standing-out for its simplicity. A term, $er.F = -r.F$ in a.u., describing the interaction between the external field, F, and the elementary charges (electrons and nuclei constituting the molecule) is added to the molecular Hamiltonian; at this level, the orbitals are self-consistent eigenfunctions of the one-electron field dependent Fock operator, $h(r)$:

$$h^{FF-SCF}(r) = -\frac{1}{2}\nabla^2 - \sum_A Z_A \ |r-R_A|^{-1} + \sum_j^{occ} \{2J_j(r) - K_j(r)\} - r.F$$

$$= h_0(r) - r.F$$

The matrix elements of $h(\mathbf{r})$ contain additional one-electron moment integrals, $\mathbf{M}_{pq} = \int \chi_p(\mathbf{r}) \, \mathbf{r} \, \chi_q(\mathbf{r}) \, d\mathbf{r}$; the latter integral can be fairly easily calculated at any level of approximation and has x- ($M_{pq}^x = \int \chi_p(\mathbf{r}) \, x \, \chi_q(\mathbf{r}) \, d\mathbf{r}$), y- ($M_{pq}^y = \int \chi_p(\mathbf{r}) \, y \, \chi_q(\mathbf{r}) \, d\mathbf{r}$), and z- ($M_{pq}^z = \int \chi_p(\mathbf{r}) \, z \, \chi_q(\mathbf{r}) \, d\mathbf{r}$) components. One has however to realize that the self-consistent procedure applies to the solutions of $h(\mathbf{r})$. More specifically, the elements D_{rs} of the density matrix are now field-dependent quantities, $D_{rs}(\mathbf{F})$ weighting the two-electron integrals in the molecular matrix elements:

$$h_{pq} = \int \chi_p(\mathbf{r}) \, h(\mathbf{r}) \, \chi_q(\mathbf{r}) \, d\mathbf{r}$$

$$= T_{pq} - \sum_A Z_A V_{pq|A}$$

$$- \, \mathbf{F} \, \mathbf{M}_{pq}$$

$$+ \sum_{r,s} D_{rs}(\mathbf{F}) \, \{ \, 2 \, (pq|rs) - (pr|qs) \, \}$$

The components of the (hyper)polarizability tensors are obtained from derivatives of the field-dependent dipole moment with respect to the external electric field in the limit of zero field. In practice, this is done numerically using finite symmetric difference approximations of the first and third-order derivatives of the dipole:

$$\alpha_i = \left\{ \frac{\delta \langle \mu_i \rangle}{\delta F_i} \right\}_{F=0} = \frac{1}{2F_i} [\mu_i(F_i) - \mu_i(-F_i)]$$

$$\gamma_{iiii} = \left\{ \frac{\delta^3 \langle \mu_i \rangle}{\delta F_i^3} \right\}_{F=0} = \frac{1}{48 F_i^3} [\, \mu_i(3F_i) - \mu_i(-3F_i) - 3 \, \{\mu_i(F_i) - \mu_i(-F_i)\} \,]$$

$$= \{\mu_i(3F_i) - \mu_i(-3F_i) - 3[\mu_i(F_i) - \mu_i(-F_i)]\}/48 \, F_i^3$$

Large values of the field intensity $\mathbf{F}$ may be thought to be preferable in terms of numerical precision on α (or γ) which is proportional to $\Delta\mu/F$ (or $\Delta\mu/6F^2$) where $\Delta\mu$ is an estimate of the numerical error on the dipole moment. The difference between the last two values of the dipole moment before convergence at the required level of accuracy may be taken as a rough estimate of $\Delta\mu$. However, higher values of F will increase the difference between the finite difference approximation of the dipole derivative and its actual value. In addition, at higher values of F, self-consistency will not be reached, and generally the best value for F will be that just below numerical divergence threshold. Typical values which allow for the computation of α, β, and γ with reasonable accuracy are of the order of 10^{-3} a.u.

4.2.1.2 Sum-Over-States perturbation method

An alternative method is based on a Sum-Over-States (S.O.S.) perturbation expansion of the Roothaan-Hartree-Fock (RHF) wave function (or the corresponding SCF-MO) orbitals in the absence of external field. The expressions for the components of the (hyper)polarizabilities are in terms of wave functions and state energies:

$$\alpha_{ij} = \sum_m \frac{<\Psi_o|\mu_i|\Psi_m><\Psi_m|\mu_j|\Psi_o>}{(E_m-E_o)}$$

$$\beta_{ijk} = \sum_n \sum_m \frac{<\Psi_o|\mu_i|\Psi_n><\Psi_n|\mu_j|\Psi_m><\Psi_m|\mu_k|\Psi_o>}{(E_n-E_o)(E_m-E_o)}$$

$$- \sum_n \frac{<\Psi_o|\mu_i|\Psi_o><\Psi_o|\mu_j|\Psi_n><\Psi_n|\mu_k|\Psi_o>}{(E_n-E_o)(E_n-E_o)}$$

where Ψ represents the total antisymmetric wave function and E are the state total energies of the molecules. Those relations can be approximated by corresponding equations in terms of molecular orbitals and orbital energies:

$$\alpha_{ij} = S_{ij} \sum_{ar} \frac{<a|\mu_i|r><r|\mu_j|a>}{(\varepsilon_r-\varepsilon_a)}$$

$$\beta_{ijk} = S_{ijk} \left\{ \sum_{ars} \frac{<r|\mu_i|a><a|\mu_j|s><s|\mu_k|r>}{(\varepsilon_r-\varepsilon_a)(\varepsilon_s-\varepsilon_a)} - \sum_{abr} \frac{<r|\mu_i|a><a|\mu_j|b><b|\mu_k|r>}{(\varepsilon_r-\varepsilon_a)(\varepsilon_r-\varepsilon_b)} \right\}$$

where $a,b,(r,s)$ represent occupied (unoccupied) molecular orbitals and ε_a, ε_b, ε_r, ε_s their corresponding one-electron energies and S is the index permutation operator. The "brute force" implementation of fourth-order perturbation formula required for getting the second-order hyperpolarizability shown below is not a trivial numerical task:

$$\gamma_{ijkl} = S_{ijkl} \ \Big[\sum_{arst} \frac{<a|\mu_i|r> <r|\mu_j|s> <s|\mu_k|t> <t|\mu_l|a>}{\{(\varepsilon_r-\varepsilon_a)(\varepsilon_s-\varepsilon_a)(\varepsilon_t-\varepsilon_a)\}}$$

$$+ \sum_{abcr} \frac{<a|\mu_i|b> <b|\mu_j|c> <c|\mu_k|r> <r|\mu_l|a>}{\{(\varepsilon_r-\varepsilon_a) (\varepsilon_r-\varepsilon_b) (\varepsilon_r-\varepsilon_c)\}}$$

$$- \sum_{abrs} \frac{<a|\mu_i|b> <b|\mu_j|s> <s|\mu_k|r> <r|\mu_l|a>}{\{(\varepsilon_r-\varepsilon_a) (\varepsilon_s-\varepsilon_b) (\varepsilon_r-\varepsilon_b)\}}$$

$$- \sum_{abrs} \frac{<a|\mu_i|b> <b|\mu_j|r> <r|\mu_k|s> <s|\mu_l|a>}{\{(\varepsilon_r-\varepsilon_a) (\varepsilon_s-\varepsilon_a) (\varepsilon_r-\varepsilon_b)\}}$$

$$- \sum_{abrs} \frac{<a|\mu_i|r> <r|\mu_j|b> <b|\mu_k|s> <s|\mu_l|a>}{\{(\varepsilon_r-\varepsilon_a) (\varepsilon_s-\varepsilon_a) (\varepsilon_s-\varepsilon_b)\}} \ \Big]$$

The numerical effort can be estimated as 25 N^2M^2 multiplications, 5 N^2M^2 divisions, and 5 N^2M^2 additions, where N and M are the number of occupied and virtual orbitals, respectively. Prohibitive computing times will therefore result and it is necessary to develop tractable expressions for the polarizability and the first and second hyperpolarizability tensor components by a recurrent S.O.S. method [4.4]. By such a methodology, the following computer-efficient formulas are obtained:

$$\alpha_{ij} = - 2 \ S_{ij} \sum_{ar} <a|\mu_j|r> \ D_{ar}^{(1)} (i)$$

$$\beta_{ijk} = \ S_{ijk} \sum_{ar} <a|\mu_k|r> \ D_{ar}^{(2)} (i,j)$$

$$\gamma_{ijkl} = - \frac{1}{3} \ S_{ijkl} \sum_{ar} <a|\mu_l|r> \ D_{ar}^{(3)} (i,j,k)$$

where:

$$D_{ar}^{(1)}(i) = \frac{<r|\mu_i|a>}{(\varepsilon_a - \varepsilon_r)}$$

$$D_{ar}^{(2)}(i,j) = \frac{\sum_s <r|\mu_j|s> D_{as}^{(1)}(i) - \sum_b <a|\mu_j|b> D_{rb}^{(1)}(i)}{(\varepsilon_a - \varepsilon_r)}$$

$$D_{ar}^{(3)}(i,j,k) = [\sum_s <r|\mu_k|s> D_{as}^{(2)}(i,j) - \sum_b <a|\mu_k|b> D_{br}^{(2)}(i,j)$$

$$- \sum_{bs} <b|\mu_k|s> \frac{D_{as}^{(1)}(i) D_{br}^{(1)}(j)}{(\varepsilon_a - \varepsilon_r)}]$$

These expressions have been implemented into a computer algorithm where the numerical calculation needs a reduced number of operations proportional to N*M multiplications and additions. It also involves the calculation of the matrix elements $D_{ar}^{(3)}$ which only requires (N+M+2N*M) multiplications and the availability of the lower order matrix elements $D_{ar}^{(1)}$ and $D_{ar}^{(2)}$ provided sufficient memory is available.

4.2.1.3 F.F. versus S.O.S.

The S.O.S. method uses as a starting point products (possibly antisymmetrized) of one-electron orbitals determined in the absence of the external perturbating field and is derived from the time-independent perturbation theory where the perturbation is the interaction between an external field **F** and the elementary charges of the molecule, $-\Sigma \mathbf{r}.\mathbf{F}$. Products of orbitals are not eigenfunctions of the total Hamiltonian in the absence of the external perturbating field, H_0, but instead are eigenfunctions of an approximate Hamiltonian which is the sum of one-electron operators. Thus, the nature of the reference unperturbed state is not a trivial point and is discussed in more detail below. In the case of the Hartree-Fock (HF) theory, for example, the HF wave function is not, by its approximate character, an eigenfunction of the total Hamiltonian, but instead an eigenfunction of the total HF operator (sum of the one-electron Fock operators) which differs from the total Hamiltonian by the correlation operator. In the standard S.O.S. method, the zeroth-order solution is actually the HF wave function and the unperturbed Hamiltonian is thus the Hartree-Fock total Hamiltonian. In this approximation, the S.O.S.

method neglects field-induced coupled reorganizational effects and is equivalent to an uncoupled HF scheme. Thus, the S.O.S. method is only to be considered as a first step to more correct coupled techniques.

As mentioned above, the Hartree-Fock wave function is not an eigenvector of the exact Hamiltonian H^{ex}:

$$H^{ex} = \sum_i \{T_i - \sum_A \frac{Z_A}{r_{iA}} + \frac{1}{2} \sum_{j \neq i} \frac{1}{r_{ij}}\}$$

but instead is an eigenvector of the total Hartree-Fock Hamiltonian H^{HF}:

$$H^{HF} = \sum_i h^{HF}(i) = \sum_i \{T_i - \sum_A \frac{Z_A}{r_{iA}} + \sum_n [2J_n(i) - K_n(i)]\}$$

where $\sum_i$ stands for the summation over the electrons and $\sum_n$ for the summation over the occupied molecular orbitals, respectively. Recognition of that difference results in the definition of a correlation perturbation operator V_C given by:

$$H^{ex} = H^{HF} + V_C = H^{HF} - \sum_i \sum_n \{2J_n(i) - K_n(i)\} + \frac{1}{2} \sum_i \sum_{j \neq i} \frac{1}{r_{ij}}$$

When the perturbation field is switched on, the Hartree-Fock wave function Ψ_0 is used as the reference state in a double perturbation process:

$$H^{ex} = H^{HF} + V_C + V_F\,F = H^{HF} - \sum_i \sum_n \{2J_n(i) - K_n(i)\} + \frac{1}{2} \sum_i \sum_{j \neq i} \frac{1}{r_{ij}} - \sum_i r_i\,F$$

where V_F is the dipolar perturbation operator with the corresponding n-order energies defined as:

$$E = E^{HF} + E^{(1)} + E^{(2)} + E^{(3)} + ...$$

Each in turn is given by:

$$E^{(1)} = \langle\Psi_0|V_F|\Psi_0\rangle F + \langle\Psi_0|V_C|\Psi_0\rangle$$

with μ_0, the Hartree-Fock dipole moment $= \langle\Psi_0|V_F|\Psi_0\rangle$ and $\langle\Psi_0|V_C|\Psi_0\rangle = $ a vanishing correlation correction.

$$E^{(2)} = \sum_n \frac{|\langle\Psi_0|V_F F + V_C|\Psi_n\rangle|^2}{(E_n - E_0)}$$

$$= F^2 \sum_n \frac{|\langle\Psi_0|V_F|\Psi_n\rangle|^2}{(E_n - E_0)}$$

$$+ \sum_n \frac{|\langle\Psi_0|V_C|\Psi_n\rangle|^2}{(E_n - E_0)} + 2F \sum_n \frac{\langle\Psi_0|V_F|\Psi_n\rangle \langle\Psi_0|V_C|\Psi_n\rangle}{(E_n - E_0)}$$

with α, being the Hartree-Fock polarizability $= \sum_n \frac{|\langle\Psi_0|V_F|\Psi_n\rangle|^2}{(E_n - E_0)}$,

the second-order Möller-Plesset $= \sum_n \frac{|\langle\Psi_0|V_C|\Psi_n\rangle|^2}{(E_n - E_0)}$

and the second-order correlation correction to the HF dipole moment being equal to:

$$2\sum_n \frac{\langle\Psi_0|V_F|\Psi_n\rangle \langle\Psi_0|V_C|\Psi_n\rangle}{(E_n - E_0)}$$

Similarly, $E^{(3)}$ is the sum of the β Hartree-Fock hyperpolarizability, of the third-order Möller-Plesset correlation, of the third-order correlation correction to the HF dipole moment, and of the third-order correlation correction to the β hyperpolarizability.

A possible way of improving the S.O.S. method is thus to perform double perturbation calculations: the first perturbation is the external field and the second one is the correlation energy operator.

As conclusion to this Section, it is to be noted that uncoupled HF calculations and correlated calculations of polarizabilities will necessarily involve the computation of two-electron integrals between molecular or polymeric orbitals. In polymeric calculations, it means the computation of a huge number of two-electron integrals between polymeric orbitals which belong to different bands and to different k-points in the first Brillouin zone of the polymer. Rules and computational tricks for efficiently solving this difficulty are needed. This is an open and timely field of research.

4.2.2 Semiempirical Approaches

In this Section, we deal with the early attempts of using theory as a tool to select and design molecules. Among these we feel that we should stress the results obtained by Hameka, prior to the exploding era of molecular calculations for optoelectronics: they have certainly shown the way of using quantum chemical calculations to get the trends of electric responses in a series of related molecules. The pertaining literature is given in Section 4.3.3. Similarly, Flytzanis and coworkers have contributed much to disentangle the problems relating to infinite systems and provided concepts to help identifying the dependence of the electric responses on the dimensionality of the systems (scaling laws). This is also discussed in Section 4.3.3. Other important contributions made at the π-electron level are due to Ratner [4.5], Beratan [4.6], and Wagnière [4.7] and their respective coworkers.

To our knowledge, Zyss and collaborators [4.8,4.9] were the first to rely in a systematic way on semiempirical quantum chemistry methods to assess the merits of specific molecular structures and the influence of substituents on the electric responses. They could predict the optical nonlinearities of a large variety of molecules and relate these values to other properties such as the transparency region and the degree of polarity of the excited charge transfer state. Other groups were also actively using semiempirical methods with the goal of identifying and designing materials with high electric responses: Garito [4.10], Morley [4.11], Waite [4.12], Pierce [4.13], Svendsen [4.14], Zerner [4.15], and Williams [4.16] and their coworkers.

4.2.3 *Ab Initio* Works

The works made at the *ab initio* level are conveniently divided in two classes: the highly accurate calculations mostly carried out on atomic and diatomic systems and those performed on relatively complex molecules. In the first group one can quote, among others, Bishop [4.17], Shelton [4.18], Thakkar [4.19], Oddershede [4.20], and Jørgensen [4.21] and their coworkers. In the other group are the works on more complex systems which therefore rely on less sophisticated methods and thus aim at qualitatively reliable numbers. Besides our contributions which will be summarized in the forthcoming Sections one can mention the contributions by Kirtman [4.22], Dupuis [4.23], and Prasad [4.24] and their coworkers. The list of references provided in Sections 4.2.2 and 4.2.3 is obviously too short to constitute a significant overview of the efforts made using quantum chemical calculations in this field. We have omitted many other important works and have to apologize for this; nevertheless, we hope that it conveys to the reader an adequate feeling of the importance this field has acquired.

4.2.4 Sample Calculations on Polyenes and Scaling Factor

Selected components of the electric polarizability tensor of polyenes calculated with minimal STO-3G and split-valence 4-31G bases in both the Finite Field and Sum-Over-States methodologies are compared in Table 4.1.

It is apparent from the results that the S.O.S. values underestimate by 16-19% the corresponding values obtained by the F.F. methodology. This arises because the S.O.S. method corresponds to an uncoupled Hartree-Fock scheme. Furthermore, the flexibility of the basis set has a strong effect on the actual values of the components. As expected, the change is larger for the direction where the electrons are poorly described by minimal basis sets (e.g., α_{yy} corresponds to π-electrons which are only described by a single carbon $2p_\pi$ orbitals in a minimal basis set calculation). A more extensive analysis on the ethylene molecule in the S.O.S. scheme is also given in Table 4.1 which includes results of minimal (STO-3G), extended (4-31G and 6-31G), and polarized (6-31G**) basis sets. A change from minimal to polarized basis sets yields an improvement by a factor 1.86 in the total polarizability, the effect being more marked for the α_{zz} component which improves by a factor of 4.53. It is important to note however that the "best" S.O.S. calculated value (17.45 a.u.) is still only 61% of the "experimental" value (28.48 a.u.). A better description of the carbon 1s inner shell has only a minor effect (which is however more marked in the y-direction). The inclusion of polarized orbitals has little effect on the molecular in-plane components but markedly modifies the perpendicular α_{yy} components.

Table 4.1 **Influence of basis sets and methodologies on the polarizabilities of polyenes (C_nH_{n+2})**
(results in a.u., for the definition of atomic units, see Section 4.2.6)

		α_{zz}*	α_{xx}	α_{yy}	α
C_2H_4					
	F.F.-STO-3G	17.4773	13.1964	2.5312	11.0653
	S.O.S.-STO-3G	15.1165	11.1140	1.9293	9.3866
	F.F.-4.31G	29.1906	22.7813	5.9019	19.2912
	S.O.S.-4.31G	23.7830	18.1468	5.6067	15.8455
	S.O.S.-6.31G	24.3127	18.4321	6.2169	16.3406
	S.O.S.-6.31G**	24.4237	19.1702	8.7550	17.4496
	Exp.				28.48
C_4H_6					
	F.F.-STO-3G	47.8512	24.8279	4.7491	25.8096
	S.O.S.-STO-3G	38.1210	22.1237	3.7711	21.3386
	F.F.-4.31G	73.3961	40.7345	12.1874	42.1060
	Exp.				56.7
C_6H_8					
	F.F.-STO-3G	100.4982	37.7190	6.9819	48.3997
	S.O.S.-STO-3G	73.7087	33.8750	5.6385	37.7387
C_8H_{10}					
	F.F.-STO-3G	174.5574	49.5785	9.1377	77.7579
	S.O.S.-STO-3G	114.7559	44.8826	7.4451	55.6946

* z is along the chain axis in the molecular plane, x is perpendicular to the chain axis in the molecular plane, and y is perpendicular to the molecular plane (π-direction). α is the average polarizability: $\frac{1}{3}\sum_i \alpha_{ii}$

The method of the scaling factor [4.25, 4.26] results from the previous observation. Using an STO-3G basis set, one treats the ethylene and acetylene series taking the first terms (four terms) of the series as presented in Table 4.2.

It is gratifying to find the ratio $R_n = \alpha_{zz}$ (STO-3G) / α_{zz} (4-31G) for n = 1 to 4 being roughly constant and similar in both series of compounds. The interval of variation of R_n for n >1 typically varies from 0.65 to 0.70, with R_n being close to 0.70 as the length of the oligomer increases. Therefore, when using the minimal STO-3G basis, a value of R_n = 0.70 can reasonably be chosen in studies related to unsaturated hydrocarbon chains. Assuming that the ratio α_{zz} (4-31G) / α_{zz} (exp.) = 0.9 for acetylene and ethylene calculated from data in Table 4.2 will stay reasonably constant for other unsaturated molecules, it is possible to predict data at the scale of experimental observations by applying the multiplicative constant 1.57 = (1/(0.9x0.7)) to the corresponding STO-3G values.

Table 4.2   **Comparison of STO-3G and 4-31G calculated longitudinal electric polarizabilities of two series of oligomers, H-$(CH=CH)_n$-H and H-$(C{\equiv}C)_n$-H ; n = 1, 2, 3, and 4** (results in a.u.)

	α_{zz}		$R_n = \dfrac{\alpha_{zz}(\text{STO-3G})}{\alpha_{zz}(\text{4-31G})}$
	STO-3G	4-31G	
ethylene	17.4773	29.1906	0.60
1,3-trans-butadiene	47.8512	73.3961	0.65
1,3,5-trans-hexatriene	104.1258	152.6748	0.68
1,3,5,7-trans-octatetraene	174.5574	246.1033	0.71
acetylene	18.1793	27.9280	0.65
1,3-butadiyne	50.6834	73.9655	0.69
1,3,5-hexatriyne	91.1179	131.3900	0.69
1,3,5,7-octatetrayne	145.4749	207.5293	0.70

4.2.5 Other Ways of Interest for Oligomers

Another interesting idea is based on the fact that the harmonic oscillator has an exact solution in the presence of an external electric field. The solution is a Gaussian whose centre is displaced by a quantity depending on the electric field intensity and the Gaussian exponent:

$$\Delta R_{ip} = \frac{\lambda_i \, F_i}{\alpha^2_p}$$

This has been applied in a FSGO scheme due to Frost (presented previously in Section 2.2) where the wave function is an antisymmetrized product of single Gaussians fully optimized (exponents, positions), each Gaussian describing a single electron pair. In a first step, we have determined the FSGO equilibrium geometry of six alkane molecules, in the absence of an external electric field. In a second step, the relaxation of the electronic density has been allowed for by an optimization of the positions of the FSGO basis orbitals in the presence of an external electric field keeping the angles defined by their positions relative to the C and H atoms fixed (calculation referred to as FSGO-1) or varying them (calculation referred to as FSGO-2). To assess the quality of the FSGO results, we have also evaluated the average polarizability tensor within STO-3G and 4-31G bases. The results are listed in Table 4.3. The theoretical average polarizabilities $<\alpha>$ are compared with experimental values when available [4.27, 4.28] and are also given in Table 4.3.

The FSGO polarizabilities are intermediate between the STO-3G and the 4-31G values. Obviously, the subminimal basis is not flexible enough to completely account for the distortion of the electronic cloud induced by an electric field and thus to produce fully quantitative results. However, the trends are well reproduced and the quantitative results obtained by the subminimal basis are better than those of the minimal STO-3G basis set. This is clearly an effect of the optimization of the orbital positions.

Table 4.3 **Average polarizabilities of alkanes**
(results in a.u.)

	STO-3G	FSGO-1	FSGO-2	4-31G	Exp. [4.8]	Exp. [4.7]
CH_4	6.000	8.173	10.316	12.422	17.999	17.547
C_2H_6	11.460	14.701	19.328	22.725	30.592	30.167
C_3H_8	16.830	21.113	27.719	32.720	43.644	42.449
C_4H_{10}	22.250	26.975	36.115	42.680	54.887	54.800
C_5H_{12}	27.755	34.558	45.170	52.752	67.035	67.622
C_6H_{14}	33.313	40.362	54.352	-	79.635	79.905

4.2.6 SI, esu and a.u. Units in Polarizability Calculations

The conversion coefficients of molecular polarizabilities from a.u. to esu and SI units are obtained from their relation with the energy. Dipole moments (μ) have dimensions QL of a charge (Q) multiplied by a length (L) while electric fields (F) and energies (E) have dimensions QL^{-2} and Q^2L^{-1}, respectively. We use the relations between electric polarizability (α), first electric hyperpolarizability (β), second electric hyperpolarizability (γ), dipole moment (μ), and field (F) to obtain the conversion factors:

$$\alpha = \frac{\mu}{F} = \frac{QL}{QL^{-2}} = L^3 \text{ (esu)} = \frac{Q^2L^2}{E} \text{ (SI)}$$

$$\beta = \frac{\mu}{F^2} = \frac{QL}{(QL^{-2})^2} = \frac{L^5}{Q} \text{ (esu)} = \frac{Q^3L^3}{E^2} \text{ (SI)}$$

$$\gamma = \frac{\mu}{F^3} = \frac{QL}{(QL^{-2})^3} = \frac{L^7}{Q^2} \text{ (esu)} = \frac{Q^4L^4}{E^3} \text{ (SI)}$$

We give below the data for their conversion to SI and esu units (a_0 = Bohr radius, E_a = Hartree, e = electron charge, C = Coulomb, m = meter, V = volt, J = Joule):

Electric dipole moment	ea_0	$8.4784 \cdot 10^{-30}$ C.m
Electric polarizability (α)	$e^2 a_0^2 E_a^{-1}$	$1.6488 \cdot 10^{-41}$ $C^2 \cdot m^2 \cdot J^{-1}$
		$0.14818 \cdot 10^{-24}$ cm^3
First electric hyperpolarizability (β)	$e^3 a_0^3 E_a^{-2}$	$3.2064 \cdot 10^{-53}$ $C^3 \cdot m^3 \cdot J^{-2}$
		$0.86392 \cdot 10^{-32}$ $cm^5 \cdot esu^{-1}$
Second electric hyperpolarizability (γ)	$e^3 a_0^3 E_a^{-3}$	$6.2354 \cdot 10^{-65}$ $C^4 \cdot m^4 \cdot J^{-3}$
		$0.50366 \cdot 10^{-39}$ $cm^7 \cdot esu^{-2}$
Electric field	$E_a e^{-1} a_0^{-1}$	$5.1423 \cdot 10^{11}$ $V \cdot m^{-1}$

It is important to note that other conventions are also widely used in the literature. As already pointed out in Section 4.2, they assume different relations between electric polarizability (α), first electric hyperpolarizability (β), second electric hyperpolarizability (γ), dipole moment (μ), energy (E), and field (F):

$$E = \frac{1}{2} \alpha F^2 = \frac{1}{3} \beta F^3 = \frac{1}{4} \gamma F^4$$

The conversion factors become in this case, 1 a.u. of polarizability (α) = $0.29635 \cdot 10^{-24}$ esu = $2*0.14818 \cdot 10^{-24}$ cm^3, 1 a.u. of first electric hyperpolarizability (β) = $0.025916 \cdot 10^{-30}$ esu = $3*0.86392 \cdot 10^{-32}$ $cm^5 \cdot esu^{-1}$, and 1 a.u. of second electric hyperpolarizability (γ) = $0.0020146 \cdot 10^{-36}$ esu = $4*0.50366 \cdot 10^{-39}$ $cm^7 \cdot esu^{-2}$ [4.8].

4.3 POLARIZABILITY OF INFINITE POLYMERIC CHAINS

4.3.1 The Difficulty of Calculating Polarizabilities in Infinite Polymeric Chains

In the study of the perturbation due to the switching of an external electric field, it is anticipated that the polarizability, normalized to the monomeric unit, tends to reach an asymptotic limit which should grow when the systems exhibit increased geometrical regularity (metallic situation). For complex systems, this limit will soon be out of reach when studying chains of increasing length. Thus, it would be very useful to be able to estimate this limit from calculations on infinite chains. In a first step, it might seem rather trivial to replace simply field-dependent MO's by field-dependent Bloch polymeric orbitals and assume the usual periodicity properties. However, two types of questions are raised.

300

On the one hand, as shown by Churchill and Holmstrom [4.29,4.30], serious difficulties arise in imposing realistic boundary conditions to solve the one-electron eigenvalue equation; under the periodic boundary conditions commonly used in treating the zero-field case (i.e., Born-Karman boundary conditions), this equation either leads to physically inconsistent results or, still worse, has no solution at all. This strange behaviour is a consequence of the pathological nature of the perturbing term, -$\mathbf{F.r}$, due to the external electric field F which becomes unbounded as $\mathbf{F} \rightarrow 0$ and $\mathbf{r} \rightarrow \infty$.

On the other hand, the periodic character of the perturbation is not guaranteed under the nonperiodic linear external perturbation which would rule out the use of field-perturbed Bloch orbitals. That point has been investigated by finite field calculations over a chain of 24 hydrogen atoms. The results are summarized in Figure 4.1.

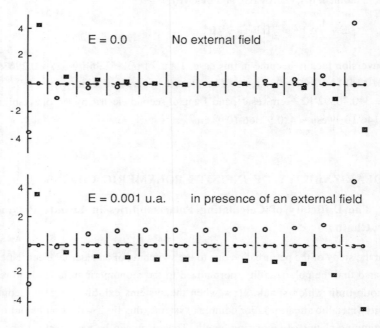

Figure 4.1   Net charges (x 10^{-2}) of a linear chain of 24 hydrogen atoms in the absence (top) and in the presence (bottom) of an external electrical field (31G calculations)

It is seen from Figure 4.1 which presents the net charge in the unperturbed and perturbed systems that the response appears at first sight to be periodic (at least in the middle of the molecule) even if the resulting potential deviates from the ideal -$\mathbf{F}.\mathbf{r}$ behaviour.

Theoretical studies of the asymptotic limit of the polarizability of infinite model polymeric chains (chains of hydrogen atoms, polyacetylene, and polydiacetylene) are described in the literature [4.31-4.34] using the S.O.S. (Summation-Over-States) perturbative methodology of Genkin & Mednis [4.35]. In this Section, we summarize for the models selected throughout this book, some results obtained with the simple Hückel method (polyacetylene chains within the π approximation) and with more complete *ab initio* techniques (chains of hydrogen atoms, polyethylene, and polysilane).

4.3.2   S.O.S. Methodology in Infinite Systems

In this Section, we are particularly interested in the S.O.S. value of the frequency-independent longitudinal polarizability introduced in Section 4.2.1.2 in terms of MO's of oligomers and polymers extending along the z-direction:

$$\alpha_{zz} = \sum_i \sum_a \frac{|\ 2 <\phi_i\ |z|\ \phi_a> |^2}{\Delta\varepsilon_{ia}}$$

The summations are extended over all the occupied {i} and unoccupied {a} MO's. The relation which defines the static dipole longitudinal polarizability per unit cell of a large oligomer:

$$\frac{\alpha_{zz}}{N} = \frac{1}{N} \sum_i \sum_a \frac{|\ 2 <\phi_i\ |z|\ \phi_a> |^2}{\varepsilon_a - \varepsilon_i}$$

has to be transformed when we are dealing with polymeric compounds. As mentioned by Callaway [4.36], it is evident that there will be difficulties associated with the application of perturbation theory because, for sufficiently large distances, the perturbation becomes arbitrarily large, no matter how weak is the field. Strictly speaking, there are no bound states when the Hamiltonian contains a term of the form of -$\mathbf{r}.\mathbf{F}$. A solution to this difficulty is given by Genkin and Mednis [4.35]. It is evident that calculations on oligomers of finite size cannot be used in order to get a reliable estimate of the polarizability per unit cell of the infinite chain. To solve this problem, we turn to the basic equations of polymer quantum chemistry (see Section 1.9.1). We recall that the LCAO Bloch form which describes the polymer orbitals extending over the whole polymer as Bloch combinations of the atomic orbitals of the polymer chain is:

$$\phi_n(k,\mathbf{r}) = \frac{1}{\sqrt{N}} \sum_j e^{ikja} \sum_p c_{np}(k) \, \chi_p(\mathbf{r}\text{-}\mathbf{P}\text{-}ja)$$

$$= e^{ikz} \frac{1}{\sqrt{N}} \sum_j e^{ik(ja\text{-}z)} \sum_p c_{np}(k) \, \chi_p(\mathbf{r}\text{-}\mathbf{P}\text{-}ja)$$

$$= e^{ikz} u_n(k,\mathbf{r})$$

where N is the (odd) number of unit cells in the polymer, n the band index, $\mathbf{r}$ the position vector and k the wave number in the first Brillouin zone ($-\pi/a \le k \le \pi/a$), $\mathbf{P}$ the position of the center of orbital χ_p in the unit cell of length a, and j the cell index; $u_n(k,\mathbf{r})$ is the periodic part of the LCAO Bloch function. For the sake of simplicity, we use the simplified notation $\chi_p(\mathbf{r}\text{-}\mathbf{P}\text{-}ja) = \chi_p^j$.

Genkin and Mednis have proposed the following formula for the calculation of the polymeric polarizability per unit cell:

$$\frac{\alpha_{zz}}{N} = \frac{2a}{\pi} \sum_i \sum_a \int_{BZ} dk \, \frac{|\,\Omega_{ia}(k)\,|^2}{\varepsilon_a(k) - \varepsilon_i(k)}$$

where:

$$\frac{\Omega_{ia}(k)}{N} = \frac{1}{a} \int_{cell} d\mathbf{r} \, u_i^*(k,\mathbf{r}) \, \frac{\delta}{\delta k} u_a(k,\mathbf{r})$$

The transformation of the matrix elements of the coordinate z of the molecular case $<\phi_i|z|\phi_a>$ or of the polymeric Bloch case $<\phi_n(k)|z|\phi_{n'}(k')>$ into the vertical matrix elements of the k-gradient $\delta/\delta k$ involving the periodic parts of the Bloch functions $u_n(k,\mathbf{r})$ is given in several books [4.36, 4.37].

4.3.3   S.O.S. Hückel Calculations of Polarizabilities of Polyacetylene Chains

4.3.3.1 Summary of Hameka's approach to the calculation of linear polarizability of polyenes of limited size

The linear polarizabilities of conjugated hydrocarbons chains (of size N, N being the number of carbon atoms) have been calculated by Davies [4.38] and recently by Risser [4.39] by conventional perturbation theory. The latter paper assumes both orthogonal and non-orthogonal Hückel atomic basis sets. Apparently, it was not recognized in these papers that the perturbation equations could be solved analytically. A full standard analytical derivation of the perturbation expansion is given in Hameka's paper in the case of an orthogonal basis set [4.40]. An alternative derivation in terms of Bessel functions is also given by Flytzanis [4.41]. The Hückel π-molecular orbitals are LCAO combinations of atomic $2p_\pi$ orbitals. In the absence of an electric field, they are determined from variational principles using a one-electron effective Hamiltonian (h), whose matrix elements with respect to the orthonormal atomic orbitals are evaluated from the standard Hückel approximation:

$h_{nn} = \alpha$ for diagonal matrix elements (sometimes improperly called Coulomb integrals),

$h_{nm} = \beta$ for the nearest-neighbour interactions (sometimes improperly called resonance integrals),

$h_{nm} = 0$ for nonnearest-neighbour interactions

The Hückel energies and molecular orbitals for finite linear polyene chains of length N are found in many textbooks [4.42-4.45]:

$$\varepsilon_i = \alpha + 2\beta\cos\left(\frac{i\pi}{N+1}\right)$$

$$\phi_i(\mathbf{r}) = \sum_{n=1}^{N} C_{in}\,\chi_n(\mathbf{r}) = \sum_{n=1}^{N} \sqrt{\frac{2}{N+1}}\ \sin\left(\frac{in\pi}{N+1}\right)\chi_n(\mathbf{r})$$

In the presence of an external electric field (F) applied along the chain axis (z), the perturbed Hamiltonian contains a supplementary term -Fz (we use the same notation as in [4.40]); Hameka has proposed to neglect the differential overlap so that the perturbation is restricted to diagonal terms which are then calibrated on the resonance integral β:

$$-F \int \chi_n(\mathbf{r}) \, z \, \chi_n(\mathbf{r}) \, d\mathbf{r} = -FM_{nn}^z = n\mu_0 F = n\lambda\beta$$

In this expression, λ is a scale parameter which defines the perturbation in β units, n is a running index starting from 1 for the diagonal matrix element of the carbon atom located at the left-hand extremity to N for the diagonal matrix element of the carbon located at the right-hand extremity. This increase over the running index n is due to the linear behaviour of the perturbation -Fz. Hameka's analytical procedure leads to the exact Hückel expression of the linear polarizability:

$$\alpha_{zz} = - \frac{2\beta\lambda^2}{F^2} E_0^{(2)} \propto E_0^{(2)}$$

4.3.3.2 Hückel calculation of linear polarizability of regular and bond alternant polyenes of limited sizes

In the Hückel approach, the lack of flexibility of the π-basis set allows us to deal only with the longitudinal component of the polarizability α_{zz}. As established in Section 4.3.3.1, the value of the z-moment integrals M_{nn}^z depends linearly on the index n of the carbon atom on which χ_n is centered. Finally, the dipole moment integrals over MO's are in units of length d (d being half the length of the translational unit cell, i.e., two carbon atoms):

$$(z)_{ai} = \sum_{n=1}^{N} n \, c_{in} \, c_{an}$$

where c_{in} (or c_{an}) are the LCAO coefficients of the MO i (or a). In terms of LCAO coefficients, the frequency-independent longitudinal polarizability is written:

$$\alpha_{zz} = 4 \sum_i \sum_a \sum_m \sum_n \frac{c_{in} \, c_{an} \, c_{im} \, c_{am} \, m \, n}{\Delta\varepsilon_{ia}}$$

The polarizabilities are expressed in $-d^2 e^2 \beta^{-1}$, since the Hückel excitation energies $\Delta\varepsilon_{ia}$ are traditionally estimated in β units. In order to generalize Hameka's approach on the polarizability of linear polyenes to bond alternant geometries (as they actually exist), we have identified the single-like bonds by an interaction β' less strong than the β interaction corresponding to the double bonds. The degree of alternation is thus characterized by the value of the bond alternation parameter $K = \beta'/\beta \leq 1$. We have used four different values of K: 1.0 (no bond alternation), 0.877 (empirical value), 0.6, and 0.1(very strong bond alternation). The empirical value $K = \beta'/\beta = 0.877$ is obtained from experimental estimates of the π-valence band width (10 eV) and the energy gap (1.4 eV) of an infinite polyacetylene chain [4.46].

In Table 4.4 and in Figure 4.2, we present the calculated Hückel longitudinal polarizabilities per unit cell (α_{zz}/N), for polyenes composed by 2 to 298 carbon atoms and 4 degrees of bond alternation (K=1.0, 0.877, 0.6, 0.1).

Table 4.4   **Hückel longitudinal polarizabilities per carbon atom (α_{zz}/N) for linear regular (K=1.0) and alternant (K= 0.877, 0.6, 0.1) polyenes** (in $-d^2 e^2 \beta^{-1}$ units). N=number of carbon atoms in the chain

N	K=1.0	K=0.877	K=0.6	K=0.1
2	0.250000	0.250000	0.250000	0.250000
6	1.416946	1.082470	0.575073	0.257142
10	3.434008	2.120716	0.755126	0.258613
26	19.987927	5.958754	0.973759	0.259970
50	70.281619	9.015633	1.042022	0.260377
98	262.541219	11.049722	1.078245	0.260593
198	1055.663589	12.139081	1.097302	0.260707
298	2379.302278	12.497352	1.103568	0.260744
polymer limit	∞	13.206728	1.115977	0.260819

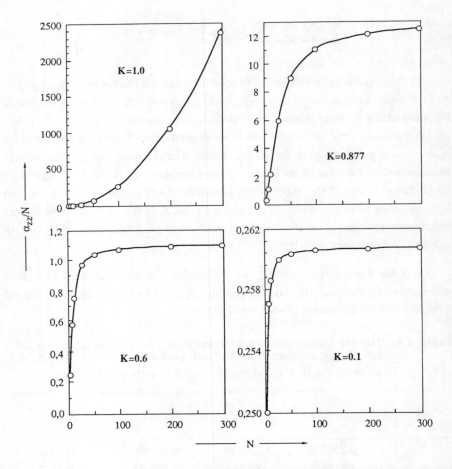

Figure 4.2 Evolution of the longitudinal polarizabilities of polyacetylene chains per carbon atom in -1/β units according to the length of the chain (N) and the degree of alternation of the single and double bonds (K)

Column 2 (K=1.0) is in full agreement with Hameka's value obtained from a completely different numerical technique. It is clear that the extrapolated polarizability per carbon atom of the infinite polymer chain is infinite in agreement with the fact that a regular chain of equidistant -(CH)- units has metallic character (degeneracy of the HOMO and the LUMO). Columns 3 (K=0.877), 4 (K=0.6), and 5 (K=0.1) corroborate the well established fact that bond alternation (or lack of regularity) decreases the polarizability [4.26, 4.43, 4.44]. This effect will be described in detail in Section 4.4.5. It is also clear that calculations of polarizability restricted to oligomers (≤ 300) are not able to give an approximation of the asymptotic limit value. It is thus important to calculate the asymptotic limit using polymeric calculations as performed in the next Section. The size effect is plotted in Figure 4.2 according to the values of K.

4.3.3.3 Hückel calculation of the linear polarizability of infinite polyacetylene chains

It is evident from Table 4.4 that, even with such a simple method as the Hückel procedure, calculations on oligomers of finite size cannot give a reliable estimate of the polarizability per unit cell of the infinite chain. To solve this problem, we turn to the LCAO Bloch form which describes the polymer orbital extending over the whole polymer as a Bloch combination of the atomic orbitals of the polymer chain. In the case of the Hückel study of polyacetylene in the π-approximation, each unit cell contains two $2p_\pi$ orbitals χ_1 and χ_2. As indicated previously in Section 1.9.3.1, the π-band structure is formed by two bands (one occupied, one unoccupied) of analytical form:

valence band:

$$\varepsilon_1(k) = \alpha + |\omega(k)|$$

$$\phi_1(k,r) = \frac{1}{\sqrt{N}} \sum_j e^{ikja} \left\{ \frac{1}{\sqrt{2}} \frac{\omega^*(k)}{|\omega(k)|} \chi_1^j + \frac{1}{\sqrt{2}} \chi_2^j \right\}$$

$$= e^{ikz} \frac{1}{\sqrt{N}} \sum_j e^{ik(ja-z)} \left\{ \frac{1}{\sqrt{2}} \frac{\omega^*(k)}{|\omega(k)|} \chi_1^j + \frac{1}{\sqrt{2}} \chi_2^j \right\}$$

$$= e^{ikz} u_1(k,r)$$

conduction band:

$$\varepsilon_2(k') = \alpha - |\omega(k')|$$

$$\phi_2(k',r) = \frac{1}{\sqrt{N}} \sum_j e^{ik'ja} \left\{ \frac{1}{\sqrt{2}} \frac{\omega^*(k')}{|\omega(k')|} \chi_1^j - \frac{1}{\sqrt{2}} \chi_2^j \right\}$$

$$= e^{ik'z} \frac{1}{\sqrt{N}} \sum_j e^{ik'(ja-z)} \left\{ \frac{1}{\sqrt{2}} \frac{\omega^*(k')}{|\omega(k')|} \chi_1^j - \frac{1}{\sqrt{2}} \chi_2^j \right\}$$

$$= e^{ik'z} u_2(k',r)$$

where $u_1(k,\mathbf{r})$ and $u_2(k',\mathbf{r})$ are the periodic parts of the Bloch functions and where:

$$\omega(k) = \beta \left(1 + K\, e^{ika}\right)$$

$$|\,\omega(k)\,| = \beta \sqrt{1 + 2K \cos(ka) + K^2}$$

$$K = \beta'/\beta$$

As seen in Section 4.3.2, the relation which defines the static dipole longitudinal polarizability per carbon atom:

$$\frac{\alpha_{zz}}{N} = \frac{1}{N} \sum_i \sum_a \frac{|\,2 <\phi_i|z|\phi_a>\,|^2}{\varepsilon_a - \varepsilon_i}$$

has to be transformed to the Genkin-Mednis formula which, in the case of polyacetylene, is written as:

$$\frac{\alpha_{zz}}{N} = \frac{a}{\pi} \int\limits_{-\pi/a}^{+\pi/a} dk\, \frac{|\,\Omega_{12}(k)\,|^2}{|\,\omega\,(k)\,|}$$

where:

$$\frac{\Omega_{12}(k)}{N} = \frac{1}{a} \int\limits_{cell} d\mathbf{r}\; u_1{}^*(k,\mathbf{r})\, \frac{\partial}{\partial k}\, u_2(k,\mathbf{r})$$

is the coupling matrix element between bands 1 and 2 (valence and conduction bands, respectively). Analogies can be made between the molecular and polymeric formulas which give the polarizability per unit cell. For the polymeric compounds, the two summations over the occupied and unoccupied states are replaced by an integration over the valence and conduction bands in the first Brillouin zone. It is important to note that the integration is restricted to the vertical terms (k=k'). This comes from the vanishing of the terms which include different k wave numbers [4.36, 4.37]. By introducing the expressions of $u_1(k,\mathbf{r})$ and $u_2(k,\mathbf{r})$ into the coupling matrix element $\Omega_{12}(k)$, we obtain:

$$\frac{\Omega_{12}(k)}{N} = \frac{1}{a} \int\limits_{cell} d\mathbf{r}\; \phi_1{}^*(k,\mathbf{r})\, \frac{\delta}{\delta k}\, \phi_2(k,\mathbf{r}) - \frac{i}{a} \int\limits_{cell} d\mathbf{r}\; \phi_1{}^*(k,\mathbf{r})\, z\, \phi_2(k,\mathbf{r})$$

We notice that the second term corresponds to the dipole integrals over a single cell. Using the definitions of $\phi_1(k,\mathbf{r})$ and $\phi_2(k,\mathbf{r})$, by performing the derivative with respect to k, and carrying out the summation over the j indices, we find the relation:

$$\Omega_{12}(k) = \frac{1}{2}\left[\frac{\omega(k)^*\,\omega(k)'}{|\omega(k)|^2} - \frac{|\omega(k)|'}{|\omega(k)|}\right] + \frac{i}{4}$$

$$= \frac{-i}{2}\left[\frac{K\cos ka + K^2}{|\omega(k)|^2}\right] + \frac{i}{4}$$

where the prime index on a term denotes the k-derivative. This expression leads to:

$$\Omega_{12}(k) = \frac{i\,a}{4}\,\frac{(1 - K^2)}{|\omega(k)|^2}$$

which gives the polarizability per unit cell of polyacetylene in $-a^2\,e^2\,\beta^{-1}$ units:

$$\frac{\alpha_{zz}}{N} = \frac{a}{16\pi}\int\limits_{-\pi/a}^{+\pi/a} dk\,\frac{(1 - K^2)^2}{|\omega(k)|^5}$$

The latter integral is solved by an equidistant numerical integration in the first Brillouin zone. The procedure is numerically very fast since it does not imply diagonalization of large matrices as needed by the extrapolation study of the oligomeric properties. The asymptotic value (polymeric limit) of the polarizability per unit cell and per carbon atom is given in Table 4.4. The asymptotic limit of the regular polyacetylene is infinite according to the metallic character of the regular chain and to the degeneracy of the energies of the valence and conduction bands at the edge of the first Brillouin zone $(k=\pm\pi/a)$.

4.3.4 S.O.S. Uncoupled Minimal Basis Set Hartree-Fock Calculations of Polarizabilities of Chains of Hydrogen Atoms

In this Section, we discuss an Hartree-Fock calculation on a model polymeric chain of hydrogen atoms [4.31-4.33]. In the model, the hydrogen molecules (with intramolecular bond length of 2 a.u.) are separated from each other by various distances (the intermolecular distances are respectively 2, 3, and 4 a.u.). In Table 4.5, we compare the polarizabilities per H_2 for the hydrogen chains obtained by S.O.S. uncoupled calculations and coupled-perturbed Finite Field Hartree-Fock (F.F.) calculations carried out using a minimal STO-3G basis set. As in the treatment of polyacetylene chains within the Hückel framework, the results show a decrease in the values of the polarizabilities as the alternance, characterized by the intermolecular distance d, increases.

Table 4.5 S.O.S. and F.F. STO-3G calculations of polarizabilities and their asymptotic limit per H_2 for hydrogen chains of different size and of different intermolecular distance, d (intramolecular bond length = 2 a.u.) (in a.u.)

Number of H_2	d = 2 a.u.		d= 3 a.u.		d = 4 a.u.	
	S.O.S.	F.F.	S.O.S.	F.F.	S.O.S.	F.F.
1	5.74	5.81	5.74	5.81	5.74	5.81
3	16.93	18.63	8.28	9.91	6.32	7.08
5	28.96	35.20	9.02	11.58	6.44	7.43
7	40.80	54.58	9.34	12.39	6.50	7.59
9	51.90	75.94	9.52	12.87	6.53	7.68
15	78.72		9.78		6.57	
polymer limit	129.42		10.22		6.63	

In all cases, the S.O.S. values are smaller than the F.F. ones. The difference increases with the size of the oligomers and with their regularity. It is a consequence of the uncoupled character of the S.O.S. methodology. By adding the reorganizational effects due to coupling, we introduce the relaxation in the excitation phenomena. Indeed, in the S.O.S. case, the denominator equals $(\varepsilon_a - \varepsilon_i)$ which corresponds to the difference between two molecular orbital energies whereas the CPHF theory allows the reorganization of the system and leads to smaller values for the denominator and thus to higher polarizabilities.

We have thus performed the calculations on the infinite polymeric hydrogen chains by using the Genkin-Mednis S.O.S. methodology. The asymptotic limits for the longitudinal static polarizabilities for the infinite hydrogen chains which present the alternances described above are also presented in Table 4.5.

4.3.5 S.O.S. Uncoupled Hartree-Fock Calculations of Polarizabilities of Polyethylene and Polysilane Chains

We complete this Section on the polarizability of infinite polymeric chains by examples on more realistic compounds; we have selected the polyethylene and polysilane chains, whose band structures have already been analyzed in Section 2.5. In Table 4.6, we present the S.O.S. polarizabilities of some alkane and silane oligomers. The x- and y-axes are perpendicular to the polymer backbone while the z-axis is longitudinal.

Table 4.6 **Polarizability components of alkane and silane chains, longitudinal polarizability per unit cell, and their asymptotic limit** (in a.u.) (*ab initio* STO-3G calculations)

TOTAL POLARIZABILITY

Number of monomer units	Polyethylene chains			Polysilane chains		
	α_{zz}	α_{xx}	α_{yy}	α_{zz}	α_{xx}	α_{yy}
1	7.604	9.089	9.089	21.551	19.153	19.153
2	15.430	17.176	18.296	47.824	41.276	39.220
3	23.268	25.340	27.500	78.663	62.030	59.230
4	31.096	33.527	36.705	109.944	82.903	79.372

POLARIZABILITY PER UNIT CELL

Number of monomer units	Polyethylene chains α_{zz}/N	Polysilane chains α_{zz}/N
1	7.604	21.551
2	7.715	23.912
3	7.757	26.221
4	7.775	27.486
Polymer limit	7.770	28.630

We note the higher polarizabilities of the silicon compounds, already pointed out by many authors and associated with the greater volume of the silicon atom and the σ-conjugation present in the silanes (a problem we shall discuss later on). More interesting are the longitudinal polarizabilities per unit cell given in the second part of Table 4.6. The van der Waals volumes for the monomer units are 268.5 a.u.^{3} and 403.1 a.u.^{3}, respectively. The ratio (α/V multiplied by 100) is larger for polysilane (7.10) than for polyethylene (2.89). It is important to note that in the alkane chains, the effect of the chain size turns out to be almost strictly additive. In the silane chains, on the other hand, there is a saturation pattern and the asymptotic limit is not reached after the first few terms, a clear indication of a σ-conjugation effect. From a methodological point of view, it again indicates the need of developing accurate techniques of calculation for the asymptotic limit of the longitudinal polarizability. The asymptotic limit of the longitudinal polarizability per unit cell is obtained by using the S.O.S.- Genkin-Mednis approach (see Section 4.3.2). It is calculated from the *ab initio* band structure. A correct band indexing and band reordering (described in Section 2.4.4) is needed in order to get significant results.

A complete analysis of the band structures of polyethylene and polysilane in terms of minimal valence basis sets (1s for hydrogen, 2s, 2px, 2py, and 2pz for carbon, plus 3s, 3px, 3py, and 3pz for silicon) and in terms of bond orbitals has been given in Section 2.5.

As mentioned in Section 4.3.2, the polarizability per unit cell can be separated into contributions from each occupied band i, due to the structure of the summations of the Genkin-Mednis formula. An interesting feature is, thus, to relate the topology of each band and their contribution to the electric polarizability. In Table 4.7, we report the contributions of each occupied Bloch valence orbital of polyethylene and polysilane according to the labelling of Section 2.5 (see Figure 2.6). The core bands (bands 1 and 2 of polyethylene, and bands 1 to 10 of polysilane) contribute very weakly.

Particularly interesting are the values of the two upper bands. In both polymers, these two bands contribute most to the longitudinal polarizability (94.0% in the case of polyethylene and 97.7% in the case of polysilane). Bands 7 and 8 of polyethylene (15 and 16 of polysilane) are antibonding combinations of the C-H (or the Si-H) orbital and of the C-C (or Si-Si) bond orbital. Band 7 of polyethylene (band 15 of polysilane) is dominated at k=0 by the C-H (or Si-H) orbitals. Recalling that band 7 (from k=0 $\rightarrow$ π/a) and band 8 (from k=π/a $\rightarrow$ k=0) of polyethylene (bands 15 and 16 of polysilane) are the two branches of the same valence band within the asymmetric cell -(CH$_2$)$_n$- (or -(SiH$_2$)$_n$-), the analysis of Section 2.5 has shown that they gradually transform into an antisymmetric (with respect to the plane of the zigzag chain) combination of the C-C (or Si-Si) bond orbitals. At k = 0, the HOMO, (band 8 of polyethylene or band 16 of polysilane) has no contribution from the C-H (or the Si-H) orbitals. The latter statement is important since it is generally believed that the HOMO of the tetrahedrally σ-bonded chains has mainly an X-H character. It is now clear that it is mainly an X-X bond orbital with nodes on each atom of the chain. This result is fully confirmed by inspection of the LCAO coefficients obtained in the minimal basis set calculation; the HOMO orbitals have only C_{2pz} contributions in polyethylene or Si_{3pz} contributions in the case of polysilane (remember that z is the chain axis). This topology, added to the large orbital polarizability of the 3p orbitals of silicon, is probably responsible for the σ-conjugation of polysilanes. This leads to a larger slope of the B_1 band near k=0 for polysilane and is also in agreement with the ratio of the effective masses (m^*): polysilane / polyethylene = 1.50 (see Section 2.5).

Table 4.7   **Contributions of the valence occupied bands to the longitudinal asymptotic polarizability per unit cell for polyethylene and polysilane chains** (in a.u.) (*ab initio* STO-3G calculations)

Polyethylene		Polysilane	
Band	Contribution	Band	Contribution
3 (A_1)	0.011	11 (A_1)	0.159
4 (B_1)	0.012	12 (B_1)	0.094
5 (B_2)	0.065	13 (B_2)	0.190
6 (A_2)	0.288	14 (A_2)	0.190
7 (A_1)	2.938	15 (A_1)	6.562
8 (B_1)	4.363	16 (B_1)	21.410

4.4 THEORETICAL DESIGN OF ACTIVE MAIN CHAIN OLIGOMERS AND POLYMERS

4.4.1 Introduction

As pointed out in the introduction to this Chapter, simple chemical intuition is valuable but seldom sufficient to provide really useful information on prospective structures for a given optoelectronic property. This is why we have undertaken systematic quantum chemical calculations on various chemical structures with polymers as our ultimate goal.

Polymers are presently considered very promising for nonlinear optics applications and constitute our main subject of interest; they can be divided in two categories:

(a) Active main chain polymers such as polyacetylene, $[-CR_1=CR_2-C\equiv C-]_x$, where the electric susceptibility originates from the response of the π-electrons distributed along the polymer backbone. Such systems are usually considered for devices based on third-order nonlinear effects and ultrafast responses.

(b) Polymers whose electric susceptibility is due to polar active organic molecules grafted on the backbone. In this case, the backbone itself is usually inactive and is mainly there to hold the active organic molecules accountable for the nonlinear response based on the first hyperpolarizability β. The required noncentrosymmetry is imposed by a poling process usually followed by several stabilization steps (cross-linking, etc.) to maintain the preferred orientation of the polar groups.

Many efforts are made towards the design of active main chain polymers. However, some of the results obtained on the monomeric units are also valuable for the design of conjugated molecules either to be used in molecular and liquid crystals, in Langmuir-Blodgett films, or as pendant groups of nonactive main chain polymers. We now report on our *ab initio* calculations of the static electric polarizability as a way to assess the relative merits of prospective conjugated structures which could be incorporated into active main chain polymers. The concepts of conjugation and delocalization are pervasive in this Chapter because there is a direct relation between the importance of the electric response and the effective delocalization length.

4.4.2 Methodological Questions

We have restricted our attention to the linear response coefficient α because, on the one hand, the first hyperpolarizability as such is of limited interest for the design of active main chain polymers, and, on the other hand, reliable calculations of the second hyperpolarizability are so far not possible for molecules of the complexity that is of interest to us. Fortunately, measured hyperpolarizabilities usually parallel the trends (enhancement or attenuation) observed for polarizability, and models to relate hyperpolarizabilities to α have been proposed on the basis of these observations [4.47, 4.48]. Similarly, test calculations on conjugated chains indicate that these assumptions are indeed reasonable [4.49, 4.50]. Accordingly, we base the investigations reported in this Chapter on the postulate that the variations obtained for the mean polarizability, $<\alpha> = \frac{1}{3}[\alpha_{xx} + \alpha_{yy} + \alpha_{zz}]$, will be indicative of the trends for the average second hyperpolarizability $<\gamma> = \frac{1}{5}[\gamma_{xxxx} + \gamma_{yyyy} + \gamma_{zzzz} + 2\gamma_{xxyy} + 2\gamma_{xxzz} + 2\gamma_{yyzz}]$.

An important aspect of *ab initio* calculations is the choice of the basis set which describes the electron distribution reorganization resulting from the external perturbation. It determines both the quality and the cost of the results. Because of the size of the systems treated and the need for results obtained at the same level of accuracy for comparison purposes, the minimal STO-3G basis has been used for all the molecules considered in the present Chapter. In spite of its obvious limitations, the STO-3G basis predicts molecular structures reasonably well in the sense that errors on bond distances and angles remain fairly constant for a wide variety of molecular structures and thus fulfills the conditions sought for a theoretical design of organic molecules and polymers for optoelectronics. It must be recalled that polarizability is quite sensitive to geometry and therefore it is preferable that all geometries be optimized to avoid structure effects in the predictions. Similarly, good qualitative estimates of the polarizability of conjugated hydrocarbons as well as structures including the common heteroatomic bondings have been noted with the STO-3G basis (see also Section 4.2.4). However, the components characterizing the direction perpendicular to the conjugation plane are systematically underestimated; this is true unless very extended basis sets are used. Thus, the predictions based on the STO-3G basis cannot be better than semiquantitative, but this is already acceptable for the immediate purpose of ranking as most promising molecules of similar composition and/or structures for a particular optoelectronic property. In some cases calculations have been carried out with larger bases to check for possible basis set dependencies that could alter the conclusions qualitatively.

Reported calculations are at the Hartree-Fock level. As pointed out in the Section on methods of calculations, the finite-field version of the coupled Hartree-Fock method provides reliable and useful estimates of the components of the electric polarizability tensor α. This, because the finite-field approach takes into account, directly and in a self-consistent way, the orbital relaxations in the presence of the perturbing external electric field. The electric polarizability α_{ii}, $i = x$, y, and z, is calculated within the framework of the finite-field method of Cohen and Roothaan discussed in Section 4.2.1.1.

4.4.3 Model Systems

As already alluded to, the design of organic polymers for optoelectronics includes many aspects and questions into which insight from quantum chemistry calculations can be obtained. Except for calibration purposes with known results, experimental or theoretical, isolated calculations are usually not very useful if not placed in the general context of the research. Therefore our contribution to this field has been more than just calculating the electric polarizability tensor of molecules and, from the results, searching for structures yielding the largest possible responses [4.51]. Just as in drug design where secondary effects can rule out an otherwise very efficient molecule, one must be aware of the constraints a system must meet to be a valuable material. By providing a unified and logical research proposition, it is our purpose in the last part of this Chapter to illustrate not only the variety of questions but also their dependence.

It is with this in mind that we have chosen illustrations from our ongoing research on finding the characteristics an active main chain polymer should have to exhibit high electric response. The problems arise at different levels, each being the heading of a Section in the sequel.

First, it is sensible that in order to exhibit high electric response a polymer should be made of already polarizable monomeric units. Experiment and theory, agree on the fact that structures with electronic and structural "homogeneity" are usually best for large electric responses. However, it can be important for various reasons to alter this homogeneity by inserting heteroatomic functions in a given conjugated structure, and thus it is useful to predict the influence of these alterations on the overall response. Assessing the polarizability of monomeric units will be the subject of Section 4.4.4; we will also deal with the question of the relative stability of chains.

One of the interests in active main chain polymers is the possibility to increase the delocalization of the π-electrons and thus the electric response along the backbone by connecting conjugated units. However, as we shall see, it is not necessarily true that connecting conjugated moieties will lead to enhanced response, even when structurally ideal conditions are assumed. This forms the subject of Section 4.4.5.

The organization at the molecular level is also important since sequences of conjugated units, sometimes assembled with great difficulties, do not distribute in an ordered way and the net response is reduced accordingly. Quantum chemistry can investigate the question of forcing the order, e.g., with hydrogen bonds, while not spoiling too much the net electric response. Results along these lines are reported in Section 4.4.6.

Finally, doping can force π-electron structures to relax in a way that enhanced electric responses result. Some of these systems are also considered as candidates for conducting polymers and thus are part of both fields. This constitutes the subject of the last Section 4.4.7.

4.4.4 Assessment of Repeating Units

In addition to problems concerning the synthesis, stability, etc., the question arises of assessing the intrinsic merits of the repeating units of conjugated polymers. This evaluation cannot be isolated from the other aspects of the problem. One may decide to favor the organization at the molecular level (order, packing, etc.), or pay attention to the stability of the resulting material, or even both. This often imposes a compromise between the wish of having very effective repeating units (e.g., highly polarizable units) and the need to meet other constraints inconsistent with the best available units. For example, contrary to normal intuition, the polarizability of conjugated structures does not always scale linearly with the number of π-electrons they contain. Quantum chemistry calculations can provide relatively fast and economic ways of performing reliable assessment of various molecular structures [4.52]. Furthermore, and this is one of the advantages of *ab initio* calculations over the semiempirical ones, it is possible to predict with an acceptable margin of accuracy quite different properties of rather different and sometimes unusual structures. This is illustrated in five situations, three with pure hydrocarbon systems and two where heteroatomic functions are inserted in the hydrocarbon backbone.

4.4.4.1 Molecules containing the same number of π-electrons

The longitudinal polarizability α_{zz} of four molecules, benzene (**1**), trans-1,3,5-hexatriene (**2**), vinylacetylene (**3**) and butatriene (**4**), all containing 6 π-electrons are compared in Table 4.8.

Table 4.8 *Ab initio* STO-3G longitudinal polarizability α_{zz} of four conjugated molecules containing 6 π-electrons
The polarizability is expressed in atomic units, l_π and V are measures of the conjugated pathway (in bohr) and of the molecular volume (in bohr^3), respectively.

Molecule	α_{zz}	α_{zz}/l_π	α_{zz}/V
benzene (**1**)	45.26	30.72	0.083
hexatriene (**2**)	104.13	30.55	0.172
vinylacetylene (**3**)	44.65	24.08	0.113
butatriene (**4**)	74.23	36.40	0.189

The calculations have been performed at the STO-3G level and all geometrical parameters have been optimized at that level of the theory. The corresponding structures are represented in Figure 4.3, with selected geometrical parameters.

The molecular planes coincide with the (x,z)-plane, the z-axis being along the direction of maximal extension of the molecules, and the y-axis is perpendicular to the molecular plane. The total longitudinal polarizability α_{zz} is not the best quantity to consider when comparing molecules for their electric response, but instead the polarizability divided by the scale of the molecule since what is important in a material is the density of active species. In the case of chain systems, the length could be appropriate to scale the longitudinal polarizability. As indicated in Figure 4.3, l_π is the vector distance (in bohr) between the most distant carbons, it provides a qualitative (admittedly arbitrary) measure of the molecules extension. A more appropriate quantity, however, is the molecular volume V because the molecular anisotropy is not always as important as in butatriene, for example. V is the volume enclosed in the van der Waals spheres centered on the atoms of the molecules, the volumes of the caps of the intersecting spheres being subtracted.

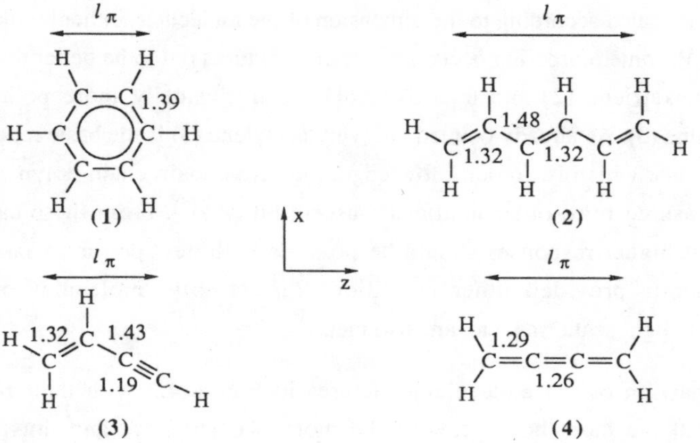

Figure 4.3 Schematic representation of molecules (1) to (4) and relevant distances (in Å, STO-3G optimized)

The numbers in Table 4.8 show important differences in the polarizability values of the structures (1) to (4) in spite of the fact that they contain the same number of π-electrons. First to point out is that vinylacetylene (3) and benzene (1) are the less polarizable of the four molecules, in absolute value and relative to both measures of their dimension, l_π and V. Hexatriene (2) is quite polarizable which points to a more effective delocalization of the 6 π-electrons along the molecular backbone. Vinylacetylene (3), the template repeating unit found in polydiacetylenes, is much less polarizable than hexatriene (4). This difference is due to a strong confinement of the four π-electrons corresponding to the C≡C bond which leads to an overall less effective delocalization. Finally, butatriene (4), a cumulenic structure, shows an appreciable polarizability value. This analysis stresses the significant dependence of the polarizability upon molecular structure. In particular, the extension of the conjugated backbone over which the π-electrons distribute is important to know. In the absence of data on the molecular structure, quantum chemistry can supplement the missing information with quite acceptable confidence [4.52]. Note that benzene (1), which has good chemical stability and tendency to induce local order due to its flat structure, is often incorporated in polymer backbones for these two reasons. Unfortunately the polarizability of isolated benzene (1) is not as favorable as these two properties.

If α_{zz} is scaled according to the dimension of the molecules, either by the length l_π or the volume V, some reordering occurs: butatriene (**4**) turns out to be better than hexatriene (**2**). Both hexatriene (**2**) and butatriene (**4**) are intrinsically more polarizable than vinylacetylene (**3**). As already pointed out, vinylacetylene (**3**) is the basic repeating unit of polydiacetylene which is, to our knowledge, the active main chain polymer having the highest measured third-order nonlinear susceptibility $\chi^{(3)}$. From these calculations it appears that higher responses should be possible with new polymers based on other repeating units provided other conditions on stability, molecular organization, transparency, light scattering, etc. are also met.

Comparison of the molecular structures in Figure 4.3 with their polarizability suggests that the more homogeneous the molecular structure and thus the electron distribution and the more polarizable is the system. As will be seen below, this seems to be quite general and could be used as a simple and convenient rule in the design of molecules for optoelectronics. However, to obtain more quantitative information on the importance of the polarizability gain with respect to geometry changes, direct calculations are required.

4.4.4.2 Dependence of the longitudinal polarizability on the molecular structure

As already mentioned, the longitudinal electric polarizability depends on the molecular structure [4.26, 4.53]. In the case of polydiacetylene crystals [4.54] large structural variations can actually be produced, e.g., by application of hydrostatic pressure, mechanical deformation or stresses induced by side groups. Thus, it is interesting to investigate in a systematic way the importance of structural modifications on α_{zz} for various model oligomers of polydiacetylenes. Here, the active part of their backbones is modelled by various geometries of the 1,5,9-decatriene-3,7-diyne (**5**) molecule represented in Figure 4.4.

The geometries are labeled by the names PDA (polydiacetylene), PTS (side group = $-CH_2OSO_2\phi CH_3$) [4.55], TCDU (side group = $- (CH_2)_4OCONH\phi$) [4.56], and PBT (polybutatriene) which, in the case of PTS and TCDU, refer to the existing polymers, but where the actual side groups have been replaced by hydrogen atoms. The geometrical parameters and the longitudinal polarizability are given in Table 4.9. To the names PDA and PBT correspond the hypothetical structures of an ideal polydiacetylene and an ideal polybutatriene, their structures have been obtained by geometry optimization of the corresponding infinite model chain [4.57].

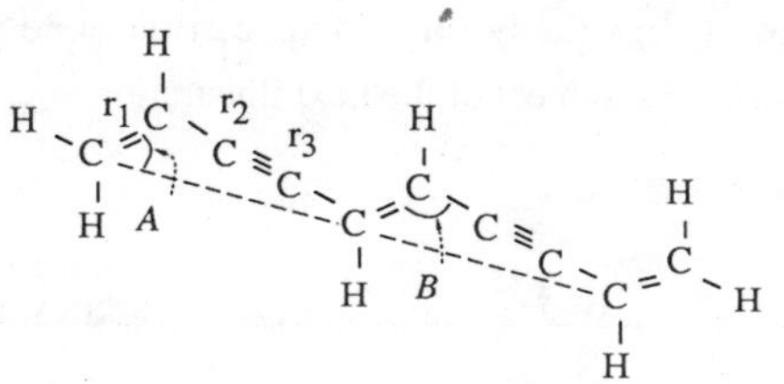

Figure 4.4 Structure of the 1,5,9-decatriene-3,7-diyne molecule (5)

PTS and TCDU polydiacetylenes exhibit different bond lengths in the polymer backbone as seen in Table 4.9; the first one adopts an acetylenic-type structure whereas the second one is closer to a butatrienic configuration. This difference in the geometry induces significant variations in the longitudinal polarizability values as can be observed in Table 4.9 where the longitudinal polarizability is shown to double from PDA to PBT.

Table 4.9 Geometries and STO-3G calculated longitudinal electric polarizability α_{zz} of 1,5,9-decatriene-3,7-diyne (5) with different geometrical parameters to model the basic repeat unit of PDA, PTS, TCDU, and PBT
(Distances in Å, angles in degrees, and polarizability in a.u.)

Oligomer	r_1	r_2	r_3	A	B	α_{zz}
PDA[a]	1.31	1.43	1.19	43.2	123.9	183.7
PTS[b]	1.36	1.43	1.21	44.5	121.9	205.5
TCDU[c]	1.42	1.38	1.24	39.3	127.7	302.7
PBT[a]	1.44	1.32	1.25	41.9	123.7	412.4

(a) optimized geometry taken from ref. [4.57]
(b) and (c) experimental geometries taken from refs. [4.55] and [4.56], respectively.

Scaling rules [4.41], which relate the linear α and the second hyperpolarizability γ, suggest that α scales with the length of conjugation l_π to the third power, whereas γ scales to the fifth power of that length. Thus, a modest increase in α can become significant when it comes to γ and controlled geometry alterations towards higher structural homogeneity to favor the π-electron delocalization may represent an interesting way to raise the electric response of organic systems. This can be done in various ways; an attempt along these lines forms the subject of the next illustration.

4.4.4.3 Structure and polarizability of acetylenic analogs of carbocyanines

As we have just seen, substantial electronic polarizations are generally obtained in systems with valence electrons highly delocalized over large distances as noted in conjugated oligomers. In designing new conjugated chains, it is important to control the molecular organization over distances at least corresponding to the so-called delocalization length l_π. Local structural defects, e.g., conformational twists, kinks, etc. can significantly reduce this delocalization length and, unfortunately, the longer the chain the more difficult it is to enforce the desired structural and electronic homogeneity.

Carbocyanines, H_2N-$(CH=CH)_n$-$CH=N^+H_2$ are characterized by a highly delocalized conjugated pathway evidenced both by a low degree of alternation between double and single bonds and a high polarizability of their backbones [4.58-4.61]. As already mentioned in Section 1.6, this idea has been already developed in the late 40's by Kuhn. It was its pioneering contribution to relate the nature of the first $\pi-\pi^*$ transition to the concept of bond alternation in conjugated chains particularly in the even-atom polyene and in the odd-atom polymethine chains. This question has been investigated in more detail [4.62] by plotting the FE transition energies of the series of oligomers $>(CH=CH)_n<$ (for $n=2,3,...6$) and $>N$-$(CH=CH)_n$-$CH=N^+<$ (for $n=1,2,...5$) as compared to the experimental values taking into account the bond end lengths as adjustable parameters. In the polymethine series, the regression line has its y-intercept at the origin of the axes. As expected, it corresponds to a zero energy gap, i.e., to a metallic character. In the polyene case, however, the y-intercept is of the order of 1.8 eV, a value in close agreement with the experimental energy gap of polymeric polyacetylene. This discussion is of crucial importance in the present discussion since in the Unsöld approximation, polarizabilities are directly related to first electronic transitions. In this way, carbocyanine based polymers should be significantly more polarizable than polyacetylenes. The question of bond alternation is thus quantitatively illustrated in Figure 4.5 where the relevant geometrical parameters (bond distances and bond angles calculated at the STO-3G level) are given for a series of four model carbocyanines, noted (6), (7), (8), and (9), of increasing length (n = 2, 3, 5, and 7). The absolute values of STO-3G geometrical parameters are known to depart somewhat from experiment, but the structural trends in a series of related molecules are usually predicted right. For instance, the bond alternation degree in polyenes is known to be exaggerated [4.26, 4.63-4.65], but the tendencies are correctly reproduced.

(6)

```
        |          |           |
   1.345 C  1.397  C  120.4   C
  N         C        C        C      N(+)
   1.382    125.2    125.0
   |        |        |        |
```

(7)

```
        |          |          |          |
   1.352 C  1.407  C  121.3   C  120.6   C
  N        C        C        C        C      N(+)
   1.372   1.388    125.2    125.2
   |        |        |        |        |
```

(8)

```
        |         |          |          |          |          |
   1.363 C  1.424 C  1.410  C  121.6   C  121.6   C  121.1   C
  N       C        C        C        C        C        C      N(+)
   1.359   1.370    1.389    125.1    125.2    125.4
   |       |        |        |        |        |        |
```

(9)

```
        |         |         |          |          |          |          |          |
   1.372 C  1.439 C  1.430 C  1.411  C  121.8   C  121.5   C  121.7   C  121.2   C
  N      C       C        C        C        C        C        C        C      N(+)
   1.348  1.355   1.370    1.389    124.3    125.3    125.4    125.5
   |      |       |        |        |        |        |        |        |
```

Figure 4.5 Ab initio STO-3G optimized bond lengths (in Å) and bond angles (in degrees) of four carbocyanines of increasing length $H_2N-(CH=CH)_n-CH=N^+H_2$ (n = 2, 3, 5 and 7)

Note that bond alternation is quickly restored as the carbocyanine chain length increases. The alternation degree of carbocyanines is classically formalized in terms of specific symmetric resonance structures which are chemically induced through the interplay of the $-NR_1R_2$ ($=N^+R_1R_2$) and ($-NR_3R_4$) $=N^+R_3R_4$ end groups; the corresponding resonance structures are shown in Figure 4.6.(a). Since the electronic polarization, in addition to the delocalization length, directly depends upon the number of active valence electrons per repeat unit, we address in this chapter the possibility of chemically deconfining π-electron rich moieties such as the -C≡C- triple bond to increase the number of electrons participating to the electronic response. The hope is that, by inserting a triple bond in a carbocyanine backbone, the -C≡C- distance will increase and lead to enhanced polarizability; the resonance forms for acetylenic carbocyanines are shown Figure 4.6.(b).

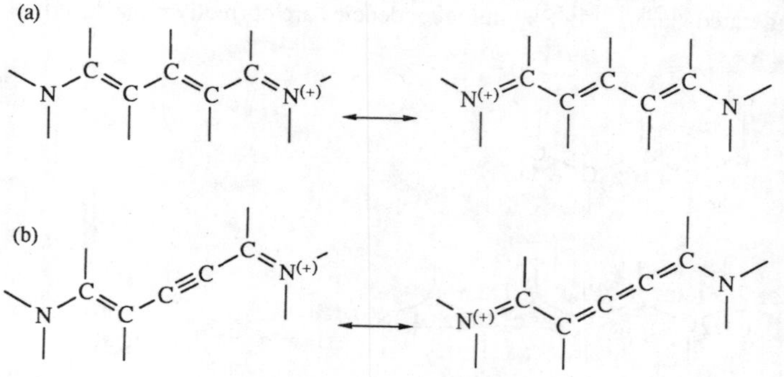

Figure 4.6   Resonance structures invoked for: (a) classical carbocyanines and (b) their acetylenic analogs

To our knowledge, the first synthesis of an acetylenic cyanine has been made by Mee et al. [4.66-4.68]. They noted an hypsochromic shift of the first optical transition as compared to the carbocyanine analogs (i.e., containing the same number of carbon atoms). They explained this shift on the basis of the asymmetric electron distribution of the acetylenic cyanines. This can also be related to the resonance structures shown in Figure 4.6.(b) which are not equivalent as opposed to the carbocyanine case, Figure 4.4.(a). In Figure 4.6.(b), one of the resonance structures is reminiscent of the monomer unit of polydiacetylene, while the second is typical of the butatrienic form. Butatriene belongs to the family of cumulenes which seem to be the most polarizable of the conjugated chains investigated so far [4.50, 4.65, 4.69-4.71]. It is thus interesting to determine which of these two limiting forms is dominant in the representation of the actual structure. A series of five molecules, denoted (10), (11), (12), (13), and (14) shown in Figure 4.7 has been considered. Molecule (10) is the simplest of the acetylenic cyanines, molecules (11) to (13) are isoelectronic to (10) and differ only by the nature of their chain ends. The general formula is R_1-(CH=CH)-C≡C-CH=R_2. For comparison purposes, the results for carbocyanine (**14**), H_2N-CH=CH-CH=CH-CH=N^+H_2, have been added to the series. Both the equilibrium geometry and polarizability have been computed for these molecules using the same conditions as described above. The optimized geometries have been used as input for static dipole polarizability calculations.

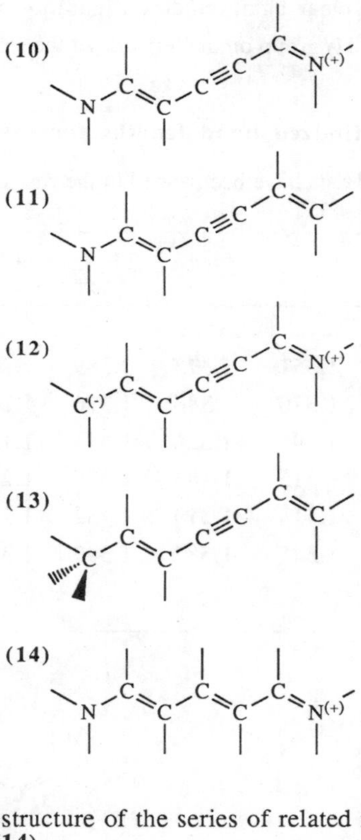

Figure 4.7   Molecular structure of the series of related molecules (**10**), (**11**), (**12**), (**13**), and (**14**)

Equilibrium geometry parameters are listed in Table 4.10 according to the convention shown in Figure 4.8. The results show that the carbon framework, -(CH=CH)-C≡C-CH=, common to all structures is quite sensitive to the nature of the chain ends. Note first the satisfactory agreement between experimental and theoretical predictions for (**10**).

Molecule (10) has a structure intermediate between acetylenic and butatrienic forms. Only r_4 is substantially shorter (1.208 Å) than the other C-C bonds in the molecule; a typical triple bond length at the STO-3G level is 1.170 Å [4.32]. This indicates a net increase of the -C≡C- bond length due to its incorporation in a cyanine type structure. Molecules (11) and (13) exhibit values typical of an acetylenic bond length and, from our previous experience [4.26,4.53], it is anticipated that their electric polarizability will be smaller than for molecule (10). Molecule (12) has a clear butatrienic configuration which originates from the constraining effect of the =CH_2 group on the left side of the structure.

Table 4.10 STO-3G optimized bond lengths for the five molecules (10) to (14) (in Å)
[Experimental data have been added in the case of molecule (10)]

	R_1	R_2	r_1	r_2	r_3	r_4	r_5	r_6
(10)	NH_2	NH_2^+	1.351	1.373	1.389	1.208	1.371	1.335
exp.			1.370	1.386	1.376	1.212	1.394	1.356
(11)	NH_2	CH_2	1.395	1.329	1.444	1.178	1.451	1.316
(12)	CH_2	NH_2^+	1.315	1.483	1.309	1.246	1.308	1.400
(13)	CH_3	CH_2	1.519	1.319	1.452	1.177	1.453	1.316
(14)	NH_2	NH_2^+	1.345	1.382	1.397	1.397	1.382	1.345

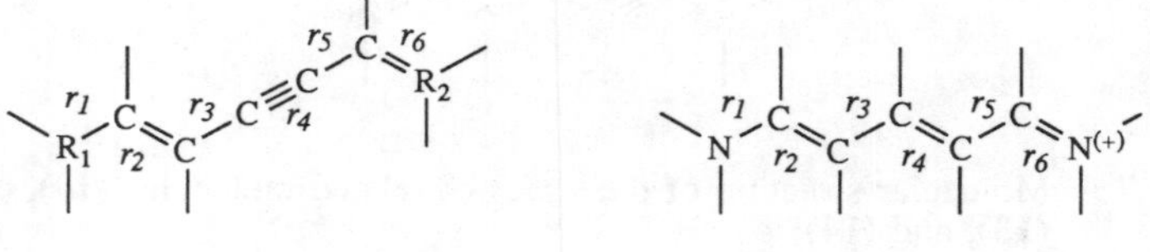

Figure 4.8   Numbering system used to denote the bonds in molecules (10) to (14)

Absolute values of the average polarizability for molecules (10) to (14) are given in Table 4.11. However, to make a fair comparison between the five systems, it is more appropriate to consider the average polarizability divided by the total number of electrons in each molecule.

Table 4.11 Average polarizability $<\alpha>$ and average polarizability divided by the number of electrons n_e in the molecule ($<\alpha>/n_e$) (in a.u.)

	$<\alpha>$	n_e	$<\alpha>/n_e$
(10)	68.25	50	1.36
(11)	46.36	50	0.93
(12)	63.67	50	1.27
(13)	46.07	50	0.92
(14)	73.30	52	1.41

The largest polarizability is obtained for carbocyanine (**14**), followed by the acetylenic cyanine (**10**) which has a geometrical structure intermediate between the alternating acetylenic structure and the more regular butatrienic form. Then comes molecule (**12**) which includes a butatrienic structure in its skeleton. Finally, molecules (**11**) and (**13**) have the smallest values of polarizability in good agreement with the fact that the -C≡C- triple bond has not been affected.

Thus, as hoped, the acetylenic cyanine exhibits a net polarizability enhancement over isoelectronic molecules (**11**) and (**13**) for which the r_4 distance is basically a -C≡C- triple bond of which the 4 π-electrons remain strongly confined. The most polarizable system is still the classical carbocyanine (**14**) which also has the maximum equalization of bond distances and thus the best chances for electronic delocalization of the structures considered in this work. Also interesting to note is the fact that even though molecule (**12**) incorporates a butatrienic structure, it is still less polarizable than (**10**), which has a structure intermediate between the vinylacetylene and butatriene forms. This is due to the fact that delocalization in (**10**) takes place over the entire molecule while in (**12**),which exhibits a butatriene-like structure known to be a highly polarizable fragment, the electronic delocalization is interrupted at r_2 (1.483 Å = the longest C-C bond distance found in the molecules (**10**) to (**14**)).

4.4.4.4 Polarizability and relative stability of octatetraene analogs

In the preparation of new materials it can be beneficial, but also difficult to either introduce or remove specific linkages and/or terminal groups in a conjugated backbone. It is therefore useful, before taking the effort to make such changes, to know the dependence on the polarizability of heteroatomic moieties incorporated in conjugated chains. In the following, we analyze the effect on the longitudinal polarizability of replacing a carbon-carbon double bond in trans-1,3,5,7-octatetraene (15) by heteroatomic moieties: -CH=O, -CH=NH, -N=NH, -N=O, -CH=N-, and -N=N- [4.72]. The choice of these heteroatomic moieties has been made in such a way as to keep the delocalized pathway l_π as similar as possible for the molecules to be compared. In a first part, the polarizability values are compared to that of the trans-1,3,5,7-octatetraene (15) which exhibits the highest electrical polarizability among the nine isoelectronic compounds. In the second, part we deal with stability questions, an aspect that is frequently overlooked in this type of prospective studies. The compounds studied here are shown in Figure 4.9 and are the all-trans conformers of 1,3,5,7-octatetraene (15), 1-aza-1,3,5,7-octatetraene (16), 4-aza-1,3,5,7-octatetraene (17), hepta 2,4,6-trienal (18), 1,2-diaza-1,3,5,7-octatetraene (19), 3,4-diaza-1,3,5,7-octatetraene (20), 1,8-diaza-1,3,5,7-octatetraene (21), 1-nitroso-1,3,5-hexatriene (22), and hexa 2,4-diene-1,6-dial (23).

In a preliminary study [4.73], the minimal STO-3G basis set has been used to determine the equilibrium geometry of the compounds shown in Figure 4.9. To assess possible basis set influences, STO-3G and 4-31G longitudinal electric polarizabilities were compared using the STO-3G equilibrium geometry as input parameters. The results on the electric polarizability were qualitatively similar in both bases and the derivatives including -CH=N- and -N=N- linkages were found to exhibit polarizabilities comparable to that of octatetraene (15). That these derivatives compare favorably with octatetraene (15) is encouraging because conjugated hydrocarbon chains are presently considered as promising organic candidates for optoelectronics. Also, the fact that polarizability estimates with STO-3G and 4-31G bases follow the same trend supports the assumptions made about the capacity of the STO-3G basis at providing qualitatively correct polarizabilities for molecules containing first row atoms. These results called further study on the relative stability of the compounds.

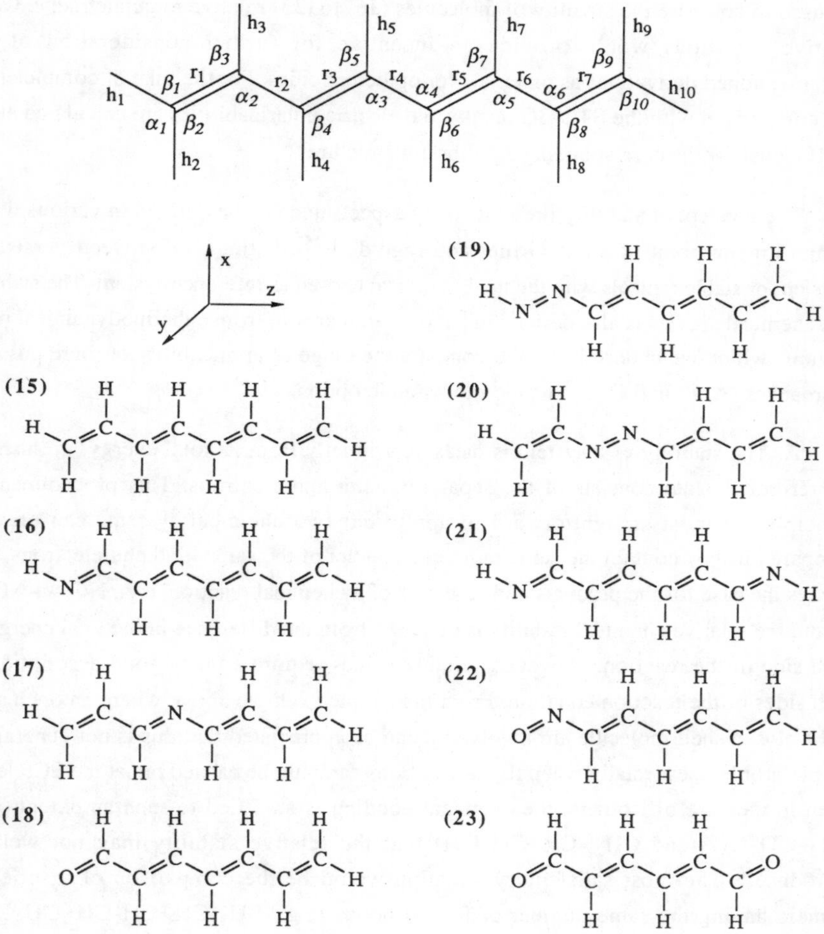

Figure 4.9 Convention used to identify the bond lengths and angles in the molecules considered in this Section

As pointed out above, the previous results on the equilibrium geometries of the compounds were obtained with the minimal STO-3G basis set. Unfortunately, minimal basis set results are usually not recommended for evaluating thermochemical properties, e.g., hydrogenation energies, etc. [4.74]. Accordingly, equilibrium geometries and total energies have been calculated with the split-valence 4-31G basis. These new calculations

330

are used to compare the stability of molecules (**15**) to (**23**) relative to octatetraene. Good relative stabilities would provide an incentive for further consideration of the aforementioned derivatives as materials for optoelectronics. For the sake of completeness and comparison with the STO-3G results, the electric polarizabilities are calculated at the 4-31G level for the corresponding equilibrium structures.

The concept of stability presents many aspects and can be defined in various ways; from a kinetic point of view, Griller and Ingold [4.75] distinguish between persistent, transient or stable radicals with the methyl radical chosen as reference system. The stability of a chemical species is also defined in a more familiar way from a thermodynamical point of view, which we choose here. We compare the range of applicability of three possible approaches (**A**, **B**, and **C**) of this thermodynamic option.

A. The stability of a system is measured directly from its total energy; in this case the reference state consists of the separate nuclei and electrons. This procedure is in principle the most straightforward to apply, but two chemical systems can only be compared if they contain the same number of nuclei of the same kind and electrons. It is always the case for the products and reactants of a chemical reaction, e.g., NO --> NO$^+$ + e$^-$, and the relative chemical stability is obtained from the difference in the total energy of each side of the reaction. However, reliable values require a fairly good description of both sides of the reaction equation. From the simple example above where an open shell and a closed shell molecule are involved it can be appreciated that this is not generally a simple problem, especially when the calculations can only be carried out at the SCF level. Even in the case of isomers, the chemical bonding is modified to such an extent (e.g., CH_3-CO-CH_3 and CH_2=CH-CH_2OH) that the relative stability may not well be reproduced. The most satisfactory situation would be the comparison of a series of isomers having the same number of formal bonds (e.g., CH_3-CH=CH-CH=CH-CH_2-CH_3 and CH_3-CH=CH-CH_2-CH=CH-CH_3). Thus, the direct use of total energies is not suitable for our purpose of comparing the relative stabilities of molecules (**15**) to (**23**).

B. Heats of atomization may also be used to measure the stabilization energy, but here the reference state is that of the separate atoms. In this case, chemical systems can only be compared if they possess the same number of atoms of the same nature. Again, this is valid for reactants and products of a chemical reaction, e.g., CH_2=CH_2 + H_2 --> CH_3-CH_3. For example, the heat of this reaction measures the stability of ethane relative to hydrogen and ethylene. Obviously, the relative stability of compounds measured via this approach will be best if the reaction schemes leading to the compounds to be compared involve the same reference molecules. Strictly speaking this is only possible for molecules (**19**) and (**20**) via an isodesmic process [4.76].

The stability measures obtained from A and B yield reliable values when highly accurate energies can be obtained. This is obviously a very restrictive condition and the number of compounds to which these approaches can be applied is somewhat limited. A more general scheme, possibly "less rigorous", is thus needed. We have chosen to use the definition of the thermodynamical stability proposed by Cox and Pilcher [4.77]. Its principles are briefly summarized below.

C. The heat of atomization of a molecule at 298.15 K is generally assumed to be equal to its binding energy [4.77], which may formally be considered as a sum of three types of contributions: bond energies, stabilization energies (for example due to delocalization), and destabilization energy (for example due to steric hindrance). Notice that the stabilization and destabilization energies cannot be obtained separately from the heats of atomization. The heat of atomization of a system with no stabilization or destabilization effects is equal to the sum of its bond energies. Cox and Pilcher have introduced the notion of conventional stabilization energy, SE, characteristic of a chemical species. It is defined as the difference between the heat of atomization and the sum of chemical bond energies:

$$SE = \Delta H_a - \Sigma N_{AB} E_{AB}$$

ΔH_a is the heat of atomization and E_{AB}'s are standard bond energy terms derived from transferable heats of atomization of reference compounds. The stabilization energy, SE, may be positive or negative depending on whether the compound is stabilized or destabilized. The atomization process transforms a molecule in the gas state to its gaseous atomic constituents: Molecule(g) ----> Atoms(g). In this work, the heat of atomization ΔH_a will always be defined with respect to the ground state of the atoms, without considering the role of valence states:

$$\Delta H_a = \Sigma \Delta H_f^0 \text{ (atoms)} - \Delta H_f^0 \text{ (molecule)}$$

332

The heats of formation of gaseous atoms needed in this work are available from the literature [4.78], but this is not the case for molecules. To solve these difficulties, Pople and coworkers [4.76] introduced the concept of bond separation reaction. Theoretical bond separation energies are obtained from the total molecular energies calculated at 0 K and corrected for zero-point vibrations. To obtain the corresponding values at room temperature, it is possible to correct the 0 K results by adding thermal translation, rotation, and vibration contributions. However, when there is an equal number of molecules on both side of a bond separation process, these contributions tend to cancel out and the calculated bond separation energies are used without any thermal adjustment for reactions at 298.15 K. We then have:

$$\Delta E^0(0 \text{ K}) \quad = \quad \Delta H^0(298.15 \text{ K}) \quad = \sum_i K_i N_i \, \Delta H_{fi}^0 \, (298.15 \text{ K})$$

where $\Delta E^0(0)$ is the theoretical bond separation energy calculated from the total energies of all partners of the reaction. If experimental heats of formation of all compounds involved in the process are known, except that of the molecule under consideration, the last equation provides a way to get an estimate of it.

A particular case of bond separation reaction is the isodesmic reaction. By definition, it is a reaction in which the number of each kind of formal chemical bond is conserved. Thus, it is a process in which only the immediate environment surrounding each bond has been altered. Isodesmic reactions will be used in this work to calculate the heats of formation of the nine molecules considered.

Molecular geometries:

Except for octatetraene (**15**) [4.78,4.79], theoretical equilibrium geometries of the compounds studied here have not been reported in an extended basis set. All compounds were assumed to be in a planar all-trans configuration during the geometry optimization process. The optimized parameters, labeled according to the convention given in Figure 4.9, are listed in Table 4.12. As before, the π-electron channel or the delocalization pathway is defined as the vector length l_π between the most distant atoms of the second period; l_π values are also included in Table 4.12.

Table **4.12** **4-31G molecular geometries** (lengths in Å, angles in degrees)

	(15)	(16)	(17)	(18)	(19)	(20)	(21)	(22)	(23)
r_1	1.3234	1.2630	1.3218	1.2143	1.2324	1.3170	1.2621	1.2107	1.2200
r_2	1.4561	1.4545	1.4550	1.4552	1.4054	1.4089	1.4558	1.4046	1.4610
r_3	1.3301	1.3286	1.2653	1.3298	1.3253	1.2362	1.3280	1.3291	1.3270
r_4	1.4515	1.4616	1.3965	1.4494	1.4495	1.4016	1.4523	1.4453	1.4526
r_5	1.3301	1.3300	1.3253	1.3305	1.3304	1.3251	1.3280	1.3315	1.3270
r_6	1.4561	1.4561	1.4550	1.4556	1.4560	1.4546	1.4558	1.4547	1.4610
r_7	1.3234	1.3232	1.3231	1.3231	1.3231	1.3230	1.2621	1.3233	1.2200
h_1	1.0713	1.0049	1.0709	-	1.0097	1.0703	1.0050	-	-
h_2	1.0735	-	1.0739	-	-	1.0708	-	-	-
h_3	1.0757	1.0824	1.0723	1.0865	-	1.0724	1.0819	-	1.0849
h_4	1.0762	1.0730	1.0840	1.0727	1.0729	-	1.0731	1.0719	1.0729
h_5	1.0764	1.0770	-	1.0773	1.0740	-	1.0763	1.0745	1.0760
h_6	1.0764	1.0756	1.0795	1.0748	1.0754	1.0737	1.0763	1.0744	1.0760
h_7	1.0762	1.0763	1.0733	1.0763	1.0761	1.0737	1.0731	1.0761	1.0729
h_8	1.0757	1.0754	1.0757	1.0749	1.0752	1.0749	1.0819	1.0747	1.0849
h_9	1.0735	1.0735	1.0733	1.0734	1.0734	1.0733	-	1.0736	-
h_{10}	1.0713	1.0712	1.0713	1.0710	1.0711	1.0711	1.0050	1.0707	-
α_1	124.51	121.33	123.33	124.14	115.86	119.85	121.05	115.70	123.44
α_2	124.20	123.14	121.53	121.53	119.12	115.43	123.16	117.61	121.70
α_3	124.38	124.53	121.47	125.03	124.31	115.86	124.20	124.83	124.11
α_4	124.38	124.02	121.08	123.57	123.66	119.48	124.20	123.02	124.11
α_5	124.20	124.25	123.93	124.36	124.30	124.00	123.16	124.38	121.70
α_6	124.51	124.30	124.21	124.02	124.16	123.84	121.05	123.76	123.44
β_1	121.85	115.60	121.85	-	110.21	121.45	115.71	-	-
β_2	121.88	-	121.96	-	-	120.79	-	-	-
β_3	119.37	122.76	121.40	119.86	-	123.75	122.90	-	120.29
β_4	119.38	121.41	121.68	121.84	124.06	-	121.42	125.04	122.00
β_5	119.19	119.20	-	118.88	117.92	-	119.53	117.46	119.54
β_6	119.19	119.46	121.00	119.76	119.55	123.66	119.53	119.83	119.54
β_7	119.38	119.40	118.14	119.36	119.35	118.21	121.42	119.27	122.00
β_8	119.37	119.48	119.38	119.67	119.56	119.60	122.90	119.80	120.29
β_9	121.88	121.91	124.84	121.93	121.91	121.88	-	121.91	-
β_{10}	121.85	121.83	121.79	121.78	121.79	121.76	115.71	121.80	-
l_π	8.55	8.47	8.38	8.44	8.32	8.21	8.37	8.28	8.32

In general, the 4-31G molecular parameters remain qualitatively similar to the STO-3G results. Systematic changes can nevertheless be noted: the C=C bond lengths are quite generally longer than the STO-3G values by roughly 0.07Å. On the contrary, the C-C,

C-N, C=O, C=N- and N=O bond distances are shorter. The valence angles between heavy atoms do not change by more than one degree, except for the angles NNC (**19**) and (**20**) and ONC (**22**). These variations in bond lengths and valence angles affect the π-channel length l_π. For all compounds, the l_π value decreases at the 4-31G level; the most important variations are observed in compounds containing heteroatomic functions such as C=O and N=O (captor groups): 0.09 Å in (**20**) and 0.13 Å in (**23**).

Single bond lengths C-H and N-H are shorter, compared to STO-3G results; one notes that when a carbon atom is attached to a terminal nitrogen or oxygen atom, its corresponding C-H length is shorter: 1.082 Å in (**16**) and 1.087 Å in (**18**) at the 4-31G level instead of 1.092 Å and 1.104 Å, respectively, at the STO-3G level.

Electric polarizabilities:

Longitudinal electric polarizabilities α_{zz} of the isoelectronic series studied in this work have already been calculated in minimal STO-3G and in extended 4-31G bases with optimized STO-3G equilibrium geometries (referred as 4-31G//STO-3G)). We have recalculated the electric polarizabilities with the geometries optimized at 4-31G level (referred to as 4-31G//4-31G) to estimate possible structural influences on this property. The computed values of longitudinal electric polarizabilities α_{zz} are listed in Table 4.13; the last two columns also contain the components α_{xx} and α_{yy}.

In general, computed longitudinal electric polarizabilities α_{zz} (4-31G//4-31G) are larger than α_{zz} (4-31G//STO-3G). The only exception is molecule (**21**). One observes an increase of this component by about 12 a.u. (6 percent) for octatetraene which exhibits the higher "homogeneity" in its geometrical structure. This effect is also quite large in (**22**) and it is generally less marked in the aza-compounds (3 to 6 a.u.). Systems which include heteroatomic links in the interior of the chain have higher polarizabilities than those for which the heteroatomic links are located at the ends of the chain: (**16**) and (**17**), (**19**) and (**20**). The polarizability per unit length is equal to 25.67 a.u. for octatetraene; it is close to the value of 25 predicted for the aza-compounds (**17**), (**19**) and (**20**).

Components α_{zz} and α_{yy} show the same trends as the longitudinal electric polarizability, the larger values are observed for octatetraene followed by the aza-compounds. Molecules with the heteroatomic link C=O (**18**) and (**23**) are less polarizable.

Table 4.13 Electric polarizabilities for compounds (15) to (23) (in a.u.) and chain lengths (Å)

	A	B	C	D	E	F	G	H
(15)	207.26	219.48	8.59	8.55	24.13	25.67	68.59	32.68
(16)	195.15	202.13	8.53	8.47	22.88	23.74	62.41	31.72
(17)	200.04	206.51	8.43	8.38	23.72	24.61	62.63	31.72
(18)	177.51	185.72	8.52	8.44	20.83	22.00	59.56	30.18
(19)	198.23	201.71	8.37	8.32	23.68	24.44	58.13	30.35
(20)	199.59	203.26	8.30	8.21	24.05	24.75	58.25	31.01
(21)	185.36	184.43	8.46	8.37	21.91	22.03	56.59	30.55
(22)	183.20	192.08	8.37	8.28	21.89	23.19	56.12	29.03
(23)	145.04	150.07	8.45	8.32	17.16	18.03	50.51	27.56

A: α_{zz}(4-31G//STO-3G) ref.[4.73]   E: ratio A/C

B: α_{zz}(4-31G//4-31G)   F: ratio B/D

C: l_{π} (4-31G//STO-3G) ref. [4.73]   G: α_{xx}(4-31G//4-31G)

D: l_{π} (4-31G//4-31G)   H: α_{yy}(4-31G//4-31G)

Semiempirical heats of formation:

Isodesmic reactions have been used to estimate semiempirical heats of formation of compounds (15) to (23). The details for the isodesmic scheme chosen for all-trans octatetraene (C_8H_{10}) are given below to serve as an illustration of the computation of the semiempirical heats of formation for the other compounds:

$$CH_2=CH-CH=CH-CH=CH-CH=CH_2 + 6\ CH_4 \ ---------> 3\ CH_3-CH_3 + 4\ CH_2=CH_2$$

If the heats of formation for the reactants and products are available, then the heat of reaction is obtained from the equation below:

$$\Delta H^0(298.15\ K) = 3\ \Delta H_f^0\ (CH_3-CH_3) + 4\ \Delta H_f^0\ (CH_2=CH_2)$$

$$- 6\ \Delta H_f^0\ (CH_4) - \Delta H_f^0\ (C_8H_{10})$$

As mentioned before, seven molecules are present on both sides of this reaction. If $\Delta E^0(0)$ is the theoretical isodesmic energy calculated from the total energies of methane,

ethane, ethylene and octatetraene, and with the premise that $\Delta E^0(0) = \Delta H^0(298.15\ K)$, the semiempirical heat of formation of octatetraene may then be deduced:

$$\Delta H_f^0\ (C_8H_{10})\text{semiemp.} = 3\ \Delta H_f^0\ (CH_3\text{-}CH_3) + 4\ \Delta H_f^0\ (CH_2\text{=}CH_2)$$

$$- 6\ \Delta H_f^0\ (CH_4) - \Delta E^0(0)$$

In the reaction [octatetraene + methane], methane, ethane and ethylene are taken as reference compounds. The choice depends on the type of chemical bonds present in the molecule involved in the isodesmic process. In Table 4.14 are collected the total energies obtained at the 4-31G level and the experimental heats of formation of the reference compounds used in this work. Total energies of all-trans octatetraene and related isoelectronic compounds are also given in this table.

Table 4.14 Total energies (in a.u.) **and heats of formation** (in kcal/mol)

Reference compounds[a]	E_t	ΔH_f^0	Studied compounds	E_t
CH_4	- 40.13977	- 17.9 [c]	(15)	- 308.25765
NH_3	- 56.10669	- 11.0 [d]	(16)	- 324.22331
$CH_2\text{=}CH_2$	- 77.92216	12.5 [c]	(17)	- 324.21526
$CH_3\text{-}CH_3$	- 79.11593	- 20.2 [c]	(18)	- 344.03902
$CH_2\text{=}NH$	- 93.88257	26.2 [b]	(19)	- 340.14521
$CH_3\text{-}NH_2$	- 95.07166	- 5.5 [c]	(20)	- 340.14042
$HN\text{=}NH$	- 109.81269	50.7 [c]	(21)	- 340.18848
$H_2C\text{=}O$	- 113.69262	- 26.0 [c]	(22)	- 359.92010
$HN\text{=}O$	- 129.57922	23.8 [e]	(23)	- 379.81573

[a] total energies E_t taken from reference [4.80], except for CH_2NH.
[b] estimated value given in reference [4.74].
[c] ref. [4.77], [d] ref. [4.81], [e] ref.[4.82].

Isodesmic energies and calculated heats of formation obtained with the method previously described are listed in Table 4.15.

Table 4.15 Isodesmic energies and calculated heats of formation (kcal/mol)

Isodesmic reactions	ΔE^0 theor.	ΔH_f^0 calc.
CH_2=CH-CH=CH-CH=CH-CH=CH_2 + 6 CH_4 ----------> 3 CH_3-CH_3 + 4 CH_2=CH_2	37.50	59.30
NH=CH-CH=CH-CH=CH-CH=CH_2 + 6 CH_4 ----------> 3 CH_3-CH_3 + 3 CH_2=CH_2 + CH_2=NH	40.47	69.71
CH_2=CH-CH=N-CH=CH-CH=CH_2 + 5 CH_4 + NH_3 ----------> 2 CH_3-CH_3 + 3 CH_2=CH_2 + CH_2=NH + CH_3-NH_2	42.77	75.53
O=CH-CH=CH-CH=CH-CH=CH_2 + 6 CH_4 ----------> 3 CH_3-CH_3 + 3 CH_2=CH_2 + CH_2=O	44.35	13.98
NH=N-CH=CH-CH=CH-CH=CH_2 + 5 CH_4 + NH_3 ----------> 2 CH_3-CH_3 + 3 CH_2=CH_2 + HN=NH + CH_3-NH_2	42.69	100.11
CH_2=CH-N=N-CH=CH-CH=CH_2 + 4 CH_4 + 2 NH_3 ----------> CH_3-CH_3 + 3 CH_2=CH_2 + HN=NH + 2 CH_3-NH_2	46.71	103.89
NH=CH-CH=CH-CH=CH-CH=NH + 6 CH_4 ----------> 3 CH_3-CH_3 + 2 CH_2=CH_2 + 2 CH_2=NH	43.77	80.43
O=N-CH=CH-CH=CH-CH=CH_2 + 5 CH_4 + NH_3 ----------> 2 CH_3-CH_3 + 3 CH_2=CH_2 + HN=O + CH_3-NH_2	47.89	68.01
O=CH-CH=CH-CH=CH-CH=O + 6 CH_4 ----------> 3 CH_3-CH_3 + 2 CH_2=CH_2 + 2 CH_2=O	48.27	-28.47

Bond separation energies for molecules with three "heavy" atoms have already been rationalized in terms of π-electron transfer [4.74] (scheme 1)

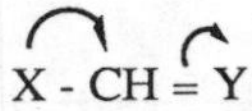

$$\overset{\frown}{X} - CH \overset{\frown}{=} Y$$

Scheme 1

The highest values of bond separation energies are found when X is a strong π-donor and Y a strong π-acceptor, while lowest values are found when X is a weak π-donor and Y a weak π-acceptor. This rule is indeed verified in our isoelectronic series containing eight heavy atoms. For example, molecules labeled (15), (16), and (18) may be written in the same manner with X equal to $CH_2=CH-CH=CH-CH=CH$ and Y equal to CH_2, NH, and O, respectively. The 4-31G π-electron distribution measured by the contributions of the $2p_y$ atomic orbitals to the gross orbital populations is shown below (scheme 2):

0.996 0.980 1.024	0.955 0.880 1.165	0.920 0.741 1.339
X - CH = CH$_2$	X - CH = NH	X - CH = O
(15)	(16)	(18)

Scheme 2

The π-acceptor character of Y increases from CH_2 group to oxygen atom; this is confirmed in Table 4.15 by the corresponding increase in the isodesmic energies, ΔE^0: 37.50, 40.47 and 44.35 kcal/mol for molecules (15), (16) and (18), respectively.

In absence of thermochemical measurements on compounds (15) to (23), it is unfortunately not possible to compare our predicted heats of formation to experimental values. However, with this procedure and the same basis set, the bond separation energies of 21 molecules [4.74] containing three heavy atoms show a mean absolute error of 3.1 kcal/mol. Thus we expect that our predicted values will fall in the same range of error. One word of caution is however necessary for those values obtained using the estimated heat of formation of methylenimine. So far no definite value can be proposed for that molecule as indicated by the large differences in the reported data for the methylenimine: 20.0 kcal/mol

[4.83] and 26.4 kcal/mol [4.84]. Thus, as indicated in Table 4.14, we have used the value of 26.2 kcal/mol originally proposed in reference [4.74].

Stabilization energies:

Bond separation energies are sometimes used as a measure of stabilization energy. The values corresponding to our series of molecules are given in Table 4.15, they are all positive. We notice in passing that our 4-31G result for all-trans octatetraene, 37.5 kcal/mol, is in good agreement with the value 37.3 kcal/mol obtained at the 6-31G level [4.79]. It may be of interest to add that in the case of polyene, a linear dependence of isodesmic energies on the number of double bond in the chain has been established [4.78].

Since it makes no relation to reference molecules, the calculated stabilization energy SE is characteristic of the chemical species considered. To obtain SE, the heat of atomization of the species under consideration and the bond energies are needed. The heat of atomization is calculated from the semiempirical heats of formation listed in Table 4.16 and the heats of formation of gaseous atoms in their ground states at 298.15 K [4.85]:

$$\Delta H_f^0 (H) = 52.1, \ \Delta H_f^0 (C) = 170.9, \ \Delta H_f^0 (N) = 113.0, \ \Delta H_f^0 (O) = 59.6 \text{ kcal/mol}.$$

Table 4.16 Bond energies and specific increments (in kcal/mol) [a]

$E(C - C)$	85.05	$E(C - N)$	76.61	$E(C_{CO} - H)_2$	100.24
$E(C = C)$	136.77	$E(N = O)$	119.35	$E(C_{CO} - H)_1$	101.33
$E(C_d - C)$	88.92	$E(C - H)_p$	98.25	$E(N - H)_2$	91.31
$E(C = O)$	160.17	$E(C_d - H)_2$	100.24	$E(N - H)_1$	90.29
$E(C_{CO} - C)$	93.68	$E(C_d - H)_1$	99.72	$\Delta(C_d - C)$	3.87
$E(N = N)$	107.68 [b]	$E(C = N)$	130.29 [b]	$\Delta(C_{CO} - C)$	8.63

[a] see notation in ref.[4.86]; [b] ref.[4.87].

Bond energies and specific increments for the calculation of stabilization energy of chemical species have been proposed by Leroy [4.86], the relevant values are collected in

Table 4.16. The contributions leading to ΔH_a and $\Sigma N_{AB}E_{AB}$ for all-trans octatetraene C_8H_{10} are:

$$\Delta H_a = 8 \Delta H_f^0 (C) + 10 \Delta H_f^0 (H) - \Delta H_f^0 (C_8H_{10})_{semiemp.}$$

$$\Sigma N_{AB}E_{AB} = 4 E(C=C) + 3 E(C-C) + 4 E(C-H)_2 + 6 E(C-H)_1$$

$$+ 8 \Delta(C_d - C)$$

Applying the data yields a stabilization energy, SE, of 4.17 kcal/mol for octatetraene which is due to delocalization effects. If the specific increments $\Delta(C_d\text{-}C)$ are not included in the sum of bond energy terms, i.e.,:

$$\Sigma N_{AB}E_{AB} = 4 E(C=C) + 3 E(C-C) + 4 E(C-H)_2 + 6 E(C-H)_1$$

the resulting stabilization energy will then be equal to 27.39 kcal/mol and will be due to both the hyperconjugation and electron delocalization effects. As the increments $\Delta(C_d\text{-}N_{N=N})$ and $\Delta(C_d\text{-}C_{C=N})$ are not available, the stabilization energies that we choose to concentrate on for the isoelectronic series of compounds (15) to (23) will be related to both the hyperconjugation and electron delocalization effects, i.e., without using the specific increments. Our aim is to estimate the changes induced in the stabilization energy by substituting isoelectronically various heteroatomic linkages in the octatetraene backbone and ultimately search for possible relations with electric polarizabilities. Table 4.17 collects the heats of atomization, the sums of bond energies, and the stabilization energies for the reference compounds and those of the studied isoelectronic series.

We observe that the reference molecules themselves are either stabilized, or destabilized. Although to a different extent, all the molecules of our series are stabilized by hyperconjugation and delocalization effects. It is interesting to notice that molecules (23) and (18) which exhibit quite low values of the longitudinal electric polarizability, 150.07 and 185.72 a.u. in Table 4.13, are 7.83 and 5.27 kcal/mol more stable than octatetraene, respectively. All the other molecules are less stabilized than octatetraene. Since for the molecules incorporating the C=O group the specific increments $\Delta(C_d\text{-}C)$ and $\Delta(C_{C=O}\text{-}C)$ are available from Table 4.16, it is possible to compute (as indicated above for octatetraene) the stabilization energy relative to the delocalization effects only. In this case the computed stabilization energies are 4.17, 4.66, and 2.88 kcal/mol, for (15), (18) and

(23), respectively. Thus (15) is more stable than (23) relative to delocalization effects and hyperconjugation effects are expected to be important in conjugated molecules including C=O groups.

Table 4.17 Stabilization energies of molecules (in kcal/mol)

Molecule	ΔH_a	$\Sigma N_{AB}E_{AB}$	SE	Molecule	ΔH_a	$\Sigma N_{AB}E_{AB}$	SE
CH_4	397.20	393.0	4.20	(15)	1828.90	1801.51	27.39
NH_3	280.30	273.93	6.37	(16)	1708.49	1684.84	23.65
CH_2CH_2	537.70	537.73	-0.03	(17)	1702.67	1686.87	15.80
CH_3CH_3	674.60	674.55	0.05	(18)	1658.68	1626.02	32.66
CH_2NH	420.00	421.06	-1.06	(19)	1568.09	1554.08	14.01
CH_3NH_2	549.90	553.98	-4.08	(20)	1564.31	1556.11	8.20
HNNH	279.50	288.26	-8.76	(21)	1587.77	1568.17	19.60
H_2CO	360.66	360.65	0.01	(22)	1494.65	1475.45	19.20
HNO	200.86	209.64	-8.78	(23)	1485.59	1450.57	35.22

Confronting the values of the longitudinal electric polarizability of molecules (15) to (23) with the corresponding stabilization energies is also quite informative. For example, molecules (17) and (20) both have a -CH=CH- replaced respectively by -CH=N- and -N=N- in the same position of the parent structure, but molecule (17) has both its electric polarizability and stabilization energy higher than (20). This suggests that if, for practical purposes or due to synthetic constraints, it is needed to substitute a -CH=CH- unit, a -CH=N- would be preferable to -N=N- at least in a polyenic structure. It would be valuable to scan theoretically (the) other environments in which both -CH=N- and -N=N- groups can be incorporated and determine which of these two groups corresponds to larger electric polarizability and relative stability. The comparison of the three isoelectronic molecules containing two nitrogen atoms, (19), (20), and (21), shows that from a stability view point it is more advantageous to have the nitrogen atoms separated. Also, a slight increase in polarizability from (19) to (20), 201.71 to 203.26 a.u., is obtained at the expense of a substantial destabilization, 5.81 kcal/mol.

To conclude this Section on the stabilization energy we would like to make a few remarks on the comparison of values obtained by different schemes. If the results above are compared to those contained in Table 4.14, we note that, in terms of total energies, isomer (16) is more stable than (17) by about 5.05 kcal/mol; this trend is also obtained for the stabilization energies of the two compounds, 23.65 and 15.80 kcal/mol, respectively. A similar situation holds for the three isomers (19), (20) and (21); (21) is more stable than (19) by about 27.15 kcal/mol; in turn (19) is more stable than (20) by 3.0 kcal/mol; stabilization energies, SE, in Table 4.17 are qualitatively in the same order but quantitatively different. These large discrepancies again occur for systems in which C=N bonds exist, (19), (20) and (21). Notice that the C=N bond energy, 130.29 kcal/mol, used here is significantly different from the one proposed by Benson [4.85], 154 ± 10 kcal/mol for the C=N bond energy in $CH_2=NH$.

The origin of the differences observed between isodesmic energies, ΔE^0, in Table 4.15 and stabilization energies, SE, in Table 4.17 can be analyzed more closely in the case of isomeric compounds. Molecules (19) and (20), for example, have the same reference molecules and their heats of formation have been reliably determined; the differences between their ΔE^0 and SE are:

$$\Delta (\Delta E^0(0)) = -4.02 \text{ kcal/mol} \quad \text{and} \quad \Delta (SE) = 5.81 \text{ kcal/mol}$$

$\Delta (SE)$ is written:

$$\Delta (SE) = \Delta (\Delta H_a) - \Delta (\Sigma N_{AB} E_{AB})$$

It follows that:

$$\Delta (\Delta H_a) = \Delta (\Delta H_f^0) = \Delta (\Delta E^0(0)) = \Delta SE$$

which is fulfilled only if $\Delta (\Sigma N_{AB} E_{AB}) = 0$ which would mean that the sum of bond energies in the two isomers does not depend on the arrangement of the atoms. This would possibly be the case if the molecules possess the same kind and the same number of chemical bonds, for example as in CH_3-CH=CH-CH=CH-CH_2-CH_3 and CH_3-CH=CH-CH_2-CH=CH-CH_3 or in cis-trans isomers. In the case of molecules (19) and (20) which contain one and two C-N bonds, respectively, we have:

$$\Delta (\Delta H_a) = 3.78 \text{ kcal/mol}; \quad \Delta (\Sigma N_{AB} E_{AB}) = -2.03 \text{ kcal/mol}$$

This simple analysis shows quite convincingly that the differences between stabilization energies, SE, and the differences between isodesmic energies are two quantities which in general are not similar and thus should not be compared numerically. The results given in this Section suggest that it is possible to use the stabilization energy concept in connection with the design of new highly polarizable molecules for optoelectronics. For instance, the design of a new molecular structure could be made in such a way as to minimize the incidence of the hyperconjugation contributions while maximizing the delocalization terms. In these attempts, the calculation of SE would provide a way to quantify these effects.

4.4.4.5. Peptide groups in conjugated hydrocarbon chains

Another example of altering the structural homogeneity of conjugated hydrocarbon chains is provided by the incorporation of a peptide or amide bond, -NH-CO-, into a butadiene backbone. As shall be discussed in Section 4.4.6, peptide bonds can be used as chemical tools to force the organization of active molecules and/or polymers in the bulk. Therefore it is interesting to estimate their influence on the resulting polarizability of a conjugated system that incorporates peptide groups. In Figure 4.10 are shown two molecules, N-vinylacrylamide (**24**) and N-butadienyl formamide (**25**), containing two C=C double bonds and an amide group. In the first system (**24**), the C=C double bonds are separated by the -NH-CO- group, while in (**25**) these C=C bonds are directly connected. This procedure of comparison has been chosen to eliminate, as much as possible, the deficiencies due to the limited basis used. In Table 4.18 are given the longitudinal polarizability α_{zz}, the average polarizability $\langle \alpha \rangle$, and the delocalization distance l_π. Polarizability and fully optimized geometries for molecules constrained to remain planar have been obtained with the STO-3G basis set. The relevant bond distances are given in Figure 4.10. Both in Table 4.18 and in Figure 4.10 have been added the corresponding results for the trans-1,3,5,- hexatriene (**2**) for comparison.

On the basis of the longitudinal and the average polarizability values given in Table 4.18, the amide group -NH-CO-, even with its 4 π-electrons, is less polarizable than a C=C double bond with only 2 π-electrons. In a peptide group the C-N bond has some degree of double-bond character resulting from the delocalization of the nitrogen lone pair into the carbonyl group; the calculated C-N bond distance is 1.42 Å in N-vinylacrylamide

(24) and N-butadienyl formamide (25). This bond length is intermediate between a true C=N bond, 1.27 Å in $H_2C=NH$, and a single C-N bond, 1.48 Å in H_3C-NH_2, both calculated at the STO-3G level [4.52]. A similar delocalization is not observed for the C-C bond adjacent to the carbonyl group; in that case the bond distance is 1.52 Å which is typical of a single C-C bond. Thus, in N-vinylacrylamide (24) the C=C double bond is in a way isolated from the rest of the conjugated framework of the molecule. This is not the case in N-butadienyl formamide (25) where the carbonyl group is located at the end of the molecule. Even in this case and in spite of the 4 π-electrons brought by the peptide group, the overall polarizability is less than in trans-1,3,5-hexatriene (2). Notice also that N-butadienyl formamide (25) is slightly more stable than N-vinylacrylamide (24); at the STO-3G level the difference of stability is of the order of 1 kcal/mol. Thus peptide bonds, when incorporated in a conjugated hydrocarbon chain of the polyene type, disrupt significantly the conjugation and should not be used unless if they can serve other important purposes such as inducing local order. It must be kept in mind however that, when located at the end of a molecule, peptide groups do not spoil the polarizability as much as when they are incorporated in the conjugated backbone. Therefore as we shall see later, they could be used as terminal groups to enhance the organization of the active species in the bulk.

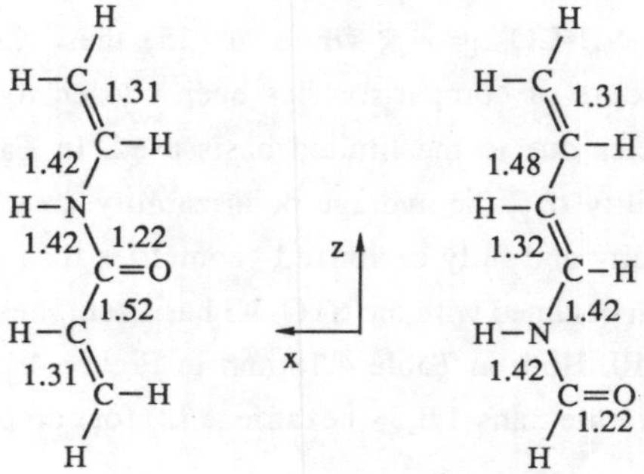

Figure 4.10 Relevant STO-3G optimized bond distances (in Å) of N-vinylacrylamide (24), N-butadienyl formamide (25)

Table 4.18 Longitudinal polarizability α_{zz}, average polarizability $<\alpha>$ (in a.u.) and delocalization length l_π (in Å) for molecules (2), (24), and (25)

Molecule	α_{zz}	$<\alpha>$	l_π
N-vinylacrylamide (24)	64.29	36.26	6.04
N-butadienyl formamide (25)	76.36	41.25	6.15
trans-1,3,5-hexatriene (2)	104.13	43.93	6.13

This series of examples on assessing the delocalization of π-electrons in various conjugated frameworks is far from being exhaustive, but it should convey some idea as to how quantum chemistry calculations can be used in the context of the design of highly polarizable systems. In the next Section, we consider chain length increase as a way to enhance the polarizability.

4.4.5 Dependence of the Polarizability on the Chain Length

We have already noted that it is important to scale the polarizability values according to the length of the conjugated framework l_π or some other quantity, such as the molecular volume V, to have a suitable measure of the intrinsic polarizability of a given conjugated system. In several cases, connecting conjugated units together leads to a net polarizability that is not simply additive of the constituents contributions but somewhat larger. Since large electric response is one of the main targets in the design of conjugated organic polymers for optoelectronics, this way of enhancing the polarizability is certainly worth considering. For example, in polydiacetylenes the one-dimensional delocalization that results from the polymerization of diacetylene molecules produces a dramatic enhancement of the optical nonlinearities of these compounds. The TCDU polymer [-CR=CR-C≡C-]$_x$, where R is -(CH$_2$)$_4$OCONHφ, shows a response 6×10^2 times larger than that of the TCDU monomer [4.88]. However, it can be anticipated that not all units are equally appropriate and, furthermore, one might be willing to mix units of different kinds. Therefore, model calculations carried out to assess the polarizability of various arrangements of units connected in chain as a function of the chain length are certainly useful to define the most suitable choice out of a set of different possibilities. Indeed, a lot of activities is devoted to engineer conjugated chain molecules and polymers with the best possible delocalization.

The size-dependence of the electronic polarizability of conjugated systems has been studied first for metallic polyene (polyacetylene) chains within the framework of the free-electron [4.89] and Hückel [4.38] models; they predict a dependence of the longitudinal polarizability α_{zz} proportional to l_π^3. According to these findings, the longitudinal polarizability should grow as the third power of the chain length and thus diverge in the limit of an infinite chain. On the one-hand, free-electron and Hückel theories do not take into account Coulombic interactions explicitly and, on the other hand, structural reorganizations in the chain cannot be ignored as the size increases. Bond alternation influence on the longitudinal polarizability of polyenes containing increasing number of carbon atoms has been investigated at the *ab initio* level. The quantity actually followed is the polarizability per vinylenic moiety, α_{zz}/n (plotted in Figure 4.11).

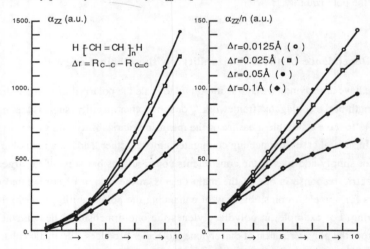

Figure 4.11 Longitudinal polarizabilities α_{zz}(total [left] and per CH=CH unit [right]) of polyenes as a function of the number (n) of CH=CH units

The polarizability per vinylenic moiety, α_{zz}/n, increases with the number of double bonds, but eventually saturates indicating that a regime of linear increase has been reached. Also seen is the fact that the value at which α_{zz}/n levels off and the number of double bonds at which the linear regime prevails is strongly related to the degree of alternation measured as the difference between the lengths of the single and double carbon-carbon bonds. Note, however, that this dependence is not trivial and it is quite unlikely that this simple measure of bond alternation will be applicable to more complex systems. Even if this would be the case, it remains to know from the start what will be the geometrical parameters connecting the units and the possible structural relaxations in the units due to these connections. Here again, quantum chemistry calculations can provide useful quantitative support as to these structural changes and their incidence on the polarizability. To illustrate the above mentioned discussion, two cases concerning the influence of the oligomerization (connecting units) on the resulting polarizability are described.

4.4.5.1 Vinylacetylene and polyene chains

In this first example, the influence of oligomerization on the longitudinal polarizability is examined by comparing 1,5-hexadiene-3-yne (26) and 1,5,9-decatriene-3,7-diyne (27), on the one hand, and trans-1,3,5-hexatriene (2) and trans-1,3,5,7-decapentaene (28), on the other hand. These molecules are schematically represented in Figure 4.12, the corresponding STO-3G values of the longitudinal polarizability α_{zz} and average polarizability $< \alpha>$ are reported in Table 4.19.

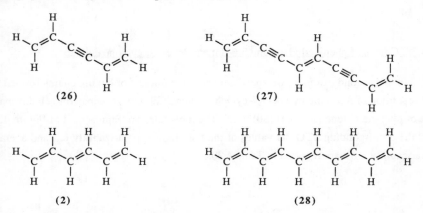

Figure 4.12 Schematic representation of 1,5-hexadiene-3-yne (26), 1,5,9-decatriene-3,7-diyne (27), trans-1,3,5-hexatriene (2) and, trans-1,3,5,7-deca-pentaene (28) molecules

Table 4.19 STO-3G longitudinal polarizability α_{zz} for molecules (26), (27), (2), and (28) (in a.u.)

Molecule	α_{zz}
1,5-hexadiene-3-yne (26)	44.65
1,5,9-decatriene-3,7-diyne (27)	85.28
trans-1,3,5-hexatriene (2)	104.13
trans-1,3,5,7-decapentaene (28)	217.25

Going from 1,5-hexadiene-3-yne (26) to 1,5,9-decatriene-3,7-diyne (27) means adding 10 more π electrons with the -C≡C-CH=CH-C≡C- moiety inserted between two vinylic groups, while going from trans-1,3,5-hexatriene (2) to trans-1,3,5,7-decapentaene (28) amounts to incorporate 3 C=C double bonds carrying a total of 6 π-electrons between the two vinylic groups. The net gain in polarizability is more substantial with the polyenic backbone, because of a more efficient delocalization of the π-electrons in this geometrical framework. As already seen in the case of the acetylenic analogs of carbocyanines, Section 4.4.4.3, triple bonds, notwithstanding the fact that they are π-electron richer than double bonds, have their electrons more confined between the nuclear centers of the bond and form less delocalized structures.

4.4.5.2 Chains of phenylethylene and 3,6-dimethylene-1,4-cyclohexadiene

In this second example, we speculate about the interest of having chains formed by the repetition of 3,6-dimethylene-1,4-cyclohexadiene (29) units compared to chains made out of phenylethylene (styrene) (30) units. The molecules are represented in Figure 4.13 and the corresponding STO-3G values of the longitudinal polarizability α_{zz} and average polarizability $<\alpha>$ are reported in Table 4.20.

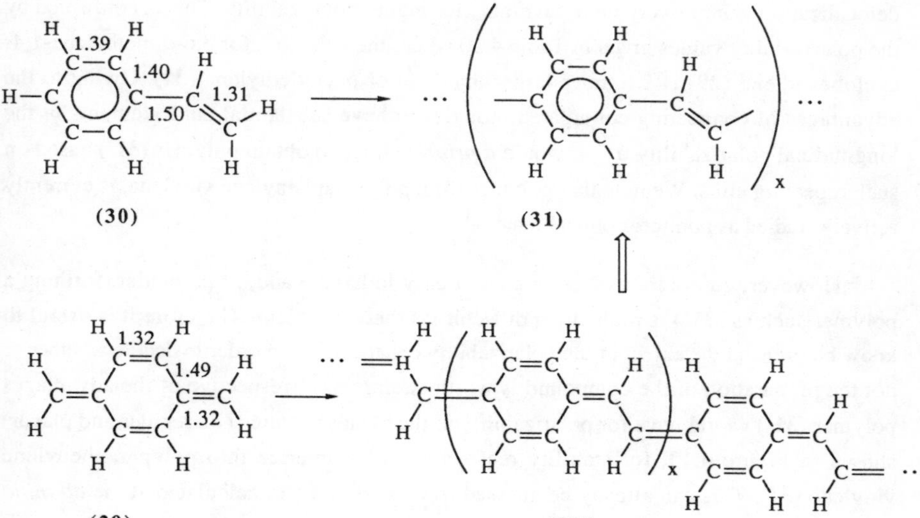

Figure 4.13 Schematic representation of 3,6-dimethylene-1,4-cyclohexadiene (**29**) and phenylethylene (styrene) (**30**) molecules and of their polymerization schemes (**31** and **31'**)

Table 4.20 STO-3G longitudinal polarizability α_{zz}, average polarizability $<\alpha>$ (in a.u.), and delocalization length l_π (in Å) for molecules (**29**) and (**30**)

Molecule	α_{zz}	$<\alpha>$	l_π
phenylethylene (**30**)	76.39	47.53	4.94
3,6-dimethylene-1,4-cyclohexadiene (**29**)	125.41	55.81	5.45

In spite of its delocalized nature benzene disrupts the conjugation and thus, as seen in Table 4.20, tends to spoil the overall polarizability of a conjugated system in which it is inserted. This can be related to its aromatic character. A possible way to weaken the aromatic character while retaining the overall geometry of benzene is to switch to a quinoidic structure as in 3,6-dimethylene-1,4-cyclohexadiene (**29**) where the delocalization is intuitively more favorable for higher polarizability. This is confirmed by the polarizability values given in Table 4.20 where the value α_{zz} for 3,6-dimethylene-1,4-cyclohexadiene (29) is 1.6 times larger than that of phenylethylene (30). Owing to the advantages of connecting conjugated moieties to have substantial enhancements in the longitudinal polarizability α_{zz}, it would *a priori* be nice to obtain polymer (31') based on such repeating units. We note that polymer (31), polyparaphenylene vinylene, is currently actively studied as nonlinear optical material.

However, quinoidic structures are not easy to handle and, in particular, forming a polymer such as (31') is probably a difficult synthetic problem. Therefore it is useful to know beforehand the extent of the polarizability enhancement in order to decide whether or not the preparation of the compound is worth attempting. Irrespective of the advantages polymer (31') would have for polarizability in the ideal structure (i.e., regular and planar) shown in Figure 4.13, for stability reasons it will isomerize into polyparaphenylene vinylene (31). This can already be guessed from total energies calculated at the *ab initio* level on the repeat units. Computed at the 6-31G level, the stability difference between 3,6-dimethylene-1,4-cyclohexadiene (29) and phenylethylene (30) is 25 kcal/mol, the latter being the more stable molecule.

4.4.5.3 Oligomers of cumulenes

As we have seen in Section 4.4.4.1, butatriene (4) is intrinsically more polarizable than hexatriene (2). Butatriene is part of the cumulene family which can be divided into two classes. The first class, $R_1R_2C[=C=]_kCR_3R_4$, is characterized by an even number k of directly connected carbons, while the second class contains those molecules for which k is an odd number. In this Section, we will only be concerned with the first class, i.e., molecules for which k is even. When such molecules are linked together to form oligomers, the two resonant structures shown in Figure 4.14 should be considered to understand the structural changes that will take place as the chain length increases.

Form (b) is more stable than form (a) and it can be expected that in the limit of large k values the actual structure will essentially correspond to form (b). Therefore, in the same line as in the previous Sections, it is interesting to know when form (b) becomes predominant over form (a), which eventually means that the intrinsic polarizability starts to level off.

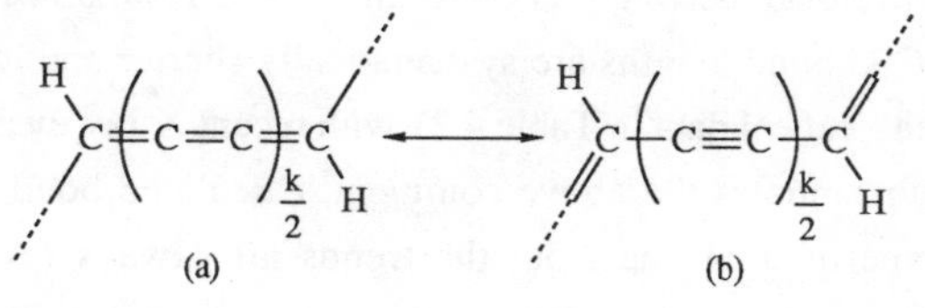

Figure 4.14 Two resonance forms of cumulene chains

Full geometry optimization for all angles and distances of the molecules shown in Figure 4.15 (forced to adopt a fully stretched configuration) has been carried out at the STO-3G level. Only the C=C and C-C bond distances are of interest to us here, they are listed in Table 4.21.

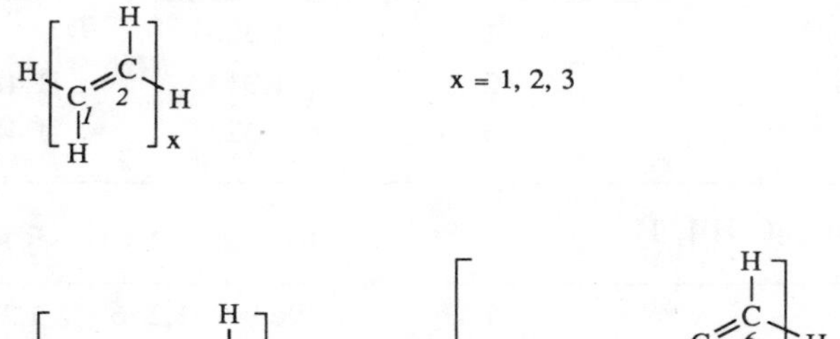

Figure 4.15 Schematic representation of the H-[-HC$_1$=C$_2$H-]$_x$-H, H-[-HC$_1$=C$_2$=C$_3$=C$_4$H-]$_x$-H and H-[-HC$_1$=C$_2$=C$_3$=C$_4$=C$_5$=C$_6$H-]$_x$-H oligomers, x= 1,2 and 3

In the dimers, the repeat units are related through inversion symmetry, and therefore only the geometrical parameters corresponding to one unit are listed. In the case of the trimers, the listed structural data have been limited to the values corresponding to the central units since they relate best to the repeat unit of a polymer. As already pointed out, the changes in the geometrical parameters of polyenes are qualitatively well predicted with the STO-3G basis as compared with more refined theoretical results. It might be useful to add that the main difference between STO-3G and 6-31G results lies in the fact that the STO-3G C=C and C-C bond lengths are systematically shorter and longer, respectively. Comparison of the theoretical data in Table 4.21 with recent X-ray measurements on small cumulenes [4.90] substantiates the above comment: calculated bond lengths are 0.02 Å shorter than the experimental ones, but the trends are always respected. Thus, it is expected that the structural changes occuring in longer oligomers will be calculated at the same level of quality.

Table 4.21 STO-3G carbon-carbon distances for three series of cumulenes with even numbers of directly connected carbons (in Å)

H-[-HC$_1$=C$_2$H-]$_x$-H	x	$r_{C1=C2}$	r_{C-C}			
ethylene (**32**)	1	1.306	-			
butadiene (**33**)	2	1.313	1.489			
hexatriene (**2**)	3	1.322	1.485			

H-[-HC$_1$=C$_2$=C$_3$=C$_4$H-]$_x$-H	x	$r_{C1=C2}$	$r_{C2=C3}$	$r_{C3=C4}$	r_{C-C}	
butatriene (**4**)	1	1.296	1.256	1.296	-	
dibutatriene (**34**)	2	1.298	1.253	1.307	1.483	
tributatriene (**35**)	3	1.311	1.247	1.311	1.479	

H-[-HC$_1$=C$_2$=C$_3$=C$_4$=C$_5$=C$_6$H-]$_x$-H	x	$r_{C1=C2}$	$r_{C2=C3}$	$r_{C3=C4}$	$r_{C4=C5}$	$r_{C5=C6}$	r_{C-C}
hexapentaene (**36**)	1	1.297	1.259	1.273	1.259	1.297	-
dihexapentaene (**37**)	2	1.299	1.257	1.277	1.253	1.312	1.477
trihexapentaene (**38**)	3	1.315	1.249	1.283	1.249	1.315	1.474

In the case of butatriene (**4**), dibutatriene (**34**), and tributatriene (**35**), the tendency for the longer chains to adopt a polyyne-like structure, (-C≡C-)$_n$, typical of the resonance form (b) is obvious and is in agreement with the calculations by Karpfen on the infinite chain [4.91]. This trend is also seen in the case of hexapentaene (36), dihexapentaene (37), and trihexapentaene (38). Not to be overlooked is the simultaneous reduction of the C-C bonds connecting the units. Thus the polarizability of these oligomers, which depends on the effectiveness of the delocalization over the entire backbone, will be controlled by two competing effects: the delocalization changes taking place in the repeating units as the number of directly connected carbon atoms increases and the parallel trend for the C-C bonds connecting the repeating units to shorten which allows for better delocalization between these units. This analysis is supported by the polarizability values listed in Table 4.22.

Table 4.22 STO-3G average polarizability $\langle\alpha\rangle$ (in a.u.) and its ratios with the number of electrons n_e, π-electrons n_π, carbon atoms n_C, and delocalization length l_π (in Å)

H-[-HC$_1$=C$_2$H-]$_x$-H

	x	$\langle\alpha\rangle$	$\langle\alpha\rangle/n_e$	$\langle\alpha\rangle/n_\pi$	$\langle\alpha\rangle/n_C$	$\langle\alpha\rangle/l_\pi$
ethylene (32)	1	10.74	0.67	5.37	5.37	8.20
butadiene (33)	2	25.15	0.84	6.29	6.29	6.85
hexatriene (2)	3	43.93	1.00	7.32	7.32	7.17

H-[-HC$_1$=C$_2$=C$_3$=C$_4$H-]$_x$-H

	x	$\langle\alpha\rangle$	$\langle\alpha\rangle/n_e$	$\langle\alpha\rangle/n_\pi$	$\langle\alpha\rangle/n_C$	$\langle\alpha\rangle/l_\pi$
butatriene (4)	1	30.19	1.08	5.03	7.55	7.84
dibutatriene (34)	2	91.11	1.69	7.59	11.39	10.57
tributatriene (35)	3	184.85	2.31	10.27	15.40	13.74

H-[-HC$_1$=C$_2$=C$_3$=C$_4$=C$_5$=C$_6$H-]$_x$-H

	x	$\langle\alpha\rangle$	$\langle\alpha\rangle/n_e$	$\langle\alpha\rangle/n_\pi$	$\langle\alpha\rangle/n_C$	$\langle\alpha\rangle/l_\pi$
hexapentaene (36)	1	65.57	1.64	6.56	10.93	10.26
dihexapentaene (37)	2	229.47	2.94	11.47	19.12	16.79
trihexapentaene (38)	3	508.57	4.38	16.95	28.25	24.24

One can relate the geometry changes discussed previously to the polarizability values listed in Table 4.22. When the number k of directly connected atoms in a cumulenic unit increases, the average polarizability $<\alpha>$ continues to increase in spite of the slight tendency for the geometries to evolve towards form (b) of the two resonance structures. It is likely that saturation in monomeric units will occur in longer structures that the ones considered in this work. In the oligomers, the C-C distances connecting the repeat units tend to shorten with the net result of improving the delocalization of the π-electrons. The overall effect is a continuous increase of the polarizability and the additivity regime is not reached for trimers. Thus, oligomers of cumulenic structures are likely to provide highly polarizable systems, which will probably be more efficient than polydiacetylenes and polyenes. However, one cannot ignore the constraints such as the stability and optical properties (transparency) of such compounds. These should be examined in detail and a combination of suitable heteroatomic linkages and cumulenic structures could provide an answer to these questions.

4.4.6 Hydrogen Bonded Systems

Controlling the structure at the molecular level, increasing the density of electroactive species and improving the stability of the material are very important aspects in the molecular engineering of new materials for optoelectronics [4.92]. In this Section, we illustrate possible contributions from quantum chemistry calculations to the question of controlling the molecular order. Experimentally, various approaches are currently considered, the most often used are the Langmuir-Blodgett film deposition, topochemical solid-state polymerization, and liquid crystal formation. More recently the approach based on hydrogen bonding has obtained increased attention. In the case of liquid crystalline systems, chain ordering by hydrogen bond formation has been demonstrated for an organic system containing carboxylic groups [4.93]. Similarly, the property of hydrogen bonding to impose specific patterns to the molecular organization has been used elegantly to engineer molecular ferromagnets [4.94]. A more systematic use of hydrogen bonding for optoelectronic materials has also been considered theoretically [4.51,4.95,4.96] and experimentally [4.97-4.99].

These examples suggest that hydrogen bonds should be considered more systematically as a tool to exert some control on the molecular assembly of conjugated organic chains. As already pointed out, free-electron and Hückel phenomenological models predict a $l_\pi{}^3$ dependence for the polarizability. In long chains however, conformational freedom can result in kinks and/or twists which disrupt the conjugation and thus spoil the expected benefit of the $l_\pi{}^3$ dependence (Figure 4.16).

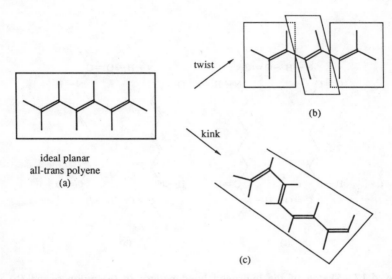

ideal planar
all-trans polyene
(a)

(b)

(c)

Figure 4.16 Schematic representation of possible structural defects that can occur in polyene chains

In this part we would like to consider from a theoretical point of view the possibility of using hydrogen bonds to prevent and/or minimize the occurrence of these undesirable effects, and more specifically illustrate the principles for a study of the intrinsic polarizability properties of some relevant hydrogen bond patterns. Before comparing the polarizability of several hydrogen bond patterns, it can be of interest to indicate how the amide linkage could be used, both directly in the active backbone or as a terminal group and explain the possible benefits of such changes.

4.4.6.1 Use of peptide linkages in the context of active main chain systems for optoelectronics

Nylons owe part of their mechanical properties to hydrogen bonds which, among other things, induce a better organization at the molecular level. Figure 4.17 provides a schematic representation of the hydrogen-bonded sheets with amide groups in nylon 6 as inferred from crystal structure determination [4.100].

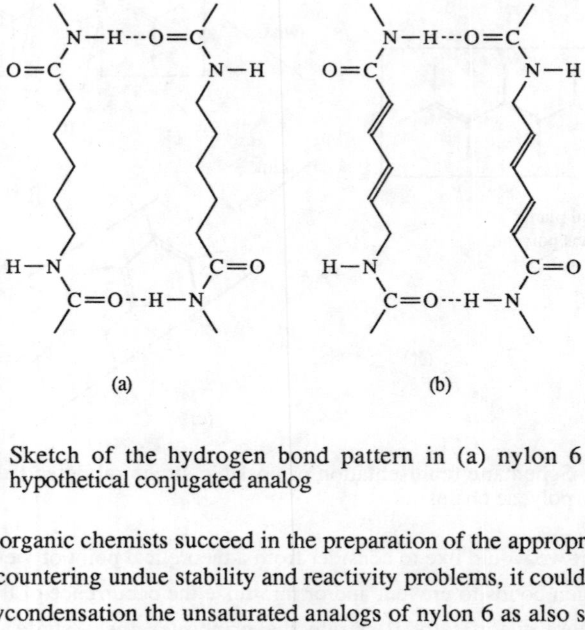

(a)   (b)

Figure 4.17 Sketch of the hydrogen bond pattern in (a) nylon 6 and (b) in its hypothetical conjugated analog

Provided organic chemists succeed in the preparation of the appropriate monomers, i.e., without encountering undue stability and reactivity problems, it could be conceivable to form by polycondensation the unsaturated analogs of nylon 6 as also shown in Figure 4.17. Again if chemistry is tractable, longer chains could be incorporated between the amide groups with the expected benefit of locking the double bond sequence in a more ordered way and thereby minimize the occurrence of kinks and twists in the conjugated backbone.

Another way of using hydrogen bonds is to increase the density of active species in a crystal and augment its cohesive interactions to force the conjugated backbone to be fully ordered. For instance, it is known that muconic acid (**39**) forms hydrogen bonds in the crystal [4.101]; these bonds force the molecule to fully align owing to the cyclic hydrogen-bonded pairs between two carboxylic groups, see Figure 4.18 (a). On the contrary, trans,trans-dimethylmuconate (**40**) - which has a structure [4.102] close that of muconic acid (**39**), but without the possibility of forming hydrogen bonds, exhibit a quite different organization in the crystal, see Figure 4.18 (b).

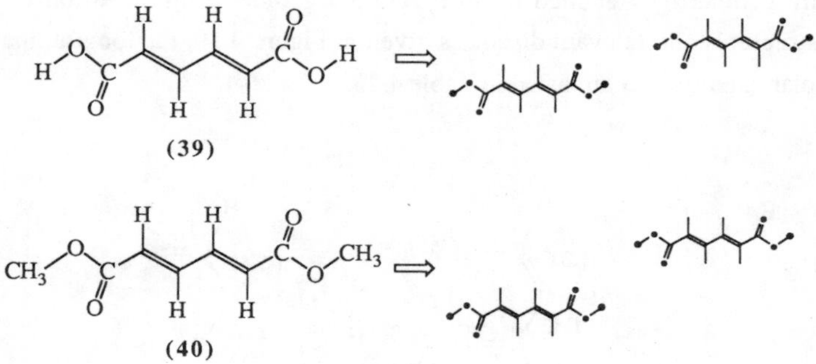

Figure 4.18   Sketch of muconic acid (**39**), of trans,trans-dimethylmuconate (**40**), and of their crystalline arrangement

There are many more examples of hydrogen bond patterns that could be used to exert some control on the molecular organization of the active species. Leiserowitz and Etter [4.103, 4.104] who have contributed much to this idea of organizing organic molecules with hydrogen bonds, have recently specialized it to the field of optoelectronics.

4.4.6.2 Dependence of the polarizability on various hydrogen-bonded patterns.

Quantum chemistry can bring useful contributions to this question by calculations of the polarizability dependence on the structure of the isolated molecules, on the one hand, and the molecules interacting via hydrogen bonds, on the other hand. Properties of hydrogen-bonded molecules are difficult to obtain correctly with small basis set calculations. Additional problems concerning the comparison of isolated and interacting systems due to the so-called superposition error come into play. Thus, the results presented in this Section should be considered with caution even though detailed calculations with larger basis sets show that from a qualitative point of view the STO-3G results are reasonable [4.96].

The first illustration considered deals with the comparison of the longitudinal polarizability of a dimer (43) of acrylamide (41) and N-vinylformamide (42) [4.51]. As usual, the polarizability calculations have been made at the STO-3G level on the equilibrium geometries obtained by full STO-3G geometry optimization of the three systems; structures and relevant distances given in Figure 4.19, the longitudinal α_{zz} and average polarizability $\langle\alpha\rangle$ are listed in Table 4.23.

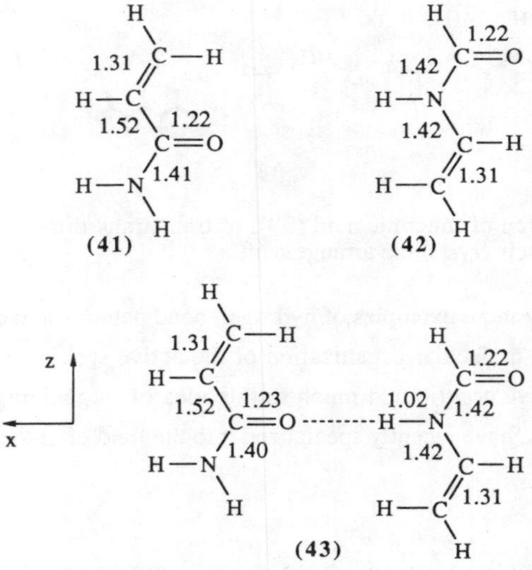

Figure 4.19 Structure and orientation of acrylamide (41), N-vinylformamide (42) and their dimer (43) (together with STO-3G optimized bond lengths)

Table 4.23 STO-3G longitudinal polarizability α_{zz}, average polarizability $<\alpha>$ (in a.u.), and delocalization length l_π for molecules (41), (42), and their dimer (43) (in Å)

Molecule	α_{zz}	$<\alpha>$	l_π
acrylamide (41)	34.40	22.02	3.68
N-vinylacrylamide (42)	38.46	23.47	3.71
dimer (43)	72.18	46.96	4.84

As seen from Table 4.23, the hydrogen bond between acrylamide (41) and N-vinylacrylamide (42) has limited influence on the resulting polarizability of the dimer as compared to the sum of the isolated molecules. This can be understood from the unchanged structure of the constituent molecules in the dimer as compared to the separate molecules. Within the present level of calculation, no significant changes occur in the C-N, C-C, and C=O bonds. Thus, the negative influence on the polarizability of a peptide group inserted in a polyene backbone is not corrected by the formation of the hydrogen bonds. Within the limits of the STO-3G calculations, it so happens that, except for its capabilities of organizing molecules, taking the pain of forming such a material for optoelectronics is probably not a good thing to do.

In the second illustration [4.96], the polarizabilities of the cyclic (44) and linear (45) hydrogen-bonded dimers of formamide (43) are compared. The STO-3G structures and relevant distances are given in Figure 4.20, the longitudinal α_{zz} and average polarizability $<\alpha>$ are listed in Table 4.24.

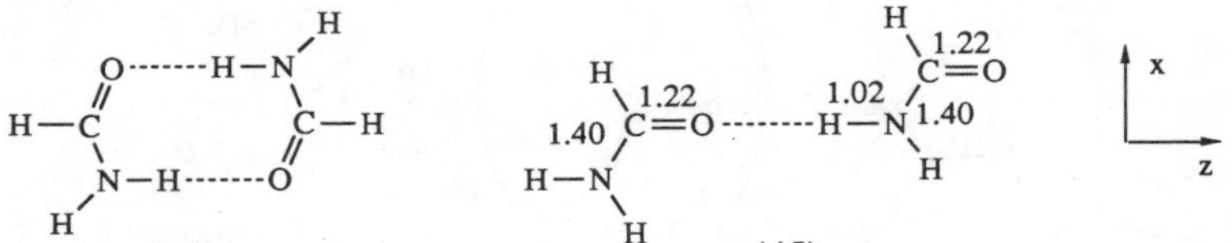

Figure 4.20 Structure and orientation of the cyclic (44) and linear (45) dimers of formamide

On the basis of the average polarizability, little polarizability enhancement is to be expected from hydrogen bond formations, at least in the case of rigid geometries. This average polarizability is barely more than twice the value of the monomer. Notice however that the longitudinal polarizability, which is of importance in highly anisotropic systems, is more important in the cyclic (less extended) dimer than in the linear dimer. This indicates that hydrogen bonding contributes significantly to the electronic polarizability of the system. However in the cyclic formamide dimer two hydrogen bonds exist instead of only one in the linear formamide dimer. Results obtained with larger basis sets confirm the above conclusion.

Table 4.24 STO-3G longitudinal polarizability α_{zz}, average polarizability $<\alpha>$ (in a.u.), and delocalization length l_π for formamide (**43**), its cyclic (**44**), and linear (**45**) dimers (in Å)

Molecule	α_{zz}	$<\alpha>$	l_π
Formamide (**43**)	17.09	10.48	3.66
Cyclic formamide dimer (**44**)	30.00	23.27	3.85
Linear formamide dimer (**45**)	38.24	22.40	6.70

It would thus be interesting to establish a saturation curve, as in the case of the polyenes, for linear chains of formamides linked together by hydrogen bonds as shown in Figure 4.21.

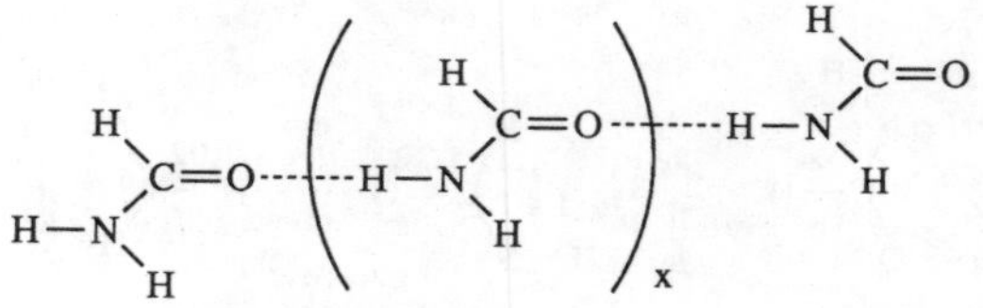

Figure 4.21 Structure of oligomer of formamide molecules held together by hydrogen bonds

It must be stressed that many more molecules capable of establishing hydrogen bonds exist and more favorable conclusions could be obtained on the role played by hydrogen bonds in organizing conjugate organic molecules for optoelectronic purposes [4.97-4.99].

4.4.7 Charge Transfer Systems

Adding electron donating (accepting) atoms and/or molecules in a medium capable of accepting (donating) the electrons usually leads to charge transfer complexes most often characterized by geometry relaxations in the partner molecules. This process is extensively used to obtain conducting polymers, as detailed in Chapter 3. Provided charge transfer can be controlled, it may serve as another way to obtain molecules characterized by better delocalized π-electron networks and thus exhibiting enhanced electric responses.

In this part, we examine [4.105] three types of chains which have been extensively studied, namely: polyparaphenylene, polypyrrole, and polythiophene. Doping induces strong geometric modification of the aromatic polymer backbone towards a more quinoid-like structure. Since a quinoidic structure is expected to be more polarizable because, contrary to benzene, there is no aromaticity driven stabilization to damp the electric response of the π-electrons, see Section 4.4.5.2, it is interesting to study the influence of doping on the electric polarizability of representative oligomers: quaterphenyl, quaterpyrryl, and quaterthienyl.

Geometry optimizations have been reported previously for these tetramers, undoped and doped with two Na atoms [4.106]. The geometry changes, when going from quaterphenyl to Na_2-doped quaterphenyl, is shown in Figure 4.22; the other tetramers follow the same trends from the point of view of the geometry modifications resulting from doping.

It is easily observed that upon doping, the inner rings adopt a quinoid-like structure, the double bonds vary from 1.35 to 1.37 Å and the single bonds vary from 1.43 to 1.46 Å [4.106].

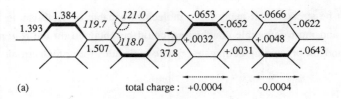

(a)   total charge :   +0.0004   -0.0004

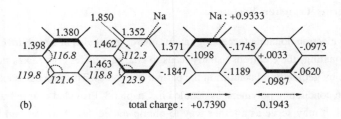

(b)   total charge :   +0.7390   -0.1943

Figure 4.22 STO-3G optimized geometry (bond lengths in Å and angles in degrees) and net Mulliken atomic charges on the carbon atoms (in units of the electronic charge) for (a) quaterphenyl for which D_2 symmetry is assumed and (b) sodium-doped quaterphenyl with C_2 symmetry assumed

The results for the average polarizability $\langle\alpha\rangle$ and its cartesian components α_{xx}, α_{yy}, and α_{zz} are given in Table 4.25 for quaterphenyl, quaterpyrryl, and quaterthienyl: (A) for the undoped oligomers, (B) for the oligomers doped with two sodium atoms, (C) for the systems in the geometry induced by the dopants, but the sodium atoms not included in the calculations, (D) for the two Na atoms only geometrically placed as in the doped situation, and (E) the sum of polarizability contributions from (C) and (D), term by term.

In all cases (A), (B), (C), and (E), quaterphenyl is found more polarizable than the other two tetramers. This is in line with the experimental polarizability of the three monomers which orders as benzene > thiophene > pyrrole [4.107].

Doping induces a strong increase in polarizability. The average value $\langle\alpha\rangle$ doubles in the case of quaterphenyl and increases by about 80 to 85% in the heteroatomic systems. Note that the average polarizability is related by a 2.5-fold increase in the longitudinal polarizability, α_{zz}, the other components being almost unaffected by the doping process.

Table 4.25 STO-3G average polarizability $<\alpha>$ and its cartesian components α_{xx}, α_{yy}, and α_{zz} for quaterphenyl, quaterpyrryl, and quaterthienyl (in a.u.)

	α_{zz}	α_{yy}	α_{xx}	$<\alpha>$
Quaterphenyl				
(A)	291.17	145.59	39.26	159.34
(B)	790.00	153.50	26.14	323.21
(C)	482.45	160.00	22.35	221.60
(D)	210.13	0.02	0.03	70.06
(E)	692.58	160.02	22.38	291.66
Quaterpyrryl				
(A)	224.65	135.09	15.94	125.23
(B)	534.15	137.62	20.03	230.60
(C)	352.42	135.34	15.97	167.91
(D)	141.07	13.25	0.02	51.45
(E)	493.49	148.49	15.99	219.36
Quaterthienyl				
(A)	268.01	140.31	19.85	142.72
(B)	608.35	138.42	23.78	256.85
(C)	473.34	145.06	19.93	212.77
(D)	160.75	12.92	0.02	57.90
(E)	534.09	157.98	19.95	270.67

(A) refers to the undoped tetramers;
(B) corresponds to the tetramers doped with two Na atoms;
(C) refers to the tetramers in the geometry of the doped oligomers, but in absence of the corresponding Na dopants;
(D) corresponds to the two Na atoms in the situation they have in the doped system, but without the tetramers;
(E) values are obtained by simply adding column wise the results (C) and (D). The z-axis is parallel to the chain axis and the y-axis is perpendicular to the average plane of the chain

From the results in situations (C) and (D), the overall polarizability increase can be explained as originating from two factors: first, the geometry modifications towards a more polarizable quinoid-type structure (this accounts for two-thirds of the total polarizability increase); second, the increase of electronic charge on the carbon backbone due to electron transfer from the sodium atoms.

The significant increase of electric polarizability resulting from doping aromatic tetramers is expected to yield an even more important enhancement in the third-order polarizability if the assumptions on the scaling laws hold true. Obviously, these prospects of using doping in the field of optoelectronics must also meet other requirements than just high electric responses; properties such as transparency, radiation damage, etc., must also be satisfied.

4.5 THEORETICAL DEVELOPMENTS NEEDED FOR ACTIVE MAIN POLYMERS

Major successes, experimental and theoretical, are noted in the first nonlinear hyperpolarizability β area where the active side chain polymers are expected to play a dominant role, but the same is far from being true for the second nonlinear hyperpolarizability γ. In this case, both theory and experiment are still in a very early stage of development where there is an obvious lack of organized studies involving experimentalists and theoreticians.

The experimental field, which shows wide and sustained activities in synthesis and characterization, is now crowded with hastily conducted works driven by the hope of being the first to obtain a good material to be incorporated in a readily marketable device. As a result, it is rather frequent that on similar compounds different conclusions are obtained from their characterization, and consistency between these results is seldom attained. Thus, more systematic and carefully designed works on the fundamental aspects of oligomers and polymers with specific γ responses are badly needed [4.108, 4.109].

The theoretical field is in no better situation [4.110]. Molecular electric responses, even the polarizability, are quite difficult to calculate *ab initio* as seen from the recent calculations carried out on very small systems with the presently most sophisticated methodologies [4.17-4.21]. In the case of "large" systems, which for quantum chemistry usually means molecules and oligomers containing up to twelve atoms of the first period of the Mendeleev table, *ab initio* calculations are almost exclusively restricted to estimates of the static molecular electric responses. However, the electronic and optical behaviour of active main chain polymers are basically related to the excited electronic states and their nature [4.111]. Thus theoretical works towards methodologies and computer programs for reliable estimates (if not highly accurate) of dynamical electric responses of chemically interesting systems are important for the field.

Finally, the complications inherent to this field are such that best chances of solving the problems require continuous contact and coordinated research plans between experimentalists and theoreticians.

References

[4.1]   N. Bloembergen, "Nonlinear Optics", Benjamin, New York (1965).

[4.2]   Y.R. Shen, "The Principles of Nonlinear Optics", J. Wiley & Sons, New York (1984).

[4.3]   H.D. Cohen and C.C.J. Roothaan, J. Chem. Phys., **43**, 534 (1965).

[4.4]   J.G. Fripiat, C. Barbier, V.P. Bodart, and J.M. André, J. Comput. Chem.,**7**, 756 (1986).

[4.5]   D. Li, M.A. Ratner, and T.J. Marks, J. Am. Chem. Soc., **110**, 1707 (1988) and references therein.

[4.6]   D.N. Beratan, J. Phys. Chem., **93**, 3915 (1989).

[4.7]   C.W. Dirk, R.J. Twieg, and G. Wagnière, J. Am. Chem. Soc., **108**, 5387 (1986).

[4.8]   J. Zyss, J. Chem. Phys., **70**, 3333 (1979); **70**, 3341 (1979); **71**, 909 (1979).

[4.9]   J. Zyss, in "Conjugated Polymeric Materials: Opportunities in Electronics, Optoelectronics and Molecular Electronics", (J.L. Brédas and R.R. Chance, Eds.), pp. 545-557, and refs. therein, Kluwer Academic Publishers, Dordrecht (1990).

[4.10] J.W. Wu, J.R. Heflin, R.A. Norwood, K.Y. Wong, O. Zamani-Khamiri, A.F. Garito, P. Kalyanaraman, and J. Sounik, J. Opt. Soc. Am. B, **6**, 707 (1989).

[4.11] J.O. Morley, V.J. Docherty, and D. Pugh, J. Mol. Electronics, **5**, 117 (1989).

[4.12] J. Waite and M.G. Papadopoulos, J. Phys. Chem., **93**, 43 (1989).

[4.13] B.M. Pierce, J. Chem. Phys., **91**, 791 (1989).

[4.14] E.N. Svendsen, Int. J. Quantum Chem. Symp., **22**, 477 (1988).

[4.15] W.A. Parkinson and M.C. Zerner, Chem. Phys. Lett., **139**, 563 (1987).

[4.16] G.R.J. Williams, J. Mol. Electronics, **6**, 99 (1990).

[4.17] D.M. Bishop, J. Pipin, and M. Rérat, J. Chem. Phys., **92**, 1902 (1990).

[4.18] D.P. Shelton, J. Opt. Soc. Am. B, **6**, 830 (1989).

[4.19] G. Maroulis and A.J. Thakkar, J. Chem. Phys., **88**, 7623 (1988).

[4.20] J. Oddershede, Adv. Chem. Phys., vol. LXIX, part II, (I. Prigogine and S.A. Rice, Eds.), pp.201 et sq., J. Wiley & Sons, New York (1987).

[4.21] H.J. Aa Jensen, H. Koch, P. Jørgensen, and J. Olsen, Chem. Phys., **19**, 297 (1988).

[4.22] B. Kirtman, Int. J. Quantum Chem., **36**, 119 (1989).

[4.23] S.P. Karna, M. Dupuis, E. Perrin, and P.N. Prasad, J. Chem. Phys., **92**, 7418 (1990).

[4.24] P.N. Prasad, in "Nonlinear Optical Effects in Organic Polymers", (J. Messier, F. Kajzar, P. Prasad, and D. Ulrich, Eds.), pp.351 et sq., Kluwer Academic Publishers, Dordrecht (1989).

[4.25] A. Chablo and A. Hinchliffe, Chem. Phys. Lett.,**72**, 149 (1980).

[4.26] V.P. Bodart, J. Delhalle, J.M. André, and J. Zyss, Can. J. Chem., **63**, 1631 (1985).

[4.27] J.P. Hermann, Ph.D.Thesis, Université de Paris-Sud, Orsay (1974).

[4.28] R.H. Boyd and L. Kesner, Macromolecules, **20**, 1802 (1987).

[4.29] J.N. Churchill and F.E. Holmstrom, Amer. J. Phys., **50**, 848 (1982).

[4.30] J.N. Churchill and F.E. Holmstrom, Physica B., **123**, 1 (1983).

[4.31] C. Barbier, Chem. Phys. Lett., **142**, 53 (1987).

368

[4.32]   C. Barbier, J. Delhalle, and J.M. André, in "Nonlinear Optical Properties of Polymers", (A.J. Heeger, J. Orenstein, and D.R. Ulrich, Eds.), p.239, Materials Research Society, Pittsburgh (1988).

[4.33]   C. Barbier, J. Delhalle, and J.M. André, J. Molec. Struct. (Theochem), **188**, 299 (1989).

[4.34]   G.P. Agrawal and C. Flytzanis, Chem. Phys. Lett., **44**, 366 (1976); C. Cojan, G.P. Agrawal, and C. Flytzanis, Phys. Rev. B, **15**, 909 (1977); G.P. Agrawal, C. Cojan, and C. Flytzanis, Phys. Rev. Lett., **38**, 711 (1977); Phys. Rev. B, **17**, 776 (1978).

[4.35]   V.M. Genkin and P.M. Mednis, Sov. Phys.-JETP **27**, 609 (1968).

[4.36]   J. Callaway in "Energy Band Theory", p.277, Academic Press, New York (1964).

[4.37]   W. Jones and N.H. March, in "Theoretical Solid Sate Physics", Vol 2, p.800, J. Wiley & Sons, London (1973).

[4.38]   P.L. Davies, Trans. Faraday Soc., **47**, 789 (1952).

[4.39]   S. Risser, S. Klemm, D.W. Allender, and M.A. Lee, Mol. Cryst. Liq. Cryst., **150**b, 631 (1987).

[4.40]   H.F. Hameka, J. Chem. Phys., **67**, 2935 (1977).

[4.41]   C. Flytzanis, in "Nonlinear Optical Properties of Organic Molecules and Crystals", (D.S. Chemla and J. Zyss, Eds.), Vol. 2, p.121, Academic Press, New York (1987).

[4.42]   W. Kauzmann, in "Quantum Chemistry", p.682, Academic Press, New York (1957).

[4.43]   F.L. Pilar, in "Elementary Quantum Chemistry", p.593, McGraw-Hill, New York (1968).

[4.44]   Th. Albright, J.K. Burdett, and M-H. Wanghbo, in "Orbital Interactions in Chemistry", p.229, J. Wiley & Sons, New York (1986).

[4.45]   T.K. Rebane, in "Methods of Quantum Chemistry", (M.G. Veselov, Ed.), p.147, Academic Press, New York (1965).

[4.46]   W.P. Su, J.R. Schrieffer, and A.J. Heeger, Phys. Rev. Lett., 42, 1698 (1979).

[4.47]   R.C. Miller, Appl. Phys. Lett., 5, 17 (1964).

[4.48]   A. Yariv, "Quantum Electronics", p.415, J. Wiley & Sons, New York (1975).

[4.49]   G.J.B. Hurst, M. Dupuis, and E. Clementi, J. Chem. Phys., 89, 385 (1988).

[4.50]   P. Chopra, L. Carlacci, H.F. King, and P.N. Prasad, J. Phys. Chem., 93, 7120 (1989).

[4.51]   J. Delhalle, M. Dory, J.G. Fripiat, and J.M. André, in "Nonlinear Optical Effects in Organic Polymers", (J. Messier, F. Kajzar, P. Prasad, and D. Ulrich, Eds.), p.13, Kluwer Academic Publishers, Dordrecht (1989).

[4.52]   W.J. Hehre, L. Radom, P. von Ragué Schleyer, and J.A. Pople, "Ab Initio Molecular Orbital Theory", J. Wiley & Sons, New York (1986).

[4.53]   V.P. Bodart, J. Delhalle, J.M. André, and J. Zyss, in "Polydiacetylenes: Synthesis, Structure and Electronic Properties", (D. Bloor and R.R. Chance, Eds.), p.125, Martinus Nijhoff Publishers, Dordrecht (1985).

[4.54]   Y. Tokura, Y. Oowaki, T. Koda, and R.H. Baughman, Chem. Phys. Lett., 88, 437 (1984).

[4.55]   D. Kobelt and E.F. Paulis, Acta Cryst., B30, 232 (1973).

[4.56]   V. Enkelmann and J.B. Lando, Acta Cryst., B34, 2352 (1978).

[4.57]   A. Karpfen, J. Phys. C, 13, 5673 (1980).

370

[4.58] V.P. Bodart, J. Delhalle and J.M. André, in "Conjugated Polymeric Materials: Opportunities in Electronics, Optoelectronics and Molecular Electronics", (J.L. Brédas and R.R. Chance, Eds.), p.509, Kluwer Academic Publishers, Dordrecht (1990).

[4.59] D.L. Smith and H.R. Luss, Acta Cryst., **B28**, 2793 (1972).

[4.60] D.L. Smith and H.R. Luss, Acta Cryst., **B31**, 402 (1975).

[4.61] R.W. Bigelow and H.J. Freund, Chem. Phys., **107**, 159 (1986).

[4.62] J.M. André, "Etude Comparative des Méthodes de Hückel et de l'Electron Libre", Ms. Sc. Thesis, Université Catholique de Louvain, Leuven-Belgium (1965).

[4.63] C.W. Bock and M.J. Trachtman, J. Mol. Struct. (Theochem), **1**, 109 (1984).

[4.64] J.L. Brédas, A.J. Heeger, and F. Wudl, J. Chem. Phys., **85**, 4673 (1986).

[4.65] V.P. Bodart, J. Delhalle, M. Dory, J.G. Fripiat, and J.M. André, J. Opt. Soc. Am., **B4**, 1047 (1987).

[4.66] J.D. Mee, J. Am. Chem. Soc., **96**, 4712 (1974).

[4.67] J.D. Mee, J. Org. Chem., **42**, 1035 (1977).

[4.68] J.D. Mee and D.M. Sturmer, J. Org. Chem., **42**, 1041 (1977).

[4.69] J. Delhalle, V.P. Bodart, M. Dory, J.M. André, and J. Zyss, Int. J. Quantum Chem. Symp., **19**, 313 (1986).

[4.70] D.N. Baratan, J.N. Onuchic, and J.W. Perry, J. Phys. Chem., **91**, 2696 (1987).

[4.71] B. Kirtman and M. Hasan, Chem. Phys. Lett., **157**, 123 (1989).

[4.72] E. Younang, J.M. André, and J. Delhalle, Int. J. Quantum Chem., in press.

[4.73] E. Younang, J. Delhalle, and J.M. André, New J. Chem., **11**, 403 (1987).

[4.74] L. Radom, W.J. Hehre, and J.A. Pople, J. Am. Chem. Soc., **93**, 289 (1971).

[4.75] D. Griller and K.U. Ingold, Acc. Chem. Res., **9**, 13 (1976).

[4.76] W.J. Hehre, R. Ditchfield, L. Radom, and J.A. Pople, J. Am. Chem. Soc., **92**, 4796 (1970).

[4.77] J.D. Cox and D. Pilcher, "Thermochemistry of Organic and Organometallic Compounds", Academic Press, New York, (1970).

[4.78] R.C. Haddon and W.H. Starness Jr., in "Advances in Chemistry Series", (D.L. Allara and W.L. Hawkins, Eds.), No 169, Chap. 27, Am. Chem. Soc., Washington DC (1978).

[4.79] C.W. Bock and M.J. Trachtman, J. Mol. Struct. (Theochem), **1**, 109 (1984).

[4.80] J.S. Binkley, J.A. Pople, and W.J. Hehre, J. Am. Chem. Soc., **102**, 939 (1980).

[4.81] S.W. Benson, F.R. Cruicshand, R.M. Golden, G.R. Haugen, H.E. O'Neal, A.S. Rodgers, R. Shaw, and R. Walsh, Chem. Rev., **69**, 279 (1969).

[4.82] S.W. Benson, "Thermochemical Kinetics", J. Wiley & Sons, New York (1971).

[4.83] H.V. Hirschhausen and K. Wenzel, Theoret. Chim. Acta (Berl.), **35**, 293 (1974), see table 1, footnote d, p. 295.

[4.84] D.J. Defrees and W.J. Hehre, J. Phys. Chem., **82**, 391 (1978).

[4.85] S.W. Benson, J. Chem. Educ., **42**, 502 (1965).

[4.86] G. Leroy, Int. J. Quantum Chem., **23**, 271 (1983).

[4.87] G. Leroy, M. Sana, C. Wilante, D. Peeters, and C. Dogimont, J. Mol. Struct. (Theochem), **153**, 249 (1987).

[4.88] C. Sauteret, J.P. Hermann, R. Frey, F. Pradere, J. Ducuing, R.H. Baughman, and R.R. Chance, Phys. Rev. Lett., **36**, 956 (1976).

[4.89] K.C. Rustagi and J. Ducuing, Opt. Commun., **10**, 258 (1974).

[4.90] H. Irngartinger and W. Götzmann, Angew. Chem. Int. Ed. Engl., **25**, 340 (1986).

[4.91] A. Karpfen, J. Phys. C, **12**, 3277 (1979).

[4.92] G.G. Roberts, Adv. Phys., **34**, 475 (1985).

[4.93] A.S. Paranjpe and V.K. Kelkar, Mol. Cryst. Liq. Cryst., **102**, 289 (1984).

[4.94] Y. Pei, M. Verdaguer, O. Kahn, J. Sletten, and J.P. Renard, J. Am. Chem. Soc., **108**, 7428 (1986).

[4.95] J. Waite and M.G. Papadopoulos, J. Chem. Phys., **83**, 4047 (1985).

[4.96] M. Dory, J. Delhalle, J.G. Fripiat, and J.M. André, Int. J. Quantum Chem. Symp., **14**, 85 (1987).

[4.97] T.W. Panunto, Z. Urbanczyk-Lipkowska, R. Johnson, and M.C. Etter, J. Am. Chem. Soc., **109**, 7786 (1987).

[4.98] M.C. Etter and G.M. Frankenbach, Chem. Materials, **1**, 10 (1989).

[4.99] E. Staab, L. Addadi, L. Leiserowitz, and M. Lahav, Adv. Mater., **2**, 3 (1990).

[4.100] J.P. Parker and P.H. Lindenmeyer, J. Appl. Polym. Sc., **21**, 821 (1977).

[4.101] J. Bernstein and L. Leiserowitz, Isr. J. Chem., **10**, 601 (1972).

[4.102] S.E. Filippakis and L. Leiserowitz, J. Chem. Soc. B, 290 (1967).

[4.103] L. Leiserowitz, Acta Cryst.., **B32**, 775 (1976).

[4.104] L. Leiserowitz and F. Nader, Acta Cryst., **B33**, 2719 (1977).

[4.105] M. Dory, V.P. Bodart, J. Delhalle, J.M. André, and J.L. Brédas, Mat. Res. Soc. Symp. Proc., **109**, 239 (1988).

[4.106] J.L. Brédas, B. Thémans, J.G. Fripiat, J.M. André, and R.R. Chance, Phys. Rev. B, **29**, 6761 (1984).

[4.107] C.G. Le Fèvre, R.J.W. Le Fèvre, B.P. Ras, and M.R. Smith, J. Chem. Soc., 1188 (1959).

[4.108] D. Bloor, "Organic Materials for Nonlinear Optics: Yesterday, Today and Tomorrow", in "Organic Materials for Nonlinear Optics", (R.Hann and D.Bloor, Eds.), The Royal Society of Chemistry, to appear (1991).

[4.109] H. J. Byrne, W. Blau, R. Giesa, and R. C. Schulz, Chem. Phys. Lett., **167**, 484 (1990).

[4.110] J. Delhalle, J. Messier, E. Orti, P. Sautet, R. Silbey, Z.G. Soos, and J.M. Toussaint, in "Conjugated Polymeric Materials: Opportunities in Electronics, Optoelectronics and Molecular Electronics", (J.L. Brédas and R.R. Chance, Eds.), p.591, Kluwer Academic Publishers, Dordrecht (1990).

[4.111] J.M. Toussaint, F. Meyers, and J.L. Brédas, in "Conjugated Polymeric Materials: Opportunities in Electronics, Optoelectronics and Molecular Electronics", (J.L. Brédas and R.R. Chance, Eds.), p.207, Kluwer Academic Publishers, Dordrecht (1990).

INDEX